Lecture Notes in Artificial Intelligence   11289

Subseries of Lecture Notes in Computer Science

LNAI Series Editors

Randy Goebel
  University of Alberta, Edmonton, Canada
Yuzuru Tanaka
  Hokkaido University, Sapporo, Japan
Wolfgang Wahlster
  DFKI and Saarland University, Saarbrücken, Germany

LNAI Founding Series Editor

Joerg Siekmann
  DFKI and Saarland University, Saarbrücken, Germany

More information about this series at http://www.springer.com/series/1244

Ildar Batyrshin
María de Lourdes Martínez-Villaseñor
Hiram Eredín Ponce Espinosa (Eds.)

Advances in Computational Intelligence

17th Mexican International Conference
on Artificial Intelligence, MICAI 2018
Guadalajara, Mexico, October 22–27, 2018
Proceedings, Part II

 Springer

Editors
Ildar Batyrshin
Instituto Politécnico Nacional
Mexico City, Mexico

María de Lourdes Martínez-Villaseñor
Universidad Panamericana
Mexico City, Mexico

Hiram Eredín Ponce Espinosa
Faculty of Engineering
Universidad Panamericana
Mexico City, Mexico

ISSN 0302-9743 ISSN 1611-3349 (electronic)
Lecture Notes in Artificial Intelligence
ISBN 978-3-030-04496-1 ISBN 978-3-030-04497-8 (eBook)
https://doi.org/10.1007/978-3-030-04497-8

Library of Congress Control Number: 2018958467

LNCS Sublibrary: SL7 – Artificial Intelligence

This Springer imprint is published by the registered company Springer Nature Switzerland AG
The registered company address is: Gewerbestrasse 11, 6330 Cham, Switzerland

Preface

The Mexican International Conference on Artificial Intelligence (MICAI) is a yearly international conference series that has been organized by the Mexican Society of Artificial Intelligence (SMIA) since 2000. MICAI is a major international artificial intelligence forum and the main event in the academic life of the country's growing artificial intelligence community.

MICAI conferences publish high-quality papers in all areas of artificial intelligence and its applications. The proceedings of the previous MICAI events have been published by Springer in its *Lecture Notes in Artificial Intelligence* series, vol. 1793, 2313, 2972, 3789, 4293, 4827, 5317, 5845, 6437, 6438, 7094, 7095, 7629, 7630, 8265, 8266, 8856, 8857, 9413, 9414, 10061, 10062, 10632, and 10633. Since its foundation in 2000, the conference has been growing in popularity and improving in quality.

The proceedings of MICAI 2018 are published in two volumes. The first volume, *Advances in Soft Computing*, contains 33 papers structured into three sections:

- Evolutionary and nature-inspired intelligence
- Machine learning
- Fuzzy logic and uncertainty management

The second volume, *Advances in Computational Intelligence*, contains 29 papers structured into three sections:

- Knowledge representation, reasoning, and optimization
- Natural language processing
- Robotics and computer vision

This two-volume set will be of interest for researchers in all areas of artificial intelligence, students specializing in related topics, and for the public in general interested in recent developments in artificial intelligence.

The conference received 149 submissions for evaluation from 23 countries: Argentina, Australia, Brazil, Canada, Colombia, Costa Rica, Cuba, Czech Republic, Finland, France, Hungary, Iran, Italy, Mexico, Morocco, Pakistan, Peru, Poland, Russia, Spain, Thailand, Turkey, and USA. Of these submissions, 62 papers were selected for publication in these two volumes after a peer-reviewing process carried out by the international Program Committee. Therefore, the acceptance rate was 41%.

The international Program Committee consisted of 113 experts from 17 countries: Azerbaijan, Brazil, Canada, Colombia, Cuba, France, Greece, India, Israel, Italy, Japan, Mexico, Portugal, Singapore, Spain, UK, and USA.

MICAI 2018 was honored by the presence of renowned experts who gave excellent keynote lectures:

- Alexander S. Poznyak Gorbatch, CINVESTAV-IPN Campus Mexico City, Mexico
- Jeff Clune, Uber AI Labs, University of Wyoming, USA

- David J. Atkinson, Silicon Valley Research & Development Center Continental AG, USA
- Gregory O'Hare, School of Computer Science, University College Dublin, Ireland
- Srinivas V. Chitiveli, Offering Manager & Master Inventor, IBM PowerAI Vision, USA

The technical program of the conference also featured 11 tutorials:

- Smart Applications with FIWARE, by Miguel Gonzalez Mendoza
- Intelligent Management of Digital Data for Law Enforcement Purposes, by Jesus Manuel Niebla Zatarain
- Introduction to Data Science: Similarity, Correlation, and Association Measures, by Ildar Batyrshin
- Brain–Computer Interface (BCI) and Machine Learning, by Javier M. Antelis, Juan Humberto Sossa Azuela, Luis G. Hernandez, and Carlos D. Virgilio
- New Models and Training Algorithms for Artificial Neural Networks, by Juan Humberto Sossa Azuela
- Intelligent Chatbots Using Google DialogFlow, by Leonardo Garrido
- Spiking Neural Models and Their Applications in Pattern Recognition: A beginner's Tutorial, by Roberto A. Vazquez
- Introduction to Quantum Computing, by Salvador Venegas
- Deep-Learning Principles and Their Applications in Facial Expression Recognition with TensorFlow-Keras, by Luis Eduardo Falcón Morales and Juan Humberto Sossa Azuela
- Introduction to Natural Language Human–Robot Interaction, by Grigori Sidorov
- Knowledge Extraction from Fuzzy Predictive Models, by Félix A. Castro Espinoza

Three workshops were held jointly with the conference:

- HIS 2018: 11th Workshop of Hybrid Intelligent Systems
- WIDSSI 2018: 4th International Workshop on Intelligent Decision Support Systems for Industry
- WILE 2018: 11th Workshop on Intelligent Learning Environments

The authors of the following papers received the Best Paper Awards based on the paper's overall quality, significance, and originality of the reported results:

- First place: "Universal Swarm Optimizer for Multi-Objective Functions," by Luis Marquez and Luis Torres Treviño, Mexico
- Second place: "Topic-Focus Articulation: A Third Pillar of Automatic Evaluation of Text Coherence," by Michal Novák, Jiří Mírovský, Kateřina Rysová and Magdaléna Rysová, Czechia
- Third place: "Combining Deep Learning and RGBD SLAM for Monocular Indoor Autonomous Flight," by José Martínez Carranza, L. Oyuki Rojas Pérez, Aldrich A. Cabrera Ponce and Roberto Munguia Silva, Mexico

The cultural program of the conference included a tour of Guadalajara and the Tequila Experience tour.

We want to thank everyone involved in the organization of this conference. In the first place, the authors of the papers published in this book: It is their research effort that gives value to the book and to the work of the organizers. We thank the track chairs for their hard work, the Program Committee members, and additional reviewers for their great effort in reviewing the submissions.

We would like to thank the Tecnológico de Monterrey Campus Guadalajara for hosting the workshops and tutorials of MICAI 2018, with special thanks to Dr. Mario Adrián Flores, Vice President of the University, and Dr. Ricardo Swain, Dean of the Engineering and Sciences School, for their generous support. We also thank Dr. José Antonio Rentería, Divisional Director of the Engineering and Sciences School, for his kind support. We also want to thank Erik Peterson from Oracle, Rodolfo Lepe and Leobardo Morales from IBM, Oscar Reyes from SinergiaSys, and Luis Carlos Garza Tamez from Grupo ABSA, for their support in organization of this conference. The entire submission, reviewing, and selection process, as well as the preparation of the proceedings, were supported free of charge by the EasyChair system (www.easychair.org). Finally, yet importantly, we are very grateful to the staff at Springer for their patience and help in the preparation of this volume.

October 2018   Ildar Batyrshin
María de Lourdes Martínez-Villaseñor
Hiram Eredín Ponce Espinosa

Conference Organization

MICAI 2018 was organized by the Mexican Society of Artificial Intelligence (SMIA, Sociedad Mexicana de Inteligencia Artificial) in collaboration with the Tecnológico de Monterrey Campus Guadalajara, the Tecnológico de Monterrey CEM, the Centro de Investigación en Computación of the Instituto Politécnico Nacional, the Facultad de Ingeniería or the Universidad Panamericana, and the Universidad Autónoma del Estado de Hidalgo.

The MICAI series website is www.MICAI.org. The website of the Mexican Society of Artificial Intelligence, SMIA, is www.SMIA.org.mx. Contact options and additional information can be found on these websites.

Conference Committee

General Chair

Miguel González Mendoza   Tecnológico de Monterrey CEM, Mexico

Program Chairs

Ildar Batyrshin   Instituto Politécnico Nacional, Mexico
María de Lourdes Martínez   Universidad Panamericana, Mexico
  Villaseñor
Hiram Eredín Ponce   Universidad Panamericana, Mexico
  Espinosa

Workshop Chairs

Obdulia Pichardo Lagunas   Instituto Politécnico Nacional, Mexico
Noé Alejandro Castro   Centro Nacional de Investigación y Desarrollo
  Sánchez   Tecnológico, Mexico
Félix Castro Espinoza   Universidad Autónoma del Estado de Hidalgo, Mexico

Tutorials Chair

Félix Castro Espinoza   Universidad Autónoma del Estado de Hidalgo, Mexico

Doctoral Consortium Chairs

Miguel Gonzalez Mendoza   Tecnológico de Monterrey CEM, Mexico
Antonio Marín Hernandez   Universidad Veracruzana, Mexico

Keynote Talks Chair

Sabino Miranda Jiménez   INFOTEC, Mexico

Publication Chair

Miguel Gonzalez Mendoza   Tecnológico de Monterrey CEM, Mexico

Financial Chair

Ildar Batyrshin   Instituto Politécnico Nacional, Mexico

Grant Chairs

Grigori Sidorov   Instituto Politécnico Nacional, Mexico
Miguel Gonzalez Mendoza   Tecnológico de Monterrey CEM, Mexico

Organizing Committee Chairs

Luis Eduardo Falcón   Tecnológico de Monterrey, Campus Guadalajara
Morales
Javier Mauricio Antelis   Tecnológico de Monterrey, Campus Guadalajara
Ortíz

Area Chairs

Machine Learning

Felix Castro Espinoza   Universidad Autónoma del Estado de Hidalgo, Mexico

Natural Language Processing

Sabino Miranda Jiménez   INFOTEC, Mexico
Esaú Villatoro   Universidad Autónoma Metropolitana Cuajimalpa,
Mexico

Evolutionary and Evolutive Algorithms

Hugo Jair Escalante   Instituto Nacional de Astrofísica, Óptica y Electrónica,
Balderas   Mexico
Hugo Terashima Marín   Tecnológico de Monterrey CM, Mexico

Fuzzy Logic

Ildar Batyrshin   Instituto Politécnico Nacional, Mexico
Oscar Castillo   Instituto Tecnológico de Tijuana, Mexico

Neural Networks

María de Lourdes Martínez   Universidad Panamericana, Mexico
Villaseñor

Hybrid Intelligent Systems

Juan Jose Flores   Universidad Michoacana, Mexico

Intelligent Applications

Gustavo Arroyo · · · · · · · · · · · Instituto Nacional de Electricidad y Energias Limpias, Mexico

Computer Vision and Robotics

José Martínez Carranza · · · · · · Instituto Nacional de Astrofísica, Óptica y Electrónica, Mexico

Daniela Moctezuma · · · · · · · · · Centro de Investigación en Ciencias de Información Geoespacial, Mexico

Program Committee

Rocío Abascal-Mena	Universidad Autonoma Metropolitana – Cuajimalpa, Mexico
Giner Alor Hernandez	Instituto Tecnologico de Orizaba, Mexico
Matias Alvarado	Centro de Investigación y de Estudios Avanzados del IPN, Mexico
Nohemi Alvarez	Centro de Investigación en Geografía y Geomática Ing. Jorge L. Tamayo A.C., Mexico
Gustavo Arechavaleta	Centro de Investigación y de Estudios Avanzados del IPN, Mexico
Gustavo Arroyo-Figueroa	Instituto Nacional de Electricidad y Energías Limpias, Mexico
Maria Lucia Barrón-Estrada	Instituto Tecnológico de Culiacán, Mexico
Rafael Batres	Tecnológico de Monterrey, Mexico
Ildar Batyrshin	CIC, Instituto Politécnico Nacional, Mexico
Davide Buscaldi	LIPN, Université Paris 13, Sorbonne Paris Cité, France
Hiram Calvo	CIC, Instituto Politécnico Nacional, Mexico
Nicoletta Calzolari	Istituto di Linguistica Computazionale, CNR, Italy
Jesus Ariel Carrasco-Ochoa	Instituto Nacional de Astrofísica, Óptica y Electrónica, Mexico
Oscar Castillo	Instituto Tecnológico de Tijuana, Mexico
Felix Castro Espinoza	CITIS, Universidad Autónoma del Estado de Hidalgo, Mexico
Noé Alejandro Castro-Sánchez	Centro Nacional de Investigación y Desarrollo Tecnológico, Mexico
Jaime Cerda Jacobo	Universidad Michoacana de San Nicolás de Hidalgo, Mexico
Ulises Cortés	Universitat Politècnica de Catalunya, Spain
Paulo Cortez	University of Minho, Portugal
Laura Cruz	Instituto Tecnologico de Cd. Madero, Mexico
Israel Cruz Vega	Instituto Nacional de Astrofísica, Óptica y Electrónica, Mexico
Andre de Carvalho	University of São Paulo, Brazil

Jorge De La Calleja	Universidad Politécnica de Puebla, Mexico
Omar Arturo Domínguez-Ramírez	CITIS, Universidad Autónoma del Estado de Hidalgo, Mexico
Leon Dozal	CentroGEO, Mexico
Hugo Jair Escalante	Instituto Nacional de Astrofísica, Óptica y Electrónica, Mexico
Bárbaro Ferro	Universidad Panamericana, Mexico
Denis Filatov	Sceptica Scientific Ltd., UK
Juan José Flores	Universidad Michoacana de San Nicolás de Hidalgo, Mexico
Anilu Franco-Árcega	CITIS, Universidad Autónoma del Estado de Hidalgo, Mexico
Sofia N. Galicia-Haro	Universidad Nacional Autónoma de México, Mexico
Cesar Garcia Jacas	Universidad Panamericana, Mexico
Milton García-Borroto	Universidad Tecnológica de la Habana José Antonio Echeverría (CUJAE), Cuba
Alexander Gelbukh	CIC, Instituto Politécnico Nacional, Mexico
Carlos Gershenson	Universidad Nacional Autónoma de México, Mexico
Eduardo Gomez-Ramirez	Dirección de Posgrado e Investigación, Universidad La Salle, Mexico
Enrique González	Pontificia Universidad Javeriana de la Compañía de Jesús, Colombia
Luis-Carlos González-Gurrola	Universidad Autónoma de Chihuahua, Mexico
Miguel Gonzalez-Mendoza	Tecnológico de Monterrey Campus Estado de México, Mexico
Mario Graff	Infotec, Centro de Investigación e Innovación en Tecnologías de la Información y Comunicación, Mexico
Fernando Gudiño	FES Cuautitlán, Universidad Nacional Autónoma de México, Mexico
Miguel Angel Guevara Lopez	Computer Graphics Center, Portugal
Andres Gutierrez	Tecnológico de Monterrey, Mexico
J. Octavio Gutierrez-Garcia	Instituto Tecnológico Autónomo de México, Mexico
Rafael Guzman Cabrera	Universidad de Guanajuato, Mexico
Yasunari Harada	Waseda University, Japan
Jorge Hermosillo	Universidad Autónoma del Estado de Hidalgo, Mexico
Yasmin Hernandez	Instituto Nacional de Electricidad y Energías Limpias, Mexico
José Alberto Hernández	Universidad Autónoma del Estado de Morelos, Mexico
Oscar Herrera	Universidad Autónoma Metropolitana – Azcapotzalco, Mexico
Pablo H. Ibarguengoytia	Instituto Nacional de Electricidad y Energías Limpias, Mexico

Sergio Gonzalo Jiménez Vargas	Instituto Caro y Cuervo, Colombia
Angel Kuri-Morales	Instituto Tecnológico Autónomo de México, Mexico
Carlos Lara-Alvarez	Centro de Investigación en Matemáticas (CIMAT), Mexico
Eugene Levner	Ashkelon Academic College, Israel
Fernando Lezama	Instituto Nacional de Astrofísica, Óptica y Electrónica, Mexico
Rodrigo Lopez Farias	CONACYT, Mexico; Consorcio CENTROMET, Mexico
Omar Jehovani López Orozco	Instituto Tecnológico Superior de Apatzingán, Mexico
Omar López-Ortega	CITIS, Universidad Autónoma del Estado de Hidalgo, Mexico
Octavio Loyola-González	Escuela de Ingeniería y Ciencias, Tecnológico de Monterrey, Mexico
Yazmin Maldonado	Instituto Tecnológico de Tijuana, Mexico
Cesar Martinez Torres	Universidad de las Américas Puebla, Mexico
María De Lourdes Martínez Villaseñor	Universidad Panamericana, Mexico
Jose Martinez-Carranza	Instituto Nacional de Astrofísica, Óptica y Electrónica, Mexico
José Fco. Martínez-Trinidad	Instituto Nacional de Astrofísica, Óptica y Electrónica, Mexico
Antonio Matus-Vargas	Instituto Nacional de Astrofísica, Óptica y Electrónica, Mexico
Patricia Melin	Instituto Tecnológico de Tijuana, Mexico
Ivan Vladimir Meza Ruiz	Instituto de Investigaciones en Matemáticas Aplicadas y en Sistemas, Universidad Nacional Autónoma de México, Mexico
Efrén Mezura-Montes	Universidad Veracruzana, Mexico
Sabino Miranda-Jiménez	Infotec, Centro de Investigación e Innovación en Tecnologías de la Información y Comunicación, Mexico
Daniela Moctezuma	CONACyT, Mexico; Centro de Investigación en Ciencias de Información Geoespacial, Mexico
Raul Monroy	Tecnológico de Monterrey Campus Estado de México, Mexico
Marco Morales	Instituto Tecnológico Autónomo de México, Mexico
Annette Morales-González	CENATAV, Cuba
Masaki Murata	Tottori University, Japan
Antonio Neme	Universidad Autónoma de la Ciudad de México, Mexico
C. Alberto Ochoa-Zezatti	Universidad Autónoma de Ciudad Juárez, Mexico
José Luis Oliveira	University of Aveiro, Portugal

Jose Ortiz Bejar	Universidad Michoacana de San Nicolás de Hidalgo, Mexico
José Carlos Ortiz-Bayliss	Tecnológico de Monterrey, Mexico
Juan Antonio Osuna Coutiño	Instituto Tecnológico de Tuxtla Gutiérrez, Mexico
Partha Pakray	National Institute of Technology Silchar, India
Leon Palafox	Universidad Panamericana, Mexico
Ivandre Paraboni	University of São Paulo, Brazil
Obdulia Pichardo-Lagunas	Unidad Profesional Interdisciplinaria en Ingeniería y Tecnologías Avanzadas, Instituto Politécnico Nacional, Mexico
Garibaldi Pineda García	Universidad Michoacana de San Nicolás de Hidalgo, Mexico; University of Manchester, UK
Hiram Eredin Ponce Espinosa	Universidad Panamericana, Mexico
Soujanya Poria	Nanyang Technological University, Singapore
Belem Priego-Sanchez	Benemérita Universidad Autónoma de Puebla, Mexico; Université Paris 13, France; Universidad Autónoma Metropolitana – Azcapotzalco, Mexico
Luis Puig	Universidad de Zaragoza, Spain
Vicenç Puig	Universitat Politècnica de Catalunya, Spain
Juan Ramirez-Quintana	Instituto Tecnológico de Chihuahua, Mexico
Patricia Rayón	Universidad Panamericana, Mexico
Juan Manuel Rendon-Mancha	Universidad Autónoma del Estado de Morelos, Mexico
Orion Reyes	University of Alberta Edmonton, Canada
José A. Reyes-Ortiz	Universidad Autónoma Metropolitana, Mexico
Noel Enrique Rodriguez Maya	Instituto Tecnológico de Zitácuaro, Mexico
Hector Rodriguez Rangel	University of Oregon, USA
Alejandro Rosales	Tecnológico de Monterrey, Mexico
Christian Sánchez-Sánchez	Universidad Autónoma Metropolitana, Mexico
Ángel Serrano	Universidad Rey Juan Carlos, Spain
Shahnaz Shahbazova	Azerbaijan Technical University, Azerbaijan
Grigori Sidorov	CIC, Instituto Politécnico Nacional, Mexico
Juan Humberto Sossa Azuela	CIC, Instituto Politécnico Nacional, Mexico
Efstathios Stamatatos	University of the Aegean, Greece
Eric S. Tellez	CONACyT, Mexico; Infotec, Mexico
Esteban Tlelo-Cuautle	Instituto Nacional de Astrofísica, Óptica y Electrónica, Mexico
Nestor Velasco-Bermeo	Tecnológico de Monterrey Campus Estado de México, Mexico
Francisco Viveros Jiménez	Efinfo, Mexico
Carlos Mario Zapata Jaramillo	Universidad Nacional de Colombia, Colombia

Saúl Zapotecas Martínez   Universidad Autónoma Metropolitana – Cuajimalpa,
                                Mexico
Ramón Zatarain   Instituto Tecnológico de Culiacán, Mexico

Additional Reviewers

David Tinoco   Kazuhiro Takeuchi
Atsushi Ito   Rafael Rivera López
Ryo Otoguro   Adan Enrique Aguilar-Justo

Organizing Committee

Local Chairs

Luis Eduardo Falcón   Tecnológico de Monterrey, Campus Guadalajara
  Morales
Javier Mauricio Antelis   Tecnológico de Monterrey, Campus Guadalajara
  Ortíz

Logistics Chairs

Olga Cecilia García Rosique Tecnológico de Monterrey, Campus Guadalajara
Edgar Gerardo Salinas   Tecnológico de Monterrey, Campus Guadalajara
  Gurrión
Omar Alejandro Robledo   Tecnológico de Monterrey, Campus Guadalajara
  Galván

Finance Chair

Mónica González Frías   Tecnológico de Monterrey, Campus Guadalajara

Contents – Part II

Robotics and Computer Vision

Best Paper Award, Third Place:

Contents – Part I

Machine Learning

Fuzzy Logic and Uncertainty Management

Knowledge Representation, Reasoning, and Optimization

Coding 3D Connected Regions with F26 Chain Code

Osvaldo A. Tapia-Dueñas[1], Hermilo Sánchez-Cruz[1(✉)], Hiram H. López[2], and Humberto Sossa[3,4]

[1] Universidad Autónoma de Aguascalientes, Centro de Ciencias Básicas, Av. Universidad 940, 20131 Aguascalientes, Ags., Mexico
`black.osvo@gmail.com, hsanchez@correo.uaa.mx`
[2] Department of Mathematical Sciences, Clemson University, Martin Hall O-2, Clemson, SC 29634-0975, USA
`hlopezv@clemson.edu`
[3] Instituto Politécnico Nacional-CIC, Av. Juan de Dios Bátiz S/N, Gustavo a Madero, 07738 Mexico City, Mexico
`hsossa@cic.ipn.mx`
[4] Tecnológico de Monterrey, Campus Guadalajara, Av. Gral. Ramón Corona 2514, 45138 Zapopan, Jal., Mexico

Abstract. There are many applications in different fields, as diverse as computer graphics, medical imaging or pattern recognition for industries, where the use of three dimensional objects is needed. By the nature of these objects, it is very important to develop thrifty methods to represent, study and store them. In this paper, a new method to encode surfaces of three-dimensional objects that are not isomorphic to the plane is developed. In the proposed method, a helical path that covers the contour is obtained and then, the Freeman F26 chain code is used to encode the helical path. In order to solve geometric problems to find optimal paths between adjacent slices, a modification of the A star algorithm was carried out. Finally, our proposed method is applied to three-dimensional objects obtained from real data.

Keywords: Voxel-based objects · Chain code
Three-dimensional objects · Helical path

1 Introduction

Today, the representation and recognition of 3D objects is a very active field in computer vision. There is a great amount of applications that require 3D images to solve real-life problems, such as medical images, where 3D imaging plays an important role in supporting experts to provide more accurate diagnostics. There are also applications in the preservation of cultural heritage, games, mechanical construction, security and surveillance, computer-aided design (*i.e.*, CAD systems) and in general, in computer vision and pattern recognition. In the literature have appeared proposals to represent a 3D object through another

© Springer Nature Switzerland AG 2018
I. Batyrshin et al. (Eds.): MICAI 2018, LNAI 11289, pp. 3–14, 2018.
https://doi.org/10.1007/978-3-030-04497-8_1

of smaller dimension, which allows an analysis and recognition. In particular, skeletonization has been used for different reasons [1–3]. Although the skeletons maintain the topological properties of the underlying object and can give qualitative results, the original object losses geometric information [1–5].

Other descriptors notably used are the chain codes, which represent movements through the contour of the object. The way in which the contour is visited and the kind of movements that can be produced by different codes, have been exploited for representation and compression, and has attracted the attention of many researchers [6–9].

For the three-dimensional case, there are important proposals for the use and exploitation of chain codes but they have not been used as heavily as in the two-dimensions, being currently a very fertile field [10–12]. In the literature, coding of three-dimensional objects of surfaces that are isomorphic to the plane is presented [13], however, in this work we address the problem of coding surfaces that are not isomorphic to the plane, taking into account the different geometries that are presented, to solve the problem of finding the shortest path that allows to optimally encode the transition from one slice to another, of the given 3D object to encode.

This paper is organized as follows. In Sect. 2, we provide some definitions. In Sect. 3, we explain how to encode a 3D object in a helical path, whereas in Sect. 4 we describe the algorithm used to find the shortest path between one slice and another. In Sect. 5, the application of our method is presented. Finally, in Sect. 6 we give some conclusions and further work.

2   Definitions

In this section we give the most important concepts and definitions used to throughout the paper.

Definition 1. A *voxel*, v, is a resolution cell of a 3D grid with Cartesian coordinates $c(x, y, z)$ and an intensity value $I_v \in \{0, 1\}$. If $I_v = 0$, we say that the voxel is a 0-*voxel*; on the contrary, we say that it is a 1-*voxel*.

Definition 2. A voxel v_0 can share its faces, edges and/or vertices, depending on the 6, 12, 18 or 26-*neighborhood*, which are defined as follows: $N_6(v_0) = \{v | d_e(v_0, v) = 1\}$, $N_{12}(v_0) = \{v | d_e(v_0, v) = \sqrt{2}\}$, $N_{18}(v_0) = \{v | d_e(v_0, v) \leq \sqrt{2}\}$ and $N_{26}(v_0) = \{v | d_e(v_0, v) \leq \sqrt{3}\}$, where d_e is the Euclidean distance between v_0 and its neighbor voxel, v.

Definition 3. A 3D object is a connected component composed of 1-voxels, which is immersed in a 3D array of columns, rows and slices. Each slice is composed by zero or more connected regions: $\mathcal{R}_0^s, \mathcal{R}_1^s, \cdots \mathcal{R}_m^s$, where s refers to the s-th slice.

Definition 4. If c_1 are the coordinates of voxel v_1, c_2 the coordinates of voxel v_2, and $b = c_2 - c_1$, then v_1 is in the vicinity of v_2 if and only if $b \in \mathcal{B} = \{(i, j, k)\}$, with $i, j, k \in \{-1, 0, 1\} \setminus \{(0, 0, 0)\}$. The set $\mathcal{B}$ is called the *grid basis*.

Definition 5. A *path* is a sequence of adjacent ordered voxels, $\mathcal{P} = \{v_1, v_2, \ldots, v_p\}$, such that v_1 is adjacent to v_2, v_2 is adjacent to $v_3, \ldots, v_{p-1}$ is adjacent to v_p. The vector set $\mathcal{P_B} = \{b_1, b_2, \ldots, b_{p-1}\} \subset \mathcal{B}$ it is called *basis of path $\mathcal{P}$*.

Definition 6. We give symbols to each of the elements of the grid base $\mathcal{B}$ as follows: $a = (1,0,0), b = (1,1,0), c = (0,1,0), d = (-1,1,0), e = (-1,0,0), f = (-1,-1,0), g = (0,-1,0), h = (1,-1,0), i = (1,0,1), j = (1,1,1), k = (0,1,1)$
$l = (-1,1,1), m = (-1,0,1), n = (-1,-1,1), o = (0,-1,1), p = (1,-1,1),$
$\quad q = (1,0,-1), r = (1,1,-1), s = (0,1,-1), t = (-1,1,-1), u = (-1,0,-1),$
$\quad\quad v = (-1,-1,-1), w = (0,-1,-1), x = (1,-1,-1), y = (0,0,1), z = (0,0,-1),$
So, the alphabet we use is F26 $= \{a,b,c,d,e,f,g,h,i,j,k,l,m,n,o,p,q,r,s,t,u, v, w, x, y, z\}$, and the coding is obtained when each vector b_k is taken from $\mathcal{P_B} = \{b_1, b_2, \ldots, b_{n-1}\}$ associating its respective symbol in F26.

3 Helical Path to Encode 3D Objects

Helical coding allows the chain codes to save starting coordinates as much as possible, while the region in the current slice is a neighbor of the previous region, there is a path to go from n-th voxel $(v_n(s))$ to the first unvisited voxel $(v_1(s+1))$, which allows to recover the shape of the object without need to know in which coordinate we start to codify.

The helical coding is carried out in the following way:

1. Do $s \to 0$ and define Z as the number of total slices of the 3D grid.
2. While $s < Z + 1$.
3. Visit the first unvisited voxel, $v_1(s)$, of current slice. If it does not exist $s \to s + 1$, go to 2.
4. Encode the contour $\mathcal{R}_k^s$ and obtain $v_n(s)$.
5. Find$(v_1(s+1))$. If it does not exist $s \to 0$, go to 3.
6. $s \to s + 1$. Go to 2.

The Find$(\cdot)$ function introduced in the previous algorithm is implemented considering the following cases, which are generated by the geometry of the object. Figure 1 shows an object composed of two slices, where dark voxels in upper slice represent the contour of the region $\mathcal{R}_1^{s+1}$.

Case 1. $v_1(s+1) \in N_{26}(v_n(s))$ (Fig. 1a).
Case 2. It is possible to draw a discrete straight line from $v_n(s)$ to $v_1(s+1)$ as the shortest path (Fig. 1b).
Case 3. It is not possible to draw a discrete straight line from $v_n(s)$ to $v_1(s+1)$, since there are 0-voxels between them, for this reason the path can not be created, unless a concavity must be surrounded. This causes that the original $v_1(s+1)$ could change, since in the i-th step, a voxel of the contour of $\mathcal{R}_{k'}^{s+1}$ has the smallest d_e with respect to the last voxel of the current path (Fig. 1c).

Case 4. There are two candidates $v_1(s+1)$, because both have the same distance with respect to $v_n(s)$ (Fig. 1d).

We must validate which $\mathcal{R}_{k'}^{s+1}$ is a neighbor of $\mathcal{R}_k^s$, selecting the correct $v_1(s+1)$ we use A* to go from $v_n(s)$ to $v_1(s+1)$. To solve these cases, algorithm A* is detailed in Sect. 4.

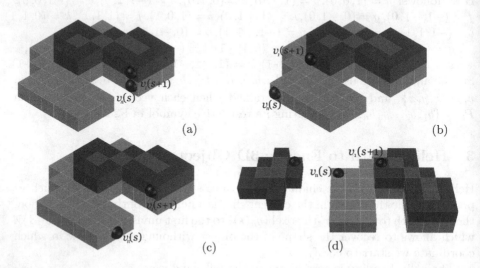

(a)   (b)

(c)   (d)

Fig. 1. Four different cases to consider for the starting and target points due to the geometry of the objects.

3.1 Used Symbols of F26

To codify our 3D object, the 26 symbols of F26 are not necessary. Part of the strategy, is to first visit each slice and encode its contours, so we can use eight different vectors for each visited region, $\mathcal{R}_k^s$, since the vectors point to one of the four faces of each voxel, plus four towards the edges. So, the symbols {a, b, c, d, e, f, g, h} are required. On the other hand, once the contour is visited, the next step is to move to the contiguous slice, i.e., from $\mathcal{R}_k^s$ to $\mathcal{R}_{k'}^{s+1}$, thus, we need vectors that point to any of its four edges or four vertices, plus one more symbol corresponding to the top face. So, the symbols {i, j, k, l, m, n, o, p, q} are used. Therefore, 17 symbols are required, at most.

4   Modifications to A* Algorithm

A* algorithm searches for the shortest path from an initial point to a target point [14]. This heuristic uses information relative to the place where the objective is located to select the next direction to be followed. The formula used to select the next point in the configuration space is: $f(v) = h(v) + g(v)$, where v is the current cell, $h(v)$ is the heuristic distance (Manhattan, Euclidean or Chebyshev) from v to the destination cell and $g(v)$ is the cumulative cost of moving from the initial state to the state v. Each adjacent point of the current one is evaluated by the formula $f(v)$. The point with the smallest value of $h(v)$ is selected as the next in the sequence [15].

The algorithm A* used for a grid configuration space is restricted to 8-connectivity. This means that we can find a path that is based on the connection between the closest cells. Due to the discretization of the plane, there may be zigzag movements to emulate the straight lines.

In the literature, the target cell does not change under any conditions. In this paper we find that the target point is the closest, and the starting cell is not in the same plane than the target cell. Taking into account these requirements, we modify the A* so that the target cell can be modified.

4.1 Conflict Zones for A*

Since we want to encode the shape of a three-dimensional object with a simple curve, *i.e.* collisions or repeated paths are not permitted, when looking for the shortest path, we avoid to go through 1-voxels that are part of the contour, since it was already visited. A drawback of doing this is that if a column or row only contains 1-voxels that are part of the contour, the A* is not able to find a viable path, so, we must validate this fact before using the algorithm. When this case is presented, we use one of two structuring elements, one to add 1-voxels to the column, and the other to add 1-voxels to the row.

Figure 2(a) shows that going from $v_n(s)$ to $v_1(s + 1)$, there is no path that satisfy our conditions. To solve this, we validate each column or/and row that exist between $v_n(s)$ and $v1(s + 1)$. If the row or column is one with less than three 1-voxels, the structuring element is used. Figure 2(b) shows the result of applying the structuring element.

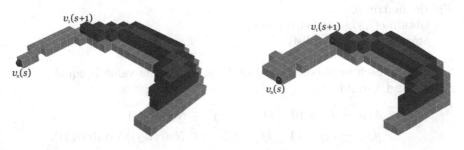

(a) Before using structuring elements   (b) After using structuring elements

Fig. 2. Example of how to use structuring elements in conflictive zone to use A*.

4.2   Pseudocode of Our A* Modified

For our proposals, we have used A* with adaptations to achieve the optimal path between $v_n(s)$ and $v_1(s+1)$. We introduce the matrices $\mathcal{M}$ and $\mathcal{E}$ composed of 1s and 0s to represent the regions, by following the next steps.

1. If there is a path to go from $v_n(s)$ to $v_1(s+1)$ in $\mathcal{R}_k^s$.
   $\mathcal{M} \to \mathcal{R}_k^s$.
   $\mathcal{E} \to \mathcal{R}_{k'}^{s+1}$.
2. else If there is a path to go from $v_n(s)$ to $v_1(s+1)$ in $\mathcal{R}_{k'}^{s+1}$.
   $\mathcal{M} \to \mathcal{R}_{k'}^{s+1}$.
   $\mathcal{E} \to \mathcal{R}_k^s$.
3. Fill the matrix h.
   Fill the matrix h with $\times$ if $c(i,j)$ of $\mathcal{M}$ is 1, which represents a 1-voxel of the contour, label with ∞ any other case. We fill the matrix g in the same way as h.
   Consider $c(i,j)$ of $v_n(s)$, and do $d_{ij} \to 0$.
   Store $c(i,j)$ in open list.
   Do
   > For each neighbor $c(i',j')$ of $N_8(c(i,j))$, if its value is equal ∞ in h and 1 in $\mathcal{M}$, calculate the Manhattan distance as following:

   $$d_{i'j'} \to d_{ij}+1 \quad if \quad c(i',j') \in N_4(c(i,j))$$
   $$d_{i'j'} \to d_{ij}+2 \quad if \quad c(i',j') \in N_8 c(i,j) \setminus N_4(c(i,j)) \tag{1}$$

   > Each $c(i',j')$ is stored in the open list.
   > Search in open list $c(i',j')$, such that $d_{i',j'}$ is the smallest and remove $c(i',j')$ from the open list and do the Eq. (2).

   $$d_{ij} \to d_{i'j'}$$
   $$c(i,j) \to c(i',j'). \tag{2}$$

   While open list is not empty.
4. Fill the matrix g.
   Obtain $c(i,j) \to v_n(s), d_{ij} \to 0$.
   Store $c(i,j)$ in open list.
   Do
   > For each neighbor $c(i',j')$ of $N_8(c(i,j))$, if its value is equal ∞ in h and 1 in $\mathcal{M}$, do:

   $$d_{i'j'} \to d_{ij}+10 \quad if \quad c(i',j') \in N_4(c(i,j))$$
   $$d_{i'j'} \to d_{ij}+14 \quad if \quad c(i',j') \in N_8(c(i,j)) \setminus N_4(c(i,j)). \tag{3}$$

   > Each $c(i',j')$ is stored in the open list.
   > Search in open list $c(i',j')$, such that the sum $d_{i'j'}$ in $h + d_{i'j'}$ in g is the smallest and remove $c(i',j')$ from the open list.
   > Obtain the Euclidean distance of 1-voxel from the contour of $\mathcal{E}$ to d_{ij}. The one with the smallest distance is now $v_1(s+1)$.

While $v_1(s + 1) \notin N_8(c(i, j))$.

5. Obtain the shortest path.

   Obtain the $c(i, j)$ of $v_1(i + 1)$.

   While $v_1(s + 1) \notin N_8(c(i, j))$

       Add $c(i, j)$ to the shortest path.

       Search $c(i', j') \in N_8(c(i, j))$, such that its $d_{i'j'}$ is the smallest in g.

       Assign $c(i, j) \rightarrow c(i', j')$.

To illustrate an example, consider the need to go from $v_n(s)$ to $v_1(s + 1)$ through an optimal path, Fig. 1(c) is used for this purpose. To achieve this, let $\mathcal{R}_k^s$ and $\mathcal{R}_{k'}^{s+1}$ the two contiguous regions. If there is a path to go from $v_n(s)$ to $v_1(s + 1)$ in $\mathcal{R}_k$, we associate $\mathcal{M}$ with $\mathcal{R}_k^s$ and $\mathcal{E}$ with $\mathcal{R}_{k'}^{s+1}$ (Fig. 3(a) and (b), respectively).

Following our modified A* method, the first step is to fill the matrix h with '×' if $c(i, j)$ of $\mathcal{M}$ is '1', which represents a 1-voxel of the contour, and we fill with ∞ in any other case (see Fig. 3(c)). We fill the matrix g in the same way as h (see Fig. 4).

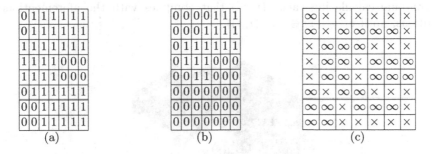

Fig. 3. Matrices (a) $\mathcal{M}$, (b) $\mathcal{E}$, and (c) h, respectively.

Next step is to assign a number greater than or equal to zero to each $c(i, j)$, such that the value of $c(i, j)$ is equal to ∞ in h and '1' in $\mathcal{M}$. This number is the Manhattan distance. We start by making $d_{ij} = 0$ in (i, j) of h, where $c(i, j)$ is the coordinate of $v_n(s)$.

The Manhattan distance is calculated as described in the algorithm of modified A* (Eq. (1)).

The coordinates $c(i', j')$ are stored in the open list to remember the cell to visit their neighbors. We search in this list the coordinates $c(i', j')$, such that $d_{i'j'}$ is the shortest. We remove $c(i', j')$ from the list and solve the assignations of Eq. (2).

We continue with the calculation of the Manhattan distance of each $c(i', j')$ in the open list, until it is empty. The matrix h becomes as in Fig. 5.

If the value of $c(i, j)$ is ∞ in g and '1' in $\mathcal{M}$, we subsequently assign to each $c(i, j)$ a number greater than or equal to zero. We start by making $d_{ij} = 0$ in the cell (i, j) of g, where $c(i, j)$ is the coordinate of $v_n(s)$. Each number is calculated by Eq. (3).

∞	×	×	×	×	×	×
∞	×	∞	∞	∞	∞	×
×	∞	∞	∞	×	×	×
×	∞	∞	×	∞	∞	∞
×	∞	∞	×	∞	∞	∞
∞	×	∞	34	×	×	×
∞	∞	×	30	20	10	0
∞	∞	×	×	×	×	×

Fig. 4. Matrix g

∞	×	×	×	×	×	×
∞	×	9	10	11	12	13
×	9	8	9	×	×	14
×	8	7	×	∞	∞	∞
×	7	6	×	∞	∞	∞
∞	×	5	4	×	×	×
∞	∞	×	3	2	1	0
∞	∞	×	×	×	×	×

Fig. 5. Matrix h.

The coordinates $c(i', j')$ are stored in the open list to remember the coordinates of the square in which we need to visit its neighbors. We search in this list the coordinates $c(i', j')$, such that the sum of $d_{i'j'}$ in h plus $d_{i'j'}$ in g is the smallest, we remove $c(i', j')$ from the list and solve the Eq. (3).

We use $\mathcal{E}$ to obtain the Euclidean distance of each 1-voxel from the contour of $\mathcal{R}_k^s$ to d_{ij}. The one with the smallest Euclidean distance is now $v_1(s+1)$. We continue calculating each $c(i', j')$ that complies with the aforementioned conditions, until $v_1(s+1) \in N_8(c(i, j))$.

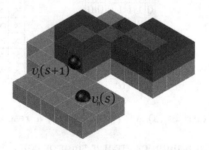

Fig. 6. Final position of $v_1(s+1)$.

On this example, it is true that when surrounding the concavity, $v_1(s+1)$ changes (Fig. 6), because there is a voxel $(v_p(s+1))$ in the contour represented in $\mathcal{E}$, which is the closest to the current voxel $(v_0(s))$. This causes $v_p(s+1)$ to be $v_1(s+1)$. As in the subsequent steps, there is no other $v_p(s+1)$ that is closer to $v_0(s)$, whereas $v_1(s+1)$ does not change until the algorithm ends.

The last step is to find the shortest path between $v_n(s)$ and $v_1(s+1)$. Assign $c(i, j) \rightarrow v_1(s+1)$. While $v_n(s) \notin N_8(c(i, j))$, we do: store $c(i, j)$ in the short path list. Search $c(i', j') \in N_8(c(i, j))$, such that its $d_{i'j'}$ is the smallest in g. Assign $c(i, j) \rightarrow c(i', j')$.

Thus, in our example the optimal path from $v_n(s)$ to the definite target found, $v_1(s+1)$, is $\mathcal{P} = \{(7, 7, 1), (7, 6, 1), (7, 5, 1), (6, 4, 1), (5, 4, 2)\}$, where $(7,7,1)$ and $(5,4,2)$ are the coordinates of $v_n(s)$ and $v_1(s+1)$, respectively.

Finally, the chain code of the helical path representing the two contours is:
$S = 771ceefeffgghgaaaaacceedcbaadefpefgahhaacceed.$

5   Experiments

To test our method, we found helical paths to a sample of 3D objects. We used surfaces directly from the voxelization implicit in a data set. These models were previously treated to simplify and eliminate noise in the data.

The models used come from different sources and helped us to test our methodology. We accessed the Stanford Computer Graphics Laboratory site: http://graphics.stanford.edu/data and the Suggestive Contour Gallery site: http://gfx.cs.princeton.edu/proj/sugcon/models.

Figure 7 presents the results of applying our method to obtain the helical path that describes the surface of the contour of each object. The figures show

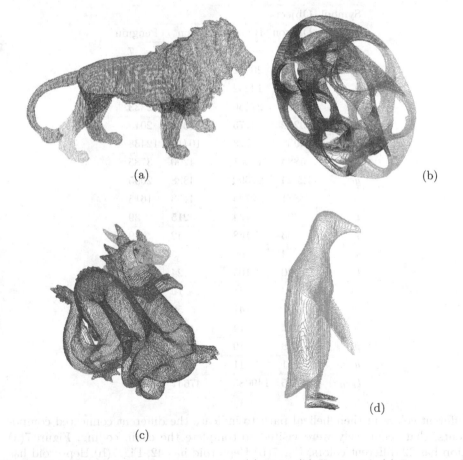

(a)

(b)

(c)

(d)

Fig. 7. Helical path found in a sample test: (a) Lion, (b) Heptoroid, (c) Dragon and (d) Penguin (Color figure online)

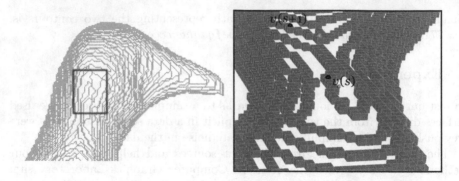

Fig. 8. On the left, the helical path of the penguin head object is shown. On the right, case 3 is exemplified with the contour voxel representation.

Table 1. Frequency of F26 symbols

Symbol	Objects			
	Dragon	Heptoroid	Lion	Penguin
cc	39	42	22	7
a	15934	20912	10414	13440
b	7332	13232	4165	3204
c	8895	27156	5079	2333
d	9252	13175	4003	2015
e	14939	20723	10419	12438
f	5884	13001	4200	3785
g	12524	27392	4828	2095
h	6871	12724	4153	1615
i	301	734	215	39
j	93	168	42	9
k	4	66	5	5
l	30	105	24	27
m	13	21	10	36
n	47	141	50	24
o	17	14	4	0
p	48	110	33	44
q	31	11	3	17
l_{F26}	82215	149685	47647	41126

different colors in their helical path to indicate the different connected components, that recursively were visited to complete the chain coding. Figure 7(a) Lion has 22 different colors, Fig. 7(b) Heptoroid has 42, Fig. 7(b) Heptoroid has 39 and Penguin just has 7.

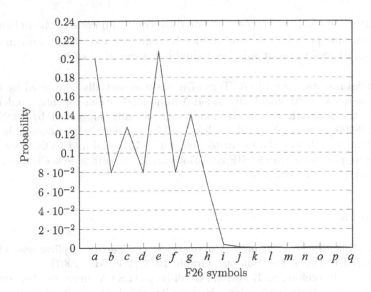

Fig. 9. Probability of occurrence of each symbol.

In Fig. 8 (left) is shown the helical path of the object called Penguin, where Case 3 of our modified A* is exemplified. In this case it is necessary to surround the concavity that exists between $v_n(s)$ and $v_1(s+1)$. This is presented in Fig. 8 (right). This representation only contains the 1-voxels that are part of the contours of coded regions.

On the other hand, Table 1, presents frequencies of the code symbols F26, as well as the length of the chain code (l_{F26}) and the number of connected components (cc) that each object has. The heptoroid object has the longest chain code, while the penguin object has the smallest. As can be seen in the Fig. 9, the presence of the first eight symbols of F26 is much greater than the rest, and of them, symbols a and b are those that appear most likely, followed by g and c. This type of information and analysis is important for dealing with recognition and compression problems.

6   Conclusions and Further Work

We have presented a new method to encode surfaces of voxel-based objects that are not isomorphic to the plane, by means of helical paths. One of the advantages of our work is that the length of the chain code is optimized by visiting the centers of the voxels that make up the contours, as well as finding the shortest trajectories between the slices, preserving the topological and geometric properties. Several applications may result from the work presented here. Information and analysis of the data obtained by the probability of the appearance of the symbols is an important topic to address problems of recognition and compression. So, a future work is to exploit this type of codification for such purposes. As another future work, it is suggested to analyze the chain code of the helical path, to find dominant points, and thus reduce to a greater extent the information of the shape-of-object without missing valuable information of it.

Acknowledgements. Osvaldo A. Tapia-Dueñas was partially supported by CONA-CyT. H. Sánchez-Cruz thanks Universidad Autónoma de Aguascalientes, under Grant PII18-8 for the support. Hiram H. López was partially supported by CONACyT, CVU no. 268999, project "Network Codes", and by Universidad Autónoma de Aguas-calientes. H. Sossa thanks the Instituto Politécnico Nacional and CONACyT for the economical support under funds: SIP 20180730 and 65 (Fronteras de la Ciencia), respectively to undertake this research.

References

1. Cornea, N.D., Silver, D., Min, P.: Curve-skeleton properties, applications, and algorithms. IEEE Trans. Vis. Comput. Graph. **13**(3), 530–548 (2007)
2. Punam, K., Borgefors, S., Borgefors, G., di Baja, G.S.: A survey on skeletonization algorithms and their applications. Pattern Recogn. Lett. **76**, 3–12 (2016). Special Issue on Skeletonization and its Application
3. Jin, D., Iyer, K.S., Chen, C., Hoffman, E.A., Saha, P.K.: A robust and efficient curve skeletonization algorithm for tree-like objects using minimum cost paths. Pattern Recogn. Lett. **76**(C), 32–40 (2016)
4. Svensson, S., Nyström, I., di Baja, G.S.: Curve skeletonization of surface-like objects in 3D images guided by voxel classification. Pattern Recogn. Lett. **23**(12), 1419–1426 (2002)
5. Arcelli, C., di Baja, G.S., Serino, L.: Distance-driven skeletonization in voxel images. IEEE Trans. Pattern Anal. Mach. Intell. **33**(4), 709–720 (2011)
6. Sánchez-Cruz, H., Rodríguez-Dagnino, R.M.: Compressing bilevel images by means of a three-bit chain code. Opt. Eng. **44**, 44–44–8 (2005)
7. Yong, K.L., Alik, B.: An efficient chain code with Huffman coding. Pattern Recogn. **38**(4), 553–557 (2005)
8. Echávarri, L., Aguinaga, R., Neri-Calderón, A., Rodriguez-Dagnino, R.M.: Compression rates comparison of entropy coding for three-bit chain codes of bilevel images. Opt. Eng. **46**, 46–46–7 (2007)
9. Yong, K.L., Wei, W., Peng, J.W., Alik, B.: Compressed vertex chain codes. Pattern Recogn. **40**(11), 2908–2913 (2007)
10. Freeman, H.: Computer processing of line-drawing images. ACM Comput. Surv. **6**(1), 57–97 (1974)
11. Bribiesca, E.: A chain code for representing 3D curves. Pattern Recogn. **33**(5), 755–765 (2000)
12. Sánchez-Cruz, H., López-Valdez, H., Cuevas, F.J.: A new relative chain code in 3D. Pattern Recogn. **47**(2), 769–788 (2014)
13. Salazar, J.M., Bribiesca, E.: Compression of three-dimensional surfaces by means of chain coding. Opt. Eng. **54**, 54–54–12 (2015)
14. Cui, S.G., Wang, H., Yang, L.: A simulation study of A-star algorithm for robot path planning, pp. 506–509, January 2012
15. Duchó, F., et al.: Path planning with modified a star algorithm for a mobile robot. Proc. Eng. **96**, 59–69 (2014). Modelling of Mechanical and Mechatronic Systems

Finding Optimal Farming Practices to Increase Crop Yield Through Global-Best Harmony Search and Predictive Models, a Data-Driven Approach

Hugo Dorado[1,2], Sylvain Delerce[2], Daniel Jimenez[2],
and Carlos Cobos[1(⊠)]

[1] Information Technology Research Group (GTI), Universidad del Cauca,
Sector Tulcán Office 422 FIET, Popayán, Colombia
ccobos@unicauca.edu.co
[2] International Center for Tropical Agriculture (CIAT),
Km 17 Recta Cali-Palmira, Apartado Aéreo 6713, Cali 763537, Colombia
{h.a.dorado, s.delerce, d.jimenez}@cgiar.org

Abstract. Increasing crops' yields to meet the world's demand for food is one of the great challenges of these times. To achieve this, farmers must make the best decisions based on the resources available for them. In this paper, we propose the use of Global-best Harmony Search (GHS) to find the optimal farming practices and increase the yields according to the local climate and soil characteristics, following the principles of site-specific agriculture. We propose to build an aptitude function based on a random forest model trained on farms' data combined with open data sources for climate and soil. The result is an optimizer that uses a data-driven approach and generates information on the optimized farming practices, allowing the farmer to harness the full potential of his land. The approach was tested on a case-study on maize in the state of Chiapas, Mexico, where the adoption of the practices suggested by our approach was estimated to increase average yield by 1.7 ton/ha, contributing to closing the yield gap. The proposal has the potential to be scaled to other locations, other response variables and other crops.

Keywords: Global-best harmony search · Machine learning · Open data
Data-driven agronomy · Optimization

1 Introduction

Sub-optimal management on farms results in yield gaps that eventually affect the global food production capacity and efficiency. These have been extensively documented and closing them is today a priority for agriculture to meet both its production and climate change mitigation goals [1–5].

Decision-making in agriculture has traditionally been based on blanket recommendations made by technicians, local knowledge or practices that are adopted as customs for generations and in many cases without detailed knowledge on the effect of

© Springer Nature Switzerland AG 2018
I. Batyrshin et al. (Eds.): MICAI 2018, LNAI 11289, pp. 15–29, 2018.
https://doi.org/10.1007/978-3-030-04497-8_2

climate, or on the nutrition requirements of the plant according to biophysical characteristics [6]. Simultaneously, generating information that helps to make site-specific decisions based on the traditional agronomical research can lead to years of experimentation [7].

A recent study conducted in 83 developing countries showed that small farms can account for more than 380 million agricultural households and produce approximately 70% of the food calories for these regions [8]. For this reason, support in implementing optimal farming practices for small-scale farmers is of great importance [9, 10]. However, due to conditions of access to farms, costs and even availability of experts, such support as technical assistance may end up covering only a small number of farmers, leaving a huge proportion of farmers with little to no support.

This generates a scenario in which there is great need to find easily accessible tools that provide support for decision making in agriculture, considering the biophysical conditions of each farmer's land and even their capacity to apply certain practices, due to economic constraints or access.

In [11], Cock et al. put forward a methodology based on data for decision making in agriculture that starts from a different approach than the traditional experiments-based research in agriculture, giving a central value to the commercial information of the producers themselves. In this approach, each harvest is considered a unique cropping event and the compilation of many cropping events enables analyses that eventually provide information for making more accurate decisions in the field. Later in [12–14], the same approach was used to analyze Andean Blackberry, lulo and plantain production systems, involving machine learning techniques such as supervised and unsupervised algorithms (Multi-Layer Perceptron, random forest, cluster and mixed models). The factors that most affected the yields of each crop were detected, and recommendations specifically tailored to each farms' climate and soil conditions were generated. In [15], Delerce et al. went on to implement a similar approach for the analysis of different varieties of rice. In that case, Conditional Inference Forest [16] was used to detect the most relevant climatic variables affecting rice yield. Using partial dependencies, the relationships between each predictor and the response variable were evaluated. Based on this, the optimum sowing periods were identified.

Up until this point, however, the machine learning models applied on observational data have mainly been used for interpretation and to find patterns and relationships between the variables. They have not yet been used for optimization, which can also help producers to find their optimal farming practices and make the most of their crops.

We find in literature metaheuristic algorithms capable of finding good solutions to complex optimization problems such as vehicle traffic, transportation, distribution of crew tasks in space flights, and construction of piping systems [17, 18]. These methods are inspired by the behavior of living organisms and their interaction with each other, among others. Metaheuristics usually follow a process in which they start with a random initial population of potential solutions. Then a fitness function is used to gauge the goodness of the solutions. Simultaneously operators are applied - for example, selection, crossover, and mutation - to obtain a new population where it is expected to keep the most promising solutions. The process is repeated until the algorithm reaches the stop criterion or the solution found meets certain criteria that allows it to be accepted or considered sufficiently close to the optimal solution [19].

Some common metaheuristics are Genetic Algorithms [20], Particle Swarm Optimization (PSO) [21, 22], Harmony Search (HS) [23], among others.

HS is a metaheuristic inspired by the process of musical improvisation, in which a musician tries different notes looking to play a harmony. HS has a number of advantages over other optimization algorithms, such as the fact that it does not require complex calculations, it avoids local optima solutions, and it can handle discrete and continuous variables [23]. The algorithm was subsequently the subject of a series of improvements. One of these was implemented by Madhavi et al. [24], with a version of the algorithm called improved harmony search (IHS), where the dynamic adjustment of two parameters was included (pitch adjustment rate and bandwidth), achieving improvements in precision and speed of convergence. In [25] Omran et al. then proposed a new variation called Global-best Harmony Search (GHS), which modifies the pitch adjustment step of HS so that the new harmony can mimic the best harmony in the memory. This concept was inspired by the PSO algorithm. The experiments carried out showed that GHS is better than IHS and HS in problems of great dimensionality and in the presence of noise. GHS is also more efficient when facing both discrete and continuous problems [25].

In this study we propose to use GHS to find an optimal solution to the combination of practices a farmer can implement according to its soil and climate conditions (site-specific), through a data-driven approach in which the aptitude (fitness) function corresponds to an empirical model that predicts crop's performance (yield). The model is built using historical data and machine learning models. The results of GHS implementation may enable the generation of personalized site-specific recommendations to assist farmers in their farming practices optimization.

The rest of the document is organized as follows: in Sect. 2 we detail the data sources, the comparison to select the best machine learning model to predict the yield, and the GHS algorithm and its representation of a solution (harmony) to the specific problem of maize crop optimization. Section 3 shows the preprocessing performed on the data, the results obtained with the machine learning models, the results of the GHS parameter tuning process and finally the results of the optimization process in maize crops, together with the discussion of the results. Finally, Sects. 4 and 5 present the conclusions and recommendations for future work in the area.

2 Materials and Methods

2.1 Data

This pilot study is based on commercial information of maize crops collected by CIMMYT's (International Center for the Improvement of Maize and Wheat) MasAgro project in the State of Chiapas in Mexico. The data used to test our method comes from a platform known as BEM (Spanish acronym for MASAGRO Electronic Field book). This platform was an initiative of CIMMYT and enables the recording of crop management and yield data from commercial fields and farms[1]. We focused on the maize

---

[1] http://conservacion.cimmyt.org/es/hubs/683 - Bitácora Electrónica MasAgro.

records for Chiapas over the 2012–2016 period. These data contain information on farming practices, geolocation, and yield. The observational unit represents one cropping event, from pre-sowing to harvest [11].

To obtain weather data in the State of Chiapas, we used the climate module of the National Institute of Forestry, Agriculture and Livestock Research (INIFAP)[2], which allows to download weather stations data and corresponds to an open data initiative of the Mexican Government. We used a similar protocol as in [14] to link weather data to the cropping events based on the coordinates and crop growth period. This assignment process consists in evaluating differences in elevation and distance between stations and fields to make an adequate match. We then calculate a series of climatic indicators given by averages, accumulated means, and standard deviations (see **Appendix 1**).

We obtained soil data from the National Institute of Statistics and Geography (INEGI) (Which is another open source data) with information on 8,526 soil sampling points throughout Mexico and a soil classification of more than 23,013 textural classes[3]. Data were combined and compiled into new layers with soil functional proprieties using Quantum GIS[4]. The association of cropping events with soil data was then performed based on coordinates.

2.2   Training Machine Learning Models

Prior to the optimization process, a fitness function was defined that predicts the maize yield (y) (see Eq. 1). Predictors related to climate $(w_1, w_2, ...)$, soil $(s_1, s_2, ...)$ and farming practices $(m_1, m_2, ...)$ were used. A description of these variables can be found in **Appendix 1**. The relationships between the predictors and the output variable are maximized by a supervised learning model (g).

$$y \approx g(w_1, w_2, \ldots, s_1, s_2, \ldots, m_1, m_2, \ldots) \tag{1}$$

Regarding the choice of the best model g, four options were tested, starting with the simplest: a Multiple Linear regression (LM), a method that seeks to maximize the linear relationships between the set of predictors and the output variable by ordinary least squares. As the second option, a non-linear regression called Multi-Layer Perceptron (MLP) was implemented [26], which comes from the use of an algorithm inspired by the way the brain learns. It is based on a network of individual units called perceptrons, which interact from weighted connections. In the training, the data is mobilized through several layers, normally 3 (input, hidden, and output), in a sequence of iterations (epochs) until reaching a limit number or convergence. Then in the process of classification/regression of a new record, the data is entered in the input layer and the result of the output layer is the response of the classifier [27].

Next, the model proposed by Breiman [28], Random Forest (RF) was also implemented. RF uses the bagging technique to combine several classification or

---

[2] http://clima.inifap.gob.mx/lnmysr/Estaciones.

[3] http://www.inegi.org.mx/geo/contenidos/recnat/edafologia/default.aspx.

[4] https://qgis.org/en/site/index.html.

regression trees, where each tree is built with observations and random predictors from the original dataset.

Also featured in the experiment was one of the improvements applied to RF, a model known as Conditional inference Forest (CF). This model uses conditional trees [29], which improves an aspect of the RF when there are presence of predictors that contain several categories, this because RF tends to highlight these predictors in the importance of variables [30].

Finally, an ensemble model (EM) [31, 32] was used, in which the predictions of the trained models were mixed by means of a greedy optimization algorithm. RF, CF and MLP were included in the ensemble. LM was not included, because it uses different rules to transform the categorical variables.

A comparison of the learning models was performed using a partition of 60% of observations for training and 40% for validation. We evaluated the goodness of the prediction based on the Root Mean Squared Error (RMSE) and the Coefficient of Determination (R^2) calculated between the predicted and real values of the test data set.

Models' parameters were optimized through a cross validation with 5 folders and 3 repetitions in the training process. The mtry parameter (the number of variables randomly sampled as candidates at each split), with possible values among 2, 19, and 37, was tested for RF and CF, and in the case of MLP, it was the number of neurons in the hidden layer (2, 18, 40). The values were selected by fixing in 3 the amount of granularity in the tuning parameters grid.

2.3 Global-Best Harmony Search

For the optimization process, the GHS algorithm was used. GHS arose from a modification made to the harmony search (HS) algorithm [23]. In the process of musical improvisation that inspired HS, musicians seek perfect harmonies based on the following actions: (1) play a familiar harmony they have learned, (2) play something like the previous harmony by adjusting it slightly to the desired pitch, and (3) compose a harmony based on his knowledge using random keys. These actions are the basis of HS and are translated into three rules: (1) use of the harmony memory, (2) pitch adjustment, and (3) randomness. HS has several advantages given its computational simplicity and few parameters [24], and has also been used in numerous applications [33–35].

GHS is a variation applied to HS in which the PSO concept is included by modifying the HS pitch adjustment step. This change consists in allowing the new harmony to imitate the best harmony in the harmony memory, reducing HS parameters (bandwidth) and adding a social component [25, 36]. The great advantage of this variation, with respect to the HS algorithm and even other metaheuristics, is that it makes it possible to work efficiently in binary, discrete and continuous problems. GHS also works efficiently when the solution representation includes a mixture of different data types, which adapts perfectly to the farming practices optimization, because we have discrete and continuous variables at the same time.

In the adaptation of the GHS algorithm to the performance predictive model, we split the dataset into two subsets of variables, some that are part of the optimization process and others that remain fixed with their current values. This is because the farmer must adapt to uncontrollable factors, such as climate and soil, and even for economic or practical reasons he may prefer to maintain some farming practices. In this case, it is assumed that farmers are willing to modify any of their farming practices and, as such, only the variables corresponding to climate and soil are fixed (see Fig. 1).

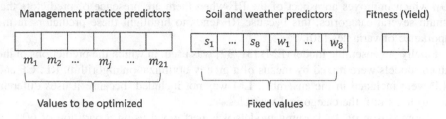

Fig. 1. Representation of a harmony for the GHS algorithm

The fitness function is equivalent to the function shown in the Eq. 1. The constraints are: (1) for each event, the soil and weather variables are statics in the GHS optimization, and only the practices are tuned, (2) the possible value for each continuous predictor should be assigned between the minimum and maximum values in the complete dataset, and (3) the possible values for each categorical predictor should be assign among all possible categories from the original dataset for each variable.

To perform the tuning of the GHS parameters, all the combinations of the values shown in Table 1 were explored, for a total of 27 combinations. The number of evaluations of the objective function (Evaluations of the Fitness Objectives, EFOs) for each GHS was fixed at 3,000 being this a compromise between computational load and the quality of fitness. Additionally a total of 30 repetitions of each experiment was performed, seeking to take the average and make the decision based on the best average of all configurations (based on the central limit theorem). The initial values for the optimized predictors were assigned randomly on the constraint space explained above. The parameter test shown in the results was applied with only data from one cropping event in the dataset. With the chosen parameters, these were then generalized for the other cropping events.

Table 1. Parameters for GHS (Recommended values from the literature (CITE))

Description	Parameter	Variation
hms	Harmony memory size	5, 10, 15
hmcr	Harmony memory considering rate	0.85, 0.9, 0.95
Par	Pitch adjustment rate	0.3, 0.35, 0.4

3 Experiments and Results

3.1 Data Processing

The initial dataset has a total of 383 observations and 48 predictor variables, of which two are related to latitude and longitude and are not considered in the final model, since both were used for the climatic and the soils assignment. Moreover, the variable variety originally had many categories because farmers use many different maize cultivars. After talking with local experts, we decided to group them into four categories (Dekalb, Criollo, P4082, Others).

The variability of the rest of the predictors (input variables) was evaluated using preprocessing rules available in the R package caret [37]: the first rule was to detect if the proportion of different values is greater than 10%. If not, the variable was removed. The second rule was: if the most prevalent value with respect to the second most frequent value is above 95/5 (frequency ratio), the variable was removed. After applying both rules, 10 predictors were discarded.

Correlation between quantitative predictors was also considered to avoid redundancy between the variables. For each pair of quantitative predictors, the Pearson correlation coefficient was calculated and, if it exceeded 0.7, only one remained. The rule for selecting which of the two should stay is based on the average of the absolute correlation that each of the predictors has with the others, maintaining the predictor with the lowest absolute correlation. At the end of this step a total of seven predictors were discarded.

Finally, a dummy transformation was implemented on the qualitative variables so that each category becomes a new binary variable, where one represents presence and zero represents absence. The table shown in **Appendix 1** describes the final set of variables where we can see that some categorical variables with more than two categories produced new variables after being transformed to binary (to convert a categorical variable into dummy variables). As a result the final dataset contains 37 predictors.

3.2 Machine Learning Models

Table 2 presents the results obtained with the test data set for each of the models. The values corresponding to the parameters that result from the tuning process carried out with cross-validation are also shown. The results show that the best fit for RF and CF was 19 variables available for splitting at each tree node (mtry = 19) and for Multi-Layer Perceptron, three neurons in the hidden layer (size = 18).

Table 2. Machine learning models for predicting yield

Ranking	Model	Parameters	RMSE	R^2
1	Random forest (RF)	mtry = 19	**0.8702**	**0.7417**
2	Ensemble (RF, CF, MLP)		0.8800	0.7370
3	Conditional inference forest (CF)	mtry = 19	0.9958	0.6584
4	Linear multiple regression (LM)		1.1314	0.5763
5	Multi-layer perceptron (MLP)	size = 18	1.1702	0.6075

The best model found was RF with a lower RMSE and higher R^2. The results of RF and the ensemble model were very close. MLP was the worst performer, even surpassed in RMSE by LM. The RF model with mtry = 19 was finally selected as the fitness function that will subsequently guide the process of GHS algorithm.

3.3   Optimization Process Based on GHS

Figure 2 shows the result of the 27 combinations of GHS parameters used on the optimization of a cropping event. The execution that used a harmony memory size (hms) of 5, a harmony memory consideration rate (hmcr) of 0.85 and a pitch adjustment rate (par) of 0.3 had the best combination of parameters to achieve the best solution in terms of yield in ton/ha. However, the difference was small with other combinations of parameters.

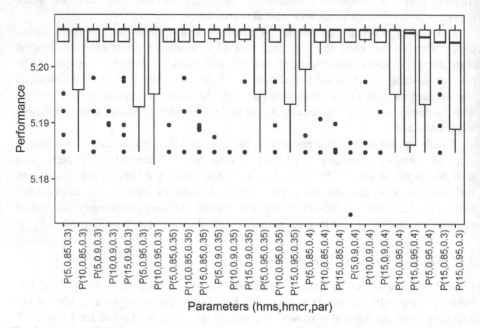

Fig. 2. Tuning of GHS parameters for the model that predicts performance using random forest as a fitness function on one cropping event. Performance values are ordered decreasingly by median

To evaluate the effect of the implementation of GHS optimized farming practices, the random forest predictive model was run for all the cropping events of the dataset, first maintaining the current practices without optimization. We obtained an estimated yield for each observation (baseline). Then the model was run again on the same cropping events but with GHS optimized farming practices. In this case, for simplicity, the same parameters were maintained for all, considering the results suggested in Fig. 2 (*hms = 5, hmcr = 0.85 y par = 0.3*).

Figure 3 shows that once the farming practices variables are optimized, average expected yield can be 1.77 ton/ha higher than with current management. The yield variability is also reduced among the group. This is because this optimization allows farmers to use all the potential of their environmental conditions, by implementing the optimal farming practices adapted to their specific conditions.

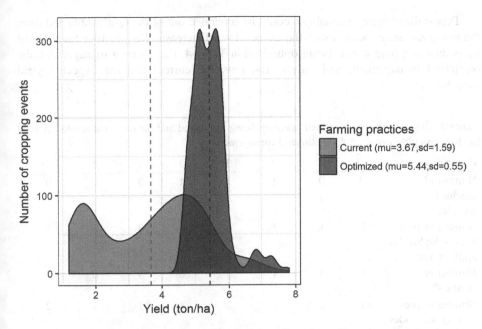

Fig. 3. Comparisons for maize yield predicted using random forest for current farming practices and productivity following the optimization of farming practices

Table 3 summarizes the results of the GHS optimization. The comparison shows that if farmers adopt the optimized farming practices, yield improvements of up to 4.28 ton/ha could be achieved, and 75% of cropping events would be expected to increase their yield by at least 0.71 ton/ha. The minimum improvement is 0.012 ton/ha. It relates to cropping events with already high yields in which cases GHS cannot deliver much improvement. The Student's t-Test suggest that the difference between the current yield and the optimized yield is significant at 0.05 level.

Table 3. Summary of the differences between current predicted yield (ton/ha) and optimized yield (ton/ha)

	Current	Optimized	Difference
Min.	1.186	4.578	0.012
1st qu.	1.881	5.076	0.711
Median	3.963	5.377	1.323
Mean	3.666	5.436	1.770
3rd qu.	4.884	5.708	3.039
Max.	7.733	7.811	4.280
Test t-student t = 20.7, degrees of freedom = 474.7 (P < 0.05)			

Personalized recommendations could be generated out of the results obtained from the optimization, associated with the expected yield increase. We give four examples of individual cropping events being optimized in Table 4, their current management, the optimized management, and on the last row, the current and the expected yield increase.

Table 4. Detail of the management practices being optimized and the expected results on yield for four cropping events. * for optimized management

Cropping event	1	1*	2	2*	3	3*	4	4*
Number of mechanical weeding	0	0	1	1	1	0	0	0
Number of post-harvest herbicides applications	0	1	1	1	0	0	0	1
Number of "rastreo"	0	3	1	3	0	3	0	3
Number of pre-sowing herbicides applications	1	2	1	2	1	2	1	2
Number of fertilizations	2	3	2	3	1	1	2	3
Number of applications of foliar fertilizers	0	0	1	0	0	2	0	2
Number of applications of bio fertilizers	0	0	0	0	1	0	1	0
Number of post-sowing herbicides applications	1	3	1	2	1	2	2	3
Number of applications of insecticides	1	3	1	3	1	3	1	3
Total amount of nitrogen applied	55.0	182.8	156.1	157.2	24.6	171.8	117.8	178.4
Total amount of phosphorus applied	23.0	106.4	46.0	170.5	0.0	162.9	0.0	92.0
Total amount of potassium applied	0.0	46.9	30.0	81.7	0.0	39.7	0.0	74.1
Cultivars' group criollo	1	0	0	0	0	0	0	0
Cultivars' group Dekalb	0	1	1	1	0	1	0	0

(*continued*)

Table 4. (*continued*)

Cropping event	1	1*	2	2*	3	3*	4	4*
Cultivars' group others	0	0	0	0	1	0	1	0
Cultivars' group P4082 W	0	0	0	0	0	0	0	1
No seed treatment	1	0	1	1	1	0	1	0
Seed treatment	0	1	0	0	0	1	0	1
Conservation agriculture	0	1	1	1	0	1	0	1
Zero or minimum tillage	1	0	0	0	0	0	1	0
Conventional tillage	0	0	0	0	1	0	0	0
Yield	1.1	5.0 355%	5.5	5.7 4%	1.8	4.8 167%	2.9	5.1 76%

4 Conclusions and Future Work

This paper proposes the implementation of a farming practices optimization method based on Global-best Harmony Search (GHS), a machine learning model and commercial agronomic data, with the aim of finding the optimal farming practices to be used in cropping events whose climatic and soil characteristics have already been characterized.

Of the models tested for yield prediction in maize in Chiapas, random forest (RF) obtained the lowest RMSE and the highest R^2, thus showing the best predictive capacity. The results indicate that in the case of the Chiapas dataset on maize, if all the famers would adopt the optimized management, an average yield increase of 1.77 tons per hectare (ton/ha) would be expected together with a reduction of 65.4% in the yield variability. Those results are very promising and will need field testing before validating the approach. Furthermore, they open the way towards the automatized generation of personalized recommendations as shown in Table 4. Indeed, such results could readily feed a web-based system that farmers could query to check the data driven recommendations for a specific field.

The information used in this study comes from open data repositories and from the dataset of the MasAgro Project, all spanning all Mexican States. Thus, the opportunity exists to scale-up the approach to other locations and crops without the need for a large investment. This new data-driven, decision-making approach will allow for quick, low-cost, easy-to-access solutions in the future so that a greater number of farmers can have access to such fine-tuned recommendations and thereby exploit the maximum potential of the resources they have access to, and therefore contributing significantly to closing of the yield gap.

For future work we propose to explore other machine learning models that could potentially improve the prediction. The opportunity also exists to use the approach involving a seasonal weather forecast in order to deliver recommendations months before the next planting. It will also be necessary to test the optimization based on multiple objectives, adding other response variables such as profitability, quality, and market access, and to automatize the process to get closer to implementation in a commercial tool.

Acknowledgements. The research has been supported by International Center for Tropical Agriculture (CIAT) and is based on data shared by the MASAGRO project lead by the International Maize and Wheat Improvement Center (CIMMYT), we also acknowledge for the open data shared by INIFAP and INEGI. We are especially grateful to Colin McLachlan for suggestions relating to the text in English.

Appendix

Appendix 1. Dataset description

Reference	Description	Classification	Scale
M1	Total amount of nitrogen applied	Practices	Continuous
M2	Total amount of phosphorus applied	Practices	Continuous
M3	Total amount of potassium applied	Practices	Continues
M4	Number of mechanical weeding	Practices	Discrete
M5	Number of post-harvest herbicides applications	Practices	Discrete
M6	Number of "rastreo"	Practices	Discrete
M7	Number of pre-sowing herbicides applications	Practices	Discrete
M8	Number of fertilizations	Practices	Discrete
M9	Number of applications of foliar fertilizers	Practices	Discrete
M10	Number of applications of biofertilizants	Practices	Discrete
M11	Number of post-sowing herbicides applications	Practices	Discrete
M12	Number of applications of insecticides	Practices	Discrete
M13	Cultivars' group criollo	Practices	Discrete
M14	Cultivars' group Dekalb	Practices	Discrete
M15	Cultivars' group others	Practices	Discrete
M16	Cultivars' group P4082 W	Practices	Discrete
M17	No seed treatment	Practices	Discrete
M18	Seed treatment	Practices	Discrete
M19	Conservation agriculture	Practices	Discrete
M20	Zero or minimum tillage	Practices	Discrete
M21	Conventional tillage	Practices	Discrete
S1	Clay content	Soil	Continuous
S2	Silt content	Soil	Continuous

(continued)

(continued)

Reference	Description	Classification	Scale
S3	Soil organic content	Soil	Continuous
S4	Cationic exchange capacity	Soil	Continuous
S5	Basis saturation	Soil	Continuous
S6	Low infiltration	Soil	Discrete
S7	Moderate infiltration	Soil	Discrete
S8	Good infiltration	Soil	Discrete
W1	Average minimum temperature	Weather	Continuous
W2	Average diurnal range	Weather	Continuous
W3	Accumulated solar energy	Weather	Continuous
W4	Frequency of days with maximum temperature above 34°C	Weather	Continuous
W5	Accumulated precipitation	Weather	Continuous
W6	Frequency of days with minimum temperature below 8°C	Weather	Continuous
W7	Average relative humidity	Weather	Continuous
W8	Standard deviation of the relative humidity	Weather	Continuous
Y	Yield	Yield	Continuous

References

1. van Ittersum, M.K., Cassman, K.G., Grassini, P., Wolf, J., Tittonell, P., Hochman, Z.: Yield gap analysis with local to global relevance—a review. Field Crops Res. **143**, 4–17 (2012)
2. Zader, A.: Food and agriculture. In: Routledge Handbook of Environment and Society in Asia, p. 237 (2014)
3. Mase, A.S., Prokopy, L.S.: Unrealized potential: a review of perceptions and use of weather and climate information in agricultural decision making. Weather. Clim. Soc. **6**(1), 47–61 (2014)
4. Tilman, D., Balzer, C., Hill, J., Befort, B.L.: Global food demand and the sustainable intensification of agriculture. Proc. Natl. Acad. Sci. **108**(50), 20260–20264 (2011)
5. Foley, J.A., et al.: Solutions for a cultivated planet. Nature **478**(7369), 337 (2011)
6. Reid, W.V., Berkes, F., Wilbanks, T.J., Capistrano, D., et al.: Bridging Scales and Knowledge Systems: Concepts and Applications in Ecosystem Assessment. Island Press, Washington (2006)
7. Rasmussen, P.E., Goulding, K.W., Brown, J.R., Grace, P.R., Janzen, H.H., Körschens, M.: Long-term agroecosystem experiments: assessing agricultural sustainability and global change. Science **282**(5390), 893–896 (1998)
8. Samberg, L.H., Gerber, J.S., Ramankutty, N., Herrero, M., West, P.C.: Subnational distribution of average farm size and smallholder contributions to global food production. Environ. Res. Lett. **11**(12), 1–12 (2016). http://iopscience.iop.org/article/10.1088/1748-9326/11/12/124010/meta

9. De Janvry, A., Sadoulet, E., Suri, T.: Field experiments in developing country agriculture. In: Handbook of Economic Field Experiments, vol. 2, pp. 427–466. Elsevier, Amsterdam (2017)

10. Leeuwis, C.: Communication for Rural Innovation: Rethinking Agricultural Extension. Wiley, Hoboken (2013)

11. Cock, J., et al.: Crop management based on field observations: case studies in sugarcane and coffee. Agric. Syst. **104**(9), 755–769 (2011)

12. Jiménez, D., et al.: From observation to information: data-driven understanding of on farm yield variation. PLoS One **11**(3), 1–20 (2016)

13. Jiménez, D., et al.: Interpretation of commercial production information: a case study of lulo (Solanum quitoense), an under-researched Andean fruit. Agric. Syst. **104**(3), 258–270 (2011)

14. Jiménez, D., et al.: Analysis of Andean blackberry (Rubus glaucus) production models obtained by means of artificial neural networks exploiting information collected by small-scale growers in Colombia and publicly available meteorological data. Comput. Electron. Agric. **69**(2), 198–208 (2009)

15. Delerce, S., et al.: Assessing weather-yield relationships in rice at local scale using data mining approaches. PLoS ONE **11**(8), e0161620 (2016)

16. Strobl, C., Boulesteix, A., Kneib, T., Augustin, T., Zeileis, A.: Conditional variable importance for random forests. BMC Bioinform. **11**, 1–11 (2008)

17. Yu, T., Davis, L., Baydar, C., Roy, R.: Evolutionary Computation in Practice. Studies in Computational Intelligence. Springer, Heidelberg (2007). https://doi.org/10.1007/978-3-540-75771-9

18. Ashlock, D.: Evolutionary Computation for Modeling and Optimization. Springer, Heidelberg (2005). https://doi.org/10.1007/0-387-31909-3

19. Yu, X., Gen, M.: Introduction to Evolutionary Algorithms. Springer, Heidelberg (2010). https://doi.org/10.1007/978-1-84996-129-5

20. Sastry, K., Goldberg, D., Kendall, G.: Genetic algorithms. In: Burke, E., Kendall, G. (eds.) Search Methodologies, pp. 93–117. Springer, Heidelberg (2014). https://doi.org/10.1007/978-1-4614-6940-7_4

21. Du, K.-L., Swamy, M.N.S.: Particle swarm optimization. In: Search and Optimization by Metaheuristics, pp. 153–173. Springer, Heidelberg (2016). https://doi.org/10.1007/978-3-319-41192-7

22. Kennedy, J., Eberhart, R.: Particle swarm optimization. In: Proceedings of IEEE International Conference on Neural Networks, vol. 4, pp. 1942–1948 (1995)

23. Geem, Z.W., Kim, J.H., Loganathan, G.V.: A New Heuristic Optimization Algorithm: Harmony Search. Simulation **76**(2), 60–68 (2001)

24. Mahdavi, M., Fesanghary, M., Damangir, E.: An improved harmony search algorithm for solving optimization problems. Appl. Math. Comput. **188**(2), 1567–1579 (2007)

25. Omran, M.G., Mahdavi, M.: Global-best harmony search. Appl. Math. Comput. **198**(2), 643–656 (2008)

26. Bishop, C.M.: Neural Networks for Pattern Recognition. Oxford University Press, Oxford (1995)

27. Hagan, M.T., Demuth, H.B., Beale, M.H., De Jesús, O., et al.: Neural Network Design, vol. 20. Pws Pub, Boston (1996)

28. Breiman, L.: Statistical modeling: the two cultures. Stat. Sci. **16**, 199–215 (2001)

29. Hothorn, T., Hornik, K., Zeileis, A.: Unbiased recursive partitioning: a conditional inference framework. J. Comput. Graph. Stat. **15**(3), 651–674 (2006)

30. Strobl, C., Boulesteix, A.-L., Zeileis, A., Hothorn, T.: Bias in random forest variable importance measures: illustrations, sources and a solution. BMC Bioinform. **8**, 25 (2007)

31. Dietterich, T.G.: Ensemble methods in machine learning. In: Kittler, J., Roli, F. (eds.) MCS 2000. LNCS, vol. 1857, pp. 1–15. Springer, Heidelberg (2000). https://doi.org/10.1007/3-540-45014-9_1

32. Al-Jarrah, O.Y., Yoo, P.D., Muhaidatc, S., Karagiannidis, G.K., Tahaa, K.: Efficient machine learning for Big Data: a review. Big Data Res. **2**(3), 87–93 (2015)

33. Gao, K.-Z., Suganthan, P.N., Pan, Q.-K., Chua, T.J., Cai, T.X., Chong, C.-S.: Discrete harmony search algorithm for flexible job shop scheduling problem with multiple objectives. J. Intell. Manuf. **27**(2), 363–374 (2016)

34. Kong, X., Gao, L., Ouyang, H., Li, S.: A simplified binary harmony search algorithm for large scale 0–1 knapsack problems. Expert Syst. Appl. **42**(12), 5337–5355 (2015)

35. Hoang, D.C., Yadav, P., Kumar, R., Panda, S.K.: Real-time implementation of a harmony search algorithm-based clustering protocol for energy-efficient wireless sensor networks. IEEE Trans. Ind. Inform. **10**(1), 774–783 (2014)

36. Assad, A., Deep, K.: Applications of harmony search algorithm in data mining: a survey. In: Pant, M., Deep, K., Bansal, J., Nagar, A., Das, K. (eds.) Proceedings of Fifth International Conference on Soft Computing for Problem Solving. Advances in Intelligent Systems and Computing, vol. 437, pp. 863–874. Springer, Singapore (2016). https://doi.org/10.1007/978-981-10-0451-3_77

37. Kuhn, M., et al.: Caret: Classification and Regression Training (2014)

On the Modelling of the Energy System of a Country for Decision Making Using Bayesian Artificial Intelligence – A Case Study for Mexico

Monica Borunda[1(✉)], Ann E. Nicholson[2], Raul Garduno[3], and Hoss Sadafi[4]

[1] Conacyt - Instituto Nacional de Electricidad y Energías Limpias, Cuernavaca, Mexico
monica.borunda@ineel.mx
[2] Faculty of Information Technology, Monash University, Melbourne, Australia
Ann.Nicholson@monash.edu
[3] Instituto Nacional de Electricidad y Energías Limpias, Cuernavaca, Mexico
rgarduno@ineel.mx
[4] Department of Mechanical and Aerospace Engineering, Monash University, Melbourne, Australia
hoss.sadafi@monash.edu

Abstract. Energy efficiency has attracted the attention of many governments around the world due to the urgent call to reduce investments in energy infrastructure, lower fossil fuel dependency, integrate renewable energies, improve consumer welfare and reduce CO_2 emissions. The conservative and smart use of energy is one of the main approaches to improve energy efficiency. However, the management of energy at the national level is a complex decision making problem involving uncertainty and therefore, Bayesian Networks are suitable paradigm to deal with this task. In this work, we present a progress report on the development of a decision making method, based on Bayesian decision networks, for the efficient use of energy as a function of the cost, efficiency and CO_2 emissions from the source of energy used.

Keywords: Energy efficiency · Decision making · Bayesian networks
Smart energy system

1 Introduction

1.1 Motivation

Climate change, damage to the environment, depletion of conventional energy sources and greenhouse gas emissions are some of the main factors that indicate the urgency of seeking for a sustainable future. Moving towards energy sustainability requires improvements in the way energy is supplied and used [1].

Energy efficiency and renewable energy are said to be the twin pillars of sustainable energy policies [2]. Recently, energy efficiency has attracted the attention of many

I. Batyrshin et al. (Eds.): MICAI 2018, LNAI 11289, pp. 30–46, 2018.
https://doi.org/10.1007/978-3-030-04497-8_3

governments around the world due to the urgent call to reduce investments in energy infrastructure, reduce dependence on fossil fuels, integrate renewable energies, improve consumer welfare and reduce energy consumption, as well as CO_2 emissions. As a result, energy efficiency policies have been implemented in many countries aiming for a better use of energy and therefore, a reduction in energy consumption.

The consumption of energy depends on many causes such as the changing climatic conditions, economic and population growth, improvement of energy efficiency, development of technology and others. The International Energy Agency (IEA) has defined indicators for energy consumption, energy efficiency and CO_2 emissions, which are then used to identify the factors that: (a) increase and restrict energy demand as well as its influence, (b) understand the differences in energy intensities between countries, and (c) quantify how the introduction of conservation of energy and the improvement in available technology can help reduce overall energy consumption [3].

In this work, we present a progress report on the development of a support decision-making system to recommend actions for the use of energy in a country, depending on the availability of resources and the energy demand. Decision making is carried out by a Bayesian Decision Network model of the country's energy system. In the following, we provide a brief summary of related work in Sect. 1. In Sect. 2, we explain the basics of artificial Bayesian intelligence and its applications to decision making. The energy system of a country is described in Sect. 3. In this section, we describe the energy sources, the energy sectors and the criteria we use to assess the use of energy. The model of an energy system is outlined in Sect. 4. We described the methodology, the Bayesian decision network (BDN) used to model the energy system of an energy sector and the BDN for a whole country. Section 5 presents a case study for Mexico. Conclusions are given in Sect. 6.

1.2 Related Work

The main related work on energy consumption worldwide and particularly in Mexico are summarized here.

In [4], it is studied the causal relationships between energy consumption and their factors such as population, gross domestic product, consumer price index and carbon dioxide emission for 30 developing countries.

In [5], some means of saving energy are presented in the residential and transport sectors in order to reduce energy consumption and CO_2 emissions.

In parallel, Yildirim [6] examined the relationship between energy consumption and economic growth for Organizations in OECD (Economic Cooperation and Development) countries for the period between 1960 to 2014. This study was used in policy making.

In [7], the causal relationship between energy consumption and real GDP is analyzed for 11 countries in Middle East and North African and reported for implementing energy conservation policies.

In the article by Almulali [8], the impact of energy consumption and CO_2 emissions in the United Arab Emirates (UAE)'s economic and financial development was studied and it was found that there was a significant seasonal relationship between energy consumption and CO_2 emissions affecting the economic and financial development indicators.

Bakirtas [9] studied the periodical relationship between energy consumption, urbanization and economic growth, in bivariate and tri-variate analyses, from 1971 to 2014 in New Emerging-Market Countries (such as Colombia, India, Indonesia, Kenya, Malaysia and Mexico). He concluded an unexpected association between all the variables in all directions.

In Mexico, Lee and Yoo [10] analyzed the unpredicted short- and long-term issues among energy consumption, CO_2 emissions and economic growth from 1971–2007 using time-series techniques. They found unidirectional causality from economic growth to energy consumption and also from CO_2 emissions to economic growth.

In Gómez publication [11], the causal link between aggregated and disaggregated levels of energy consumption and economic growth in Mexico during 1965 and 2014 was given. They find a long-run relationship between production, capital, labor and energy and linear causal links from energy consumption to economic growth.

The potential areas for reducing building energy consumption were outlined by Omer [12] as it is considered to contribute to approximately 40% of the total world annual energy consumption. They discussed patterns of future energy use and their environmental impacts.

In Lentz [13], the nexus between energy demands, population growth, and climate change in Mexico City is reported. Likewise, they formulated relationships between power consumption and maximum temperatures during the long-term scenarios and specific extreme heat events.

In the next section we present the basics of Bayesian Artificial Intelligence.

2 Decision Making Using Bayesian Artificial Intelligence

2.1 Bayesian Reasoning

In Artificial Intelligence (AI), the ability to reason probabilistically is called Bayesian reasoning, which is a way to reason under uncertainty. This is important since practical real AI systems contain uncertainty such as: (a) ignorance – limits in our knowledge about the system; (b) physical randomness or indeterminism – real systems comprehend non-deterministic factors and; (c) vagueness – many of the statement we make have at least a small degree of impreciseness [14, 15].

Bayesian networks provide a natural representation of probabilities using Bayes' theorem. This theorem represents uncertainty with probabilities through the conditional probability

$$P(h|e) = \frac{P(e|h)P(h)}{P(e)}, \tag{1}$$

which states that the probability of a hypothesis h conditioned to some evidence e is equal to its likelihood $P(e|h)$ times its probability prior to any evidence $P(h)$, normalized by $P(e)$.

A Bayesian network is a probabilistic graphical model that allows us to represent and reason about an uncertain domain. The nodes in a Bayesian network represent a set of random variables $X = X_1, \ldots X_i, \ldots X_n$, from the domain. A set of directed arcs connects pairs of nodes, $X_i \rightarrow X_j$, representing the direct dependencies between variables, given by the conditional probability distribution associated with each node, and in the case of discrete variables they correspond to a Conditional Probability Table (CPT). The CPT is constructed such that (a) each row contains the conditional probability of each node value for each possible combination of values of its parent nodes; (b) each row must sum to 1; (c) a node with no parents has one row (the prior probabilities). Figure 1 shows two nodes X_1 and X_2 with two states, $S_{*,1}$ and $S_{*,2}$.

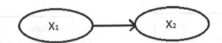

Fig. 1. Basic example of BN.

The a priori probabilities of node X_1 are defined in Table 1.

Table 1. A Priori probabilities of the node X_1.

X_1	
$S_{X_1,1}$	$P(X_1 = S_{X_1,1})$
$S_{X_1,2}$	$P(X_1 = S_{X_1,2})$

The CPT is defined by the conditional probabilities $P(X_j|X_i)$ associated to node X_j. It defines the probability distributions over the states of X_j given the states of X_i. Table 2 shows the CTP of the node X_2 given the node X_1. Once a domain and its uncertainty are represented by a BN, the BN computes the posterior probability distribution for a set of query nodes, given values for some evidence. BNs have four types of reasoning: (a) diagnostic: reasoning from symptoms to cause, (b) predictive: reasoning from new information about causes to new beliefs about effects, (c) intercausal: reasoning about mutual causes and a common effect and, (b) combined. Therefore, a BN computes the probabilities attached to a node state given the state of one or several variables, becoming a powerful modeling tool for complex systems.

Table 2. CPT of the node X_2, given the node X_1.

X_1	$S_{X_1,1}$	$S_{X_1,2}$	
X_2	$S_{X_2,1}$	$P(X_2 = S_{X_2,1}|X_1 = S_{X_1,1})$	$P(X_2 = S_{X_2,1}|X_1 = S_{X_1,2})$
	$S_{X_2,2}$	$P(X_2 = S_{X_2,2}|X_1 = S_{X_1,1})$	$P(X_2 = S_{X_2,2}|X_1 = S_{X_1,2})$

2.2 Decision Networks

BNs handle uncertainty and do probabilistic inference. This can be extended to support decision-making. Adding an explicit representation of the actions under considerations and the utility of the resultant outcomes results in a decision network [16]. Therefore, Bayesian decision networks (BDN) combine probabilistic reasoning with utilities to make decisions that maximize the expected utility. The utility function quantifies preferences, indicating the usefulness of the outcomes, by mapping them to numbers.

Figure 2 shows generic decision networks. They are formed by: (a) chance nodes: they represent random variables as in BNs; (b) decision nodes: they represent the decisions being made at a particular point in time and their values are the actions that the decision maker must choose; (c) utility nodes: they represent the utility function and have an associated utility table with one entry for each possible instantiation of its parents.

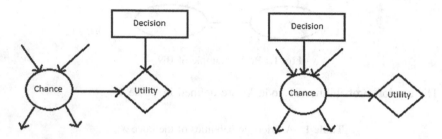

Fig. 2. Generic decision networks for (a) non-intervening and (b) intervening.

There are two main types of actions in decision problems: intervening and non-intervening. Non-intervening actions do not have a direct effect on the chance variables as shown in Fig. 2(a), contrary to intervening actions, as shown in Fig. 2(b). Therefore, the decision will affect the utility, either directly or indirectly, such that the decision influences the real world.

2.3   Knowledge Engineering with Bayesian Networks

Bayesian Networks (BN) can represent uncertain processes, make decision, perform causal modeling and optimize planning under uncertainty. Knowledge Engineering with Bayesian Networks (KEBN) is a lifecycle model, which allows for the construction of Bayesian models under a variety of circumstances.

The construction of a Bayesian network faces modeling tasks such as the definition of the variables and their values/states, the graph structure, the parameters, the available actions/decisions and their impact, and the utility nodes and their dependencies, among others.

The KEBN lifecycle model is a convenient way of introducing many aspects of the problem and involves the following steps: (a) Building the BN: defines the network structure, as well as its parameters and preferences. (b) Validation: evaluates the network with a sensitivity analysis and accuracy testing. (c) Field-testing: tests the actual use of the network either by Alpha testing (by in-house people) or by Beta testing (by a friendly-user). (d) Industrial use: implements the BN in regular use in the field. (e) Refinement: deals with requests for enhancement or fixing bugs.

In the following section, the energy system of a country is described to be able to model it in a BDN.

3 The Energy System of a Country

One of the major issues to reach a sustainable energy future is the reduction of energy consumption. World energy consumption is increasing and therefore global warming and greenhouse gas emissions. The need for renewable energies due to the depletion of conventional energy sources and the previous facts is mandatory. Institutions such as the International Energy Agency (IEA), the European Environment Agency (EEA), the U.S. Energy Information Administration (EIA), the French Environment & Energy Agency (ADEME) and the National Council of the Efficient Use of Energy (CONUEE) in Mexico, record and publish energy data. This is done in order to analyze trends and patterns to frame current energy issues and encourage movement towards collectively useful solutions. Due to the large number of countries belonging to IEA and the widely spread use of its information, in this work we follow the IEA's guidelines for collecting and publishing energy information [3]. Moreover, the structure of the data collection has an impact on the structure of the model of the energy system.

Energy consumption growth depends on the countries. For instance, in several developing countries, specifically energy consumption in Asia grows at 4%. On the other hand, energy consumption in OECD (Organization for Economic Cooperation and Development) countries was cut by 4.7%. In North America, Europe and the Commonwealth of Independent States (CIS) consumptions shrank by 4.5%, 5% and 8.5% respectively [17]. Energy consumption is loosely correlated with gross national product and climate, but the energy consumption per person varies even between developed or developing countries.

In order to study the energy consumption in a country, it is useful to model the energy system of the country. In accordance to IEA guidelines, we consider the energy system of a country as formed by: (a) the energy sources in the country and, (b) the energy consumption sectors.

3.1 Energy Sources

The main energy resources considered by IEA are coal, oil, natural gas, hydropower and renewable energies, such as solar, wind, geothermal and biomass. In 2016, 18% of the total world energy was used in the form of electricity and the rest was used for heat and transportation. This energy came from fossil fuels 80%, biofuels 10%, nuclear 5% and renewable 5% [18].

In this work, the energy system of a country is characterized by the installed capacity of each of the main available energy sources. The installed capacity refers to the amount of infrastructure able to produce usable energy and it is given in units of power [MW]. The amount of energy generated from a source being used in a year is called energy consumption and is given in units of energy [TWh].

3.2   Energy Consumption Sectors

The energy consumed by a country depends on many factors such as: (a) economy: gross domestic product, exchange rate, added value, etc.; (b) demography: population, households, dwellings, etc.; (c) climate: temperature, precipitation, etc.; (d) geography: location, ecosystems, environmental awareness, etc.; (e) technology development: air conditioners, cars, lighting, etc.

According to IEA, energy consumption can be classified by sectors. The energy consumption sectors include: transport, industry, residential, services and agriculture. In general, energy consumption by sector is different in every country.

In particular, the IEA has developed indicators of energy use, efficiency developments and CO_2 emissions to identify the factors driving and restraining the demand for energy, the differences in energy intensities amongst countries and quantify how the introduction of energy conservation and best available technology can help reduce energy use. The Energy Efficiency Indicators (EEI) provide the elements for an integrated analysis of how energy efficiency, in all end-use sectors and electricity generation, has affected energy use and CO_2 emissions.

There are aggregated and disaggregated indicators. Aggregated indicators are the energy intensity and energy consumption for different sectors. Disaggregated indicators are end-use efficiency or process efficiency indicators per sector. For the purpose of this work, we use aggregated indicators for energy consumption in each sector.

3.3   Criteria to Assess the Use of Energy

The good use of energy is important since it can result in an increase of environmental quality and prevention of future resource depletion [19]. It is related to energy costs, greenhouse emissions, energy efficiency and so on. In order to grade what is a good use of energy, meaningful criteria must be introduced. In this work, the following criteria is used:

(a) CO_2 emissions. Caring for the environment is one of the most important issues in shaping the energy policy of many countries. In particular, global warming mainly caused by greenhouse gases is a big concern. Even though carbon dioxide, methane, nitrous oxide, chlorofluorocarbons (CFCs) and hydro chlorofluorocarbons (HCFCs) are all potent greenhouse gases, as a first step we concentrate on the CO_2 emissions released for every unit of energy produced.

(b) Production cost. An important issue to take into consideration in the assessment of the use of a resource is its production cost. Production cost includes expenditures relating to the manufacturing or creation of the energy resource.

(c) Energy efficiency of the resource. The American Council for an Energy-Efficient Economy (ACEEE) defines energy efficiency as the capacity of an energy resource to yield energy. It is important to consider it together with the production cost since it provides information about the resource cost advantage of energy efficiency.

(d) Import/export. Energy demand and energy supply does not always match. As a result of a shortfall of production of resources and energy consumption, the need to import energy to meet demand becomes imminent. On the other hand, when the production of energy exceeds the demand, the export of energy is convenient.

These criteria are the basic ones to be considered in a first step to assess the use of energy resources. As it is shown in the next section, they can be incorporated into utility nodes when modeling the energy system of a country using Bayesian intelligence.

4 Modelling the Energy System of a Country

As it was described in Sect. 2, BNs are probabilistic graphical models that can model a system and graphically capture its structure in a net, with the possibility to incorporate the uncertainty of the process. In particular, as it is described in Sect. 2.3, the construction of model with Bayesian intelligence is performed with KEBN. In this section we describe the methodology used to model an energy system.

4.1 Methodology

The methodology being used to model an energy system as a BDN or influence diagram is shown in Fig. 3. The flowchart to build a BDN comprises the following steps:

1. Define the problem to be solved in a given domain of the energy system by gathering information about it and specifying what the use of the desired BDN is going to be.
2. Identify the components, functions, relations, characteristics, etc., of the energy system that are relevant to characterize and describe it in the sense of the decision making problem to be solved by the BDN.
3. Establish a correspondence between the relevant elements of the energy system identified in the previous step and the components of a BDN, i.e. nodes and arcs.
4. Embrace the know-how from the expert in energy systems, the know-how from the expert in Bayesian Intelligence, information in the technical literature, available data, missing information, intuition, creativity, etc., to propose the initial structure of the BDN.
5. Apply the mastery from the expert in energy systems, the knowledge provided in the literature, available data and missing information to propose the parameters in the elements of the BDN using an influence diagram technology software (e.g. Netica, GeNIe, Hugin Expert, etc.).
6. Validate the network operation and performance with known-data case-study first and then with what-if case-studies. Observe results for reasonability and consistency.
7. Iterate and improve the BDN as required taken into account the results of the validation step. Stop when sufficiently reasonable and consistent decisions outcome from the BDN.

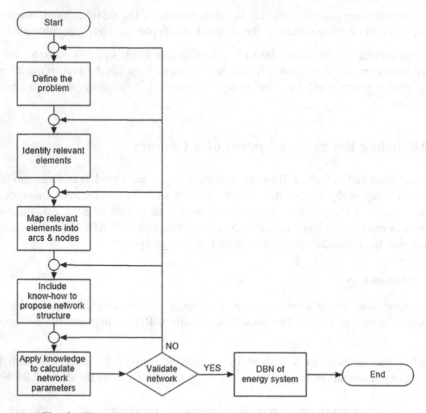

Fig. 3. Flowchart for the construction of a BDN of an energy system.

In the case of the energy system of a country, we are interested in developing a decision maker algorithm, using Bayesian intelligence, for the efficient use of energy depending on the cost, efficiency and CO_2 emissions from the source of energy. As it is described in Sect. 3, the energy system of a country is formed by energy sources and energy consumption sectors. These are mapped into chance nodes in the BDN. Since these elements may change depending on many factors, we can include this information adding causal parent nodes to them. On the other hand, the constraint that the energy consumption per sector must be covered by the energy sources, and the way this is performed, is addressed in a decision node. Finally, the criteria to assess the use of energy described in Sect. 3.3 is used to build utility functions, which provide the information that is compared in the decision making process. In the next section, we illustrate the construction of the BDN model for one energy sector.

4.2   BDN Model for One Energy Sector

The BDN model for the energy system of a country can be built from the BDN of several energy sectors, which might have the same structure. Figure 4 shows the BDN for one energy sector.

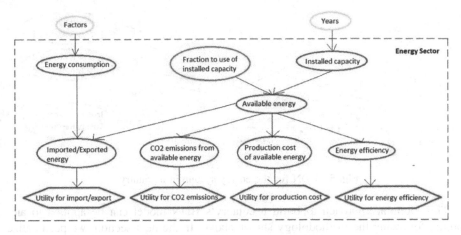

Fig. 4. BDN for one energy sector.

The first element of the energy system to be considered is the source of energy. It is denoted by the installed capacity of this kind of resource in the country, and it is represented by a chance or belief node. We also consider that the installed capacity changes over the years, but in future world one can include more factors affecting it. The second element of the energy system is the energy consumption. It is also represented by a chance node, as well as the factors affecting it. The blue node is a decision node encoding the constraint of the energy system. It corresponds to the fraction of energy from the installed capacity which is used to satisfy the energy consumption. This decision is made based on the criteria to assess the use of energy: (a) the amount of CO_2 emissions produced from the fraction of available energy, (b) the production cost of the fraction of the available energy, (c) the energy efficiency of that type of energy and, (d) the required energy to be imported or exported in order to satisfy the energy demand. The four chance nodes at the bottom of the BDN represent these criteria. Once the criteria are quantified, the utility nodes at the lowest bottom of the BDN compute utility values for energy use assessment of the sector being considered.

In the following section, we apply this model to different sectors in order to build the BDN model of the energy system of a country.

4.3 BDN Model for a Country

The BDN for one energy sector described in the previous sector is the main block of the decision model for a country. The BDN for a country is a modular system made of blocks representing the energy sectors in the country as shown in Fig. 5. The n types of energy resources available in the country are in the right hand side of the network and correspond to chance nodes. Each of them may vary depending on different factors. In this case, we simplify the scenario and consider that the installed capacities change over time, but different factors can be taken into consideration as well. The m energy sectors are represented by the dashed blocks at the bottom of the net. Each dashed block contains a BDN as shown in Fig. 4. As before, we may consider many factors that affect the energy consumption in the different sectors.

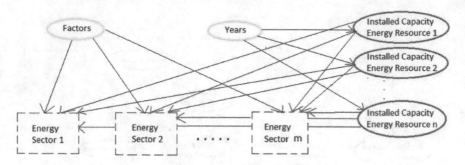

Fig. 5. BDN for the energy system in a country.

This modular approach to build a country's BDN model can be applied to any country following the methodology shown above. In the next section, we personalize the generic BDN with the energy data and conditions available for Mexico.

5   A Case Study – Mexico

5.1   Energy Sectors in Mexico

CONUEE records and publishes energy data in Mexico in accordance to IEA guidelines [3], and provides a free access online database [20]. This data shows that in the last years, the main energy sectors in Mexico have maintained a nearly constant behavior. As it is shown in Fig. 6, the share of energy from 2000 to 2015 has changed in some sectors about 5%. The share of energy consumption since 2000 has been led by the transport sector with nearly 45%, followed by the industry sector with approximately 34%, then the residential sector with closely 15% and finally the agriculture and services sectors with just about 3%.

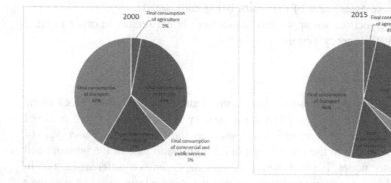

Fig. 6. Energy sectors in Mexico.

According to the data provided by CONUEE, the main energy resources used in the major energy sectors are as follows:

(a) The transport sector mainly consumes oil products like gasoline, diesel, jet fuel and fuel oil. Electricity and gas are consumed in a small amount.
(b) The industry sector consumes natural gas, electricity, oil products, coal, bagasse and solar energy.
(c) The residential sector consumes oil products, electricity, wood, gas and solar energy.

5.2 BDN Model for the Energy System in Mexico

In order to build the BDN model for the energy system in Mexico, we consider the two most important energy sectors and the main consumed resources as a first approximation. Therefore, we consider the transport and industry energy sectors, and electricity and products from oil as main consumed resources.

We place the nodes corresponding to the energy consumption of the transport and industry sectors on the left top of the network shown in Fig. 7. The consumption in these sectors depends on many factors. Demography, economy, geography, climatology, culture, technology, etc. are some of the factors driving energy consumption. In this work, we concentrate on demography and economy since they are the most influential and there is available data for them. Therefore, we consider the population in Mexico over the years 2000–2015 [thousands of inhabitants] and the exchange rate of the Mexican peso versus the US dollar.

For the sake of simplicity, we consider the reserves of crude oil [thousands of barrels per day] as the installed capacity of this resource. On the other hand, we consider the installed capacity for electric generation by power plants [MW] [21]. These main consumed resources are represented by chance nodes in the right top of the network, also shown in Fig. 7. Both of them change over the years and therefore we consider a parent node *years* to make use of this information.

To find the best way to assess the use of energy resources to satisfy energy consumption in the two sectors we introduce four decision nodes. These decision nodes are: (a) fraction of oil reserves to be used in the transport sector, (b) fraction of electricity to be used in the transport sector, (c) fraction of oil reserves to be used in the industry sector and, (d) fraction of electricity to be used in the industry sector. Each of these nodes is discretized and takes a value between 0 and 1 corresponding to the fraction of the kind of energy to be used in that energy sector.

The chance nodes on the bottom of the BDN give the criteria (production costs, CO_2 emissions, efficiency and import/export) to assess the use of energy. The utility nodes are located in the lowest bottom of the network. These nodes have utility functions, which provide values that can be used to rank a given energy consumption scenario.

The states of every node or variable in the BDN are constructed using the available data. As an example, Table 3 shows the states corresponding to the node: *total installed capacity for the generation of electricity*. This data shows that the total installed capacity in Mexico has changed from 35 GW to 42 GW in the period from 2000 through 2015, according to the Federal Commission of Electricity (CFE). Since the installed capacity changes over time, it has a parent node corresponding to the variable *years*. The discretization of the states is done in such a way that there is an increase of

0.5 GW in each state. The rest of the nodes and their states are discretized according to the available data for each corresponding variable in a similar way.

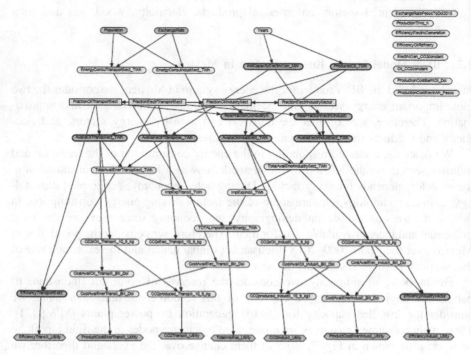

Fig. 7. Simplified BDN modeling the energy system in Mexico.

Table 3. Example of model variable, states and or data used to discretize the variable.

Model variable	Variable states	Node parent	Purpose/Usage of variable in the model
InstCapacityElectricGen_MW	35000–35500 MW 35500–36000 MW 36000–36500 MW 36500–37000 MW 37000–37500 MW 37500–38000 MW 38000–38500 MW 38500–39000 MW 39000–39500 MW 39500–40000 MW 40000–40500 MW 40500–41000 MW 41000–41500 MW 41500–42000 MW	Years	Total installed capacity for electric generation in Mexico in the last years. The evolution depends on many economic, political, resource availability, technology conditions and for the sake of simplicity, we only consider its evolution on time

5.3 Examples of the Assessment of the Use of Energy Use for Decision Making

Figure 8 illustrates the operation of the BDN shown in Fig. 7, which corresponds to Mexico's energy system. The energy system, consisting of energy sources and energy consumption sectors, is determined by given factors or initial conditions. For example, for one state of the *population* variable P_1 and one state of the *exchange rate* X_1, there is a given state of the energy consumption in both, the transport and industry sectors $S_1 S_2(P_1, X_1)$, where S_1 and S_2 refer to the energy consumption in the transport and industry sectors, respectively. Likewise, for one state of the *years* variable Y_1, there is a given state of the installed capacity for both, production of oil and electricity generation $E_1 E_2(Y_1)$, where E_1 and E_2 refer to the installed capacity for production of oil and electricity generation, respectively. Once the energy system is set in the previous way, an energy scenario can be determined by choosing the fractions $f_{E_1 S_1}, \ldots, f_{E_2 S_2}$ of each kind of energy to be used in each sector. Where the fraction of energy used in a sector is denoted by f and the indices E_1 refers to oil, E_2 refers to electricity, and S_1 and S_2 refers to the transport and industry sectors respectively. In this way, f_{E_1, S_1} corresponds to the fraction of the installed capacity of crude oil to be consumed in the transport sector, f_{E_1, S_2} corresponds to the fraction of the installed capacity of crude oil to be

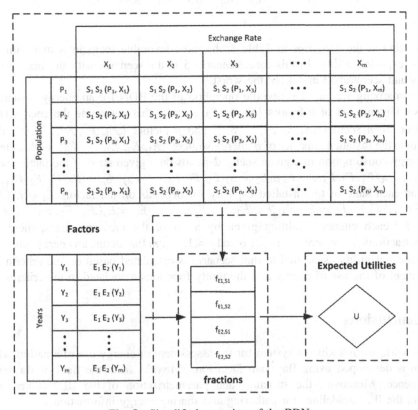

Fig. 8. Simplified operation of the BDN.

consumed in the industry sector and so on. For every energy scenario being considered, the four criteria are quantified and the utility nodes compute utility values for energy use assessment and for benchmarking amongst energy scenarios.

The BDN can work in various ways. For instance, one can set a single factor and swept the rest of them, or set all factors and look at that particular state of the system. Hence, a BDN can be able to model different energy systems and, by choosing different fractions, we generate different energy scenarios.

Consider for example an energy system with a large population (between 12 to 12.2 billion people), high exchange rate (between 15.2 and 15.9) and an installed capacity as in the year 2013. Table 4 shows examples of five different energy scenarios defined by different choices of fractions their corresponding expected utility.

Table 4. Expected utilities for different scenarios.

Scenario	f_{E_1,S_1}	f_{E_1,S_2}	f_{E_2,S_1}	f_{E_2,S_2}	U
1	0.5	0.5	0.5	0.5	3.93
2	0.7	0.7	0.3	0.3	3.81
3	0.9	0.9	0.1	0.1	3.80
4	0.1	0.1	0.9	0.9	3.80
5	0.5	0.9	0.5	0.1	3.95

Looking at the scenarios in Table 4, the most favorable scenario is that with the highest expected utility. In this case, scenario 5 is the scenario with the best use of energy and scenarios 3 and 4 are the worst.

Summarizing we can describe the operation of the BDN for an energy system as follows. Given a set of n factors $\{F_1, F_2, F_3, \ldots, F_n\}$ that determine the energy consumption in m sectors $\{S_1, S_2, S_3, \ldots, S_m\}$, and p factors $\{\mathcal{F}_1, \mathcal{F}^2, \mathcal{F}^3, \ldots, \mathcal{F}^p\}$ that determine the installed capacity of q energy sources $\{E_1, E_2, E_3, \ldots, E_q\}$, each state of the energy consumption in a given sector depends on a given state of the factors such that $S_1 = S_1(F_1, F_2, F_3, \ldots, F_n)$, $S_2 = S_2(F_1, F_2, F_3, \ldots, F_n)$, $S_m = S_m(F_1, F_2, F_3, \ldots, F_n)$ and each state of the installed capacity for depends on the factors $E_1 = E_1(\mathcal{F}_1, \mathcal{F}_2, \mathcal{F}_3, \ldots, \mathcal{F}_p)$, $E_2 = E_2(\mathcal{F}_1, \mathcal{F}_2, \mathcal{F}_3, \ldots, \mathcal{F}_p)$, $\ldots$, $E_q = E_q(\mathcal{F}_1, \mathcal{F}_2, \mathcal{F}_3, \ldots, \mathcal{F}_p)$. Then, for each energy condition given by a set of the previous states, there are $q \times m$ fractions f_{E_i,S_j} where $i = 1, \ldots, q$ and $j = 1, \ldots, m$ that define an energy scenario. The expected utility associated to that scenario is calculated based on the criteria for assessment of the use of energy and the utility functions associated to the criteria.

6 Conclusions

In this work, we introduce a system for the assessment of energy use of a nation. This system is developed using Bayesian Decision Networks from the field of Bayesian Intelligence. Moreover, the structure and parametrization of the BDN takes into account the IEA guidelines for gathering and sharing energy information.

This system is intended to provide utility values for specific scenarios of energy use. Then, the system allows us to perform benchmarking to determine the best scenario according to any given set of criteria about the use of energy. With this aim, the energy state of a country is defined by specific values of a set of factors, which includes up to now: population, currency exchange rate, and year. For any given energy state there are multiple scenarios that are defined by the share of energy resources for each energy sector. Finally, a utility value is inferred for each energy scenario. The utility values of all scenarios can be used for benchmarking to aid for decision making regarding the convenience of the way energy is employed. As a case study, we apply the system to a simplified version of Mexico's energy system obtaining encouraging results.

The results provided by the system reveal a high degree of reasonableness. However, up to now, the BDN is a simplified model and there is still much work to do to improve the BDN to describe more accurately and precisely the energy system of a nation. Future work includes the incorporation of more factors (i.e. technological, geographical, etc.), more energy sectors (i.e. residential, agricultural, etc.), and probabilistic distribution models of the states of the energy system, such that the results provided by the BDN will be more suitable for decision making. It is worth mentioning that the proposed approach to design and to build the BDN can be easily adapted to any energy system at different and larger energy scales. Finally, the current BDN, which provides a single expected utility for a given energy scenario, must be complemented with the required elements of an influence diagram to implement a truly decision system [22].

Acknowledgments. Monica Borunda wishes to thank Consejo Nacional de Ciencia y Tecnología, CONACYT, support for her Catedra Research Position with ID 71557, and to Instituto Nacional de Electricidad y Energías Limpias, INEEL, for its hospitality. She also wants to thank Australia-APEC woman in research fellowship for its grant to perform this research.

References

1. InterAcademy Council: Lighting the Way: Toward a Sustainable Energy Future, p. xvii (2007)
2. Aceee.org: The Twin Pillars of Sustainable Energy: Synergies between Energy Efficiency and Renewable Energy Technology and Policy, 11 January 2015
3. IEA. https://www.iea.org/
4. Kapusuzoglu, A., Karan, M.B.: The drivers of energy consumption in developing countries. In: Dorsman, A., Simpson, J., Westerman, W. (eds.) Energy Economics and Financial Markets, pp. 49–69. Springer, Heidelberg (2013). https://doi.org/10.1007/978-3-642-30601-3_4
5. Ngo, C., Natowitz, J.: Our Energy Future: Resources, Alternatives, and the Environment, 2nd edn (2016)
6. Yildirim, H.H.: Economic growth and energy consumption for OECD countries. In: Bilgin, M.H., Danis, H., Demir, E., Can, U. (eds.) Regional Studies on Economic Growth, Financial Economics and Management. ESBE, vol. 7, pp. 245–255. Springer, Cham (2017). https://doi.org/10.1007/978-3-319-54112-9_15

7. Ozturk, F.: Energy consumption – GDP causality in Middle East and North Africa (MENA) countries. Energy Sources Part B Econ. Plann. Policy **12**(3), 231–236 (2017)
8. Al-mulali, U., Sab, C.N.C.: Energy consumption, CO_2 emissions, and development in the UAE. Energy Sources Part B Econ. Plann. Policy **13**(4), 1–6 (2018)
9. Bakirtas, T., Akpolat, A.G.: The relationship between energy consumption, urbanization, and economic growth in new emerging-market countries. Energy **147**, 110–121 (2019)
10. Lee, S.-J., Yoo, S.H.: Energy consumption, CO_2 emission, and economic growth: evidence from Mexico. Energy Sources Part B Econ. Plann. Policy **11**(8), 711–717 (2016)
11. Gómez, M., Ciarreta, A., Zarraga, A.: Linear and nonlinear causality between energy consumption and economic growth: the case of Mexico 1965–2014. Energies **11**(4), 784 (2018)
12. Omer, A.M.: Energy, environment and sustainable development. Renew. Sustain. Energy Rev. **12**(9), 2265–2300 (2008)
13. Lentz, A.E., Angeles, M., Glenn, E., Ramírez, N., González, J.E.: On the recent climatological and energy trends in Mexico City. In: ASME 2016 10th International Conference on Energy Systainability collocated with the ASME 2016 Power Conference and the ASME 2016 14th International Conference on Fuel Cell Science, Engineering and Technology (2016)
14. Pearl, J.: Probabilistic Reasoning in Intelligent Systems: Networks of Plausible Inference. Morgan Kaufmann Publishers, San Francisco (1988)
15. Korb, K.B., Nicholson, A.E.: Bayesian Artificial Intelligence, 2nd edn. CRC Press, Boca Raton (2011)
16. Howard, R., Matheson, J.: Influence diagrams. In: Howard, R., Matheson, J. (eds.) Readings in Decision Analysis, pp. 763–771. Strategic Decisions Group, Menlo Park (1981)
17. Statistical Review of World Energy. Workbook, London (2016)
18. International Energy Statistics. Energy Information Administration. Accessed 5 June 2013
19. Top 5 reasons to be energy efficient. Alliance to Save Energy (ASE). Accessed 14 June 2016
20. CONUEE. http://www.biee-conuee.enerdata.net/
21. INEGI. http://inegifacil.com/energia
22. Garduno, R., Ibarguengoitia, P.: On the development of industrial-grade intelligent supervisory systems for power plant operation. In: IEEE Power Engineering Society General Meeting, Canada (2006)

Natural Language Processing

Enhancement of Performance of Document Clustering in the Authorship Identification Problem with a Weighted Cosine Similarity

Carolina Martín-del-Campo-Rodríguez[✉], Grigori Sidorov,
and Ildar Batyrshin

Instituto Politécnico Nacional (IPN), Centro de Investigación en Computación (CIC),
Mexico City, Mexico
cm.del.cr@gmail.com,sidorov@cic.ipn.mx,batyr1@gmail.com

Abstract. Distance and similarity measures are essential to solve many
pattern recognition problems such as classification, information retrieval
and clustering, where the use of a specific distance could led to a better
performance than others. A weighted cosine distance is proposed considering a variation in the weights of exclusive attributes of the input vectors. An agglomerative hierarchical clustering of documents was used for
the comparison between the traditional cosine similarity and the one proposed in this paper. This modified measure has outcome in an improvement in the formation of clusters.

1 Introduction

A similarity measure is a real-valued function that quantifies the similarity
between two objects, representing the inverse of the distance between such elements.

There are many distance/similarity measures encountered in different fields.
In ecology Forbes proposed in [1] a coefficient for clustering ecological related
species. In biology Jaccard applied in [2] a measure of similarity to compare the
distribution of flora in different areas. In chemistry different similarity measures
were applied in [3] and [4] for searching in chemical databases. Overview and
general methods of construction of similarity measures are considered in [5].

In linguistics, Sahu et al. proposed a modified cosine distance to cluster documents using Mahout with Hadoop [6]; for the task of grouping in the problem of
authorship identification Gómez-Adorno et al. [7] used a hierarchical clustering
analysis based on an average linkage algorithm, to join the clusters a cosine similarity was used; García-Mondeja et al. evaluated different similarity functions to
perform clustering based on a threshold [8]; in [9] Kocher et al. used the measure
SPATIUM (Latin word that means distance) to determine the clusters based on
rules.

© Springer Nature Switzerland AG 2018
I. Batyrshin et al. (Eds.): MICAI 2018, LNAI 11289, pp. 49–56, 2018.
https://doi.org/10.1007/978-3-030-04497-8_4

There have been considerable efforts in finding the appropriate measures among such choices because it is of fundamental importance to pattern classification, clustering, and information retrieval problems [10].

In this paper a weighted cosine similarity measure is proposed to reduce the values between the vectors (representing documents) where the attributes are exclusive. This for an enahancement of the clustering in the authorship identification problem. This measure differs from the soft cosine introduced in [11] where the weights are used as similarity between attributes.

2   Scientific Backgrounds

2.1   Cluster Analysis

Cluster analysis groups objects based on the information found in the data describing the objects or their relationships. The goal is that objects in the same group should be similar (or related) to another and different from (or unrelated to) the objects in other groups [12]. A clustering is considered as good when the distances between the members of each cluster are small and the distances between the clusters are large.

There are many approaches for clustering, but the main distinction is between hierarchical and partitional approaches.

Hierarchical Clustering. This clustering approach produces a nested sequence of partitions, with a single, all-inclusive cluster at the top and singleton clusters of individual points at the bottom. Generally the hierarchical clustering fall into two types:

- Agglomerative: Start with clusters containing a single point, and then merge them.
- Divisive: Start with one large cluster and then split it.

Most hierarchical algorithms involve joining two clusters or splitting a cluster into two sub-clusters. Furthermore, for these clustering types, there are different techniques, the main ones are described below [12]:

- Single link or MIN: the proximity of two clusters is defined to be minimun of the distances between any two point in the different clusters.
- Complete link or MAX: the proximity of two clusters is defined to be maximun of the distances between any two point in the different clusters.
- Group average: the proximity of two clusters is defined to be the average of the pairwise proximities between all pairs of points in the different clusters. Notice that this is an intermediate approach between MIN and MAX.

Partitional Clustering. This approach creates a one-level partitioning of the data points. If K is the desired number of clusters, then partitional approaches typically find all K clusters at once. Contrast this with traditional hierarchical schemes, which bisect a cluster to get two clusters or merge two clusters to get one.

2.2   Presence/Absence Data

In [13] Gower used the concept of presence/absence data. The author established that for two binary vectors x and y of size n, were n is the number of attributes that describe the objects (1 for presence of the attribute and 0 for absence) such that $x = (x_1, x_2, ..., x_n)$ and $y = (y_1, y_2, ..., y_n)$, the following four values can be defined:

- a as the number of attributes which value is one in both vectors, x and y.
- b as the number of attributes which value is one in the vector x but zero in vector y.
- c as the number of attributes which value is zero in the vector x but one in the vector y.
- d as the number of attributes which value is zero in both vectors, x and y.

2.3   Cosine Similarity

The cosine similarity is based on the measure of cosine of the angles between the vectors thus, if the angle is small the cosine is big, that means that the similarity between the vectors is large. In Eq. (1) this measure is shown for the vector x and y.

$$cos(x, y) = \frac{x \cdot y}{||x||\ ||y||} = \frac{\sum_{i=1}^{n} x_i\, y_i}{\sqrt{\sum_{i=1}^{n} x_i^2}\ \sqrt{\sum_{i=1}^{n} y_i^2}} \tag{1}$$

where, $x \cdot y$ is the dot product of the vectors, $||x||$ is the magnitud of the vector x and n is the dimensionality of the vectors.

3   Weighted Cosine

The basic idea behind the representation of presence/absence data for binary vectors could be applied to real-valued vectors. Consider the real-valued vectors x and y of size n, the values of presence/absence data values are defined as:

- a as the number of attributes which value is greater than zero in both vectors, x and y.
- b as the number of attributes which value is greater than zero in the vector x but zero in vector y.
- c as the number of attributes which value is zero in the vector x but is greater than zero in the vector y.
- d as the number of attributes which value is zero in both vectors, x and y.

Using the redefinition of the values of presence/absence data, we can rewrite the indexes of the sum in the Eq. (1) considering only those indexes where the value is not zero as is shown in (2).

$$cos(x, y) = \frac{\sum_{i \in I_a} x_i\, y_i}{\sqrt{\sum_{i \in I_{a \cup b}} x_i^2}\ \sqrt{\sum_{i \in I_{a \cup c}} y_i^2}}$$

$$= \frac{\sum_{i \in I_a} x_i\, y_i}{\sqrt{\sum_{i \in I_a} x_i^2 + \sum_{i \in I_b} x_i^2}\ \sqrt{\sum_{i \in I_a} y_i^2 + \sum_{i \in I_c} y_i^2}} \tag{2}$$

where i is the index of x and y such that $I_a \Rightarrow x_i > 0$ and $y_i > 0$, $I_b \Rightarrow x_i > 0$ and $y_i = 0$, $I_c \Rightarrow x_i = 0$, $y_i > 0$.

Therefore, based on the idea for visualization used by [14], we propose a constant w such that the values of x_i in I_b and I_c will be modified. Thereby, those dimensions where the values of the vectors are exclusive (I_b and I_c) will be reduced by the factor w and the cosine similarity will be more directly related with the non-exclusive (I_a) values. The Eq. (3) shows this relation,

$$cos_w(x, y) = \frac{\sum_{i \in I_a} x_i \, y_i}{\sqrt{\sum_{i \in I_a} x_i^2 + w \sum_{i \in I_b} x_i^2} \; \sqrt{\sum_{i \in I_a} y_i^2 + w \sum_{i \in I_c} y_i^2}} \tag{3}$$

where w is a constant in $[0, 1]$. As it can be seen in (3) the cos_w is still symmetric and reflexive.

The Eq. (3) can be written as a distance measure like in Eq. (4).

$$dist_w(x, y) = 1 - \frac{\sum_{i \in I_a} x_i \, y_i}{\sqrt{\sum_{i \in I_a} x_i^2 + w \sum_{i \in I_b} x_i^2} \; \sqrt{\sum_{i \in I_a} y_i^2 + w \sum_{i \in I_c} y_i^2}} \tag{4}$$

3.1   Corpus Description

Two corpora were used for testing the proposed measure, both of them belong to tasks proposed by PAN (Evaluation Lab on Digital Text Forensics). PAN is an organization that year by year proposes series of tasks related to the processing of natural language. These tasks are divided in three principal groups:

- Author Identification: determine the authorship.
- Author Profiling: determine the traits of an author.
- Author Obfuscation: hide the characteristics that determine the authorship of a document.

The two corpora used belong to the category of author identification in the subcategory of author clustering. Both corpora have a similar general description: a collection of documents is given, the goal is to identify groups of documents by the same author. All documents are single-authored, in the same language, and belong to the same genre. However, the topic or text-length of documents may vary. The number of distinct authors whose documents are included in the collection is not given [15, 16].

Base Corpus (Short Texts). The Base Corpus (BC) used is the one of the tasks of PAN 2017 [15]. The only difference between this corpus and the one of the PAN 2016 task is that this is a collection of short documents (paragraphs extracted from larger documents). These documents have an average word length of 70.

Two corpora are provided, one for training and one for testing. Both corpora are in three diferent languages: English, Dutch and Greek.

- Training Corpus: this is divided into 60 problems, each problem contains between 16 and 20 documents. These problems are divided into 20 problems per language and each problem is only in one language. The documents have a length between 15 and 345 words.
- Testing Corpus: this is divided into 120 problems, each problem contains 20 documents. These problems are divided into 40 problems per language and each problem is only in one language. The documents have a length between 13 and 388 words. The documents have a length between 13 and 388 words.

Second Corpus (Long Texts). The Second Corpus (SC) used is from the task of PAN 2016 [16]. The general description is the one given before. Worth noting that these are long documents. These documents have an average word length of 667.

Two corpora are provided, one for training and one for testing. For both corpora there are 18 problems. These problems are divided into three diferent languages: English, Dutch and Greek (6 problems per language and each problem is only in one language).

- Training Corpus: each problem contains between 50 and 100 documents. The documents have a length between 53 and 3972 words.
- Testing Corpus: each problem contains between 57 and 100 documents. The documents have a length between 50 and 3942 words.

3.2   Measures

The F-Bcubed score [17] was used to evaluate the clustering output. This measure corresponds to the harmonic mean between precision and recall. The Bcubed precision (P-Bcubed) represents the ratio of documents written by the same author in the same cluster. While the Bcubed recall (R-Bcubed) represents the ratio of documents written by an author that appear in its cluster.

4   Experiments and Results

The clustering of the documents was realized following the approach of Gómez-Adorno et al. [7] (the best approach of [15]). Gómez-Adorno et al. implemented a hierarchical cluster analysis using an agglomerative approach. To join the clusters they chose an average linkage algorithm (using cosine distance) and the Calinski-Harabaz score [18] to evaluate the clustering model. We made the same implementation considering the weighted cosine distance (4) as the measure to join the clusters with the average linkage algorithm.

The baseline is considered as the result of clustering when the value of w is 1, that means that there was no modification in the cosine. The same value of w was applied to all the problems of each corpus.

Table 1 shows the result of applying Eq. (4) to the BC corpus. As it can be seen, there are two values of w (0.6, 0.7) that improve the performance of the

Table 1. Bcubed values overall BC corpus. The results are ordered based on the value of the Bcubed F-score of the testing

w	Training			Testing		
	Bcubed recall	Bcubed precision	Bcubed F-score	Bcubed recall	Bcubed precision	Bcubed F-score
0.6	0.57611	0.64802	0.57421	0.63362	**0.61272**	**0.57569**
0.7	0.57778	0.6441	0.57386	0.63498	0.60995	0.57437
1 (baseline)	*0.57823*	*0.6444*	**0.57425**	**0.63739**	*0.60708*	*0.57316*
0.8	0.57823	0.6444	**0.57425**	0.63243	0.60993	0.57297
0.9	0.57823	0.6444	**0.57425**	0.63299	0.60947	0.57294
0.5	0.56913	**0.65072**	0.5697	0.63037	0.60888	0.57105
0.4	0.56546	0.64749	0.5643	0.63544	0.60179	0.56939
0.3	0.59373	0.62836	0.56496	0.62914	0.6058	0.56755
0.2	0.59557	0.62281	0.57302	0.63397	0.58613	0.5579
0.1	**0.60473**	0.59659	0.55545	0.62664	0.58721	0.54857
0	0.56642	0.55144	0.50962	0.58766	0.52813	0.49992

Table 2. Bcubed values overall SC corpus. The results are ordered based on the value of the Bcubed F-score of the testing

w	Training			Testing		
	Bcubed recall	Bcubed precision	Bcubed F-score	Bcubed recall	Bcubed precision	Bcubed F-score
0	0.83203	0.57964	0.53404	0.78977	**0.68256**	**0.60986**
0.2	**0.85736**	0.43357	0.43147	0.80061	0.65529	0.59454
1 (baseline)	*0.73748*	*0.77756*	*0.64261*	*0.82172*	*0.58541*	*0.53260*
0.9	0.72614	0.8249	0.67541	0.82172	0.58461	0.53131
0.7	0.72679	0.82552	0.67707	0.82092	0.5839	0.53099
0.8	0.72679	**0.82591**	**0.67779**	0.82172	0.58382	0.53095
0.4	0.76197	0.73007	0.61893	0.81004	0.59107	0.53087
0.6	0.75254	0.73927	0.6288	0.82394	0.58371	0.52978
0.3	0.80151	0.58249	0.51624	0.8139	0.55593	0.51738
0.5	0.77868	0.63123	0.5455	0.80776	0.58421	0.51509
0.1	0.85108	0.42823	0.40794	**0.83233**	0.54361	0.50317

baseline. The improvement is not truely significant, however certain values of w result in a model more robust than the classical cosine distance since there is not loss in the Bcubed values from the training to testing.

The result of applying Eq. (4) to the SC corpus is shown in Table 2. There exist two values of w that improve the performance of the baseline (0, 0.2).

Here, the improvement is more significant than the one obtained for the BC corpus. The Bcubed F-score values are still more robust than the values from the baseline.

5   Conclusions

A weighted cosine similarity was presented with the purpose of the improvement of clustering of documents in the authorship attribution problem. After several experiments, this modification showed an improvement in the clusters formation. This improvement is not so significant for short documents (with an average word length of 70); on the other hand, for long documents (with an average word length of 667) this improvement is around six percent in the testing corpus despite the diminution in the values of the Bcubed F-score in the training corpus.

For future work we would like to apply a feature selection on the current set of features (20,000 features). A smaller dimensionality would help us to avoid the redundant or irrelevant features. The distances would be based only in the characteristics that describe better the documents and this could enhance the clusters formation.

As well as, we would like to extend our investigation to others similarity measures.

Acknowledgments. This work was partially supported by the Mexican Government (CONACYT projects 240844, SNI, COFAA-IPN, SIP-IPN 20181849, 20171813, BEIFI 20181315).

References

1. Forbes, S.: On the local distribution of certain Illinois fishes: an essay in statistical ecology. In: Bulletin of the Illinois State Laboratory of Natural History, vol. 7, no. 8. Illinois State Laboratory of Natural History (1907)
2. Jaccard, P.: Étude comparative de la distribution florale dans une portion des alpes et des jura. Bulletin del la Société Vaudoise des Sciences Naturelles **37**, 547–579 (1901)
3. Willett, P., Barnard, J.M., Downs, G.M.: Chemical similarity searching. J. Chem. Inf. Comput. Sci. **38**, 983–996 (1998)
4. Arif, S.M., Holliday, J.D., Willett, P.: Comparison of chemical similarity measures using different numbers of query structures. J. Inf. Sci., 1–8 (2013)
5. Batyrshin, I.: Towards a general theory of similarity and association measures: similarity, dissimilarity and correlation functions. J. Intell. Fuzzy Syst. (2018)
6. Sahu, L., Mohan, B.R.: An improved k-means algorithm using modified cosine distance measure for document clustering using Mahout with Hadoop. In: 2014 9th International Conference on Industrial and Information Systems (ICIIS), pp. 1–5 (2014)
7. Gómez-Adorno, H., Alemán, Y., Vilariño Ayala, D., Sanchez-Perez, M., Pinto, D., Sidorov, G.: Author clustering using hierarchical clustering analysis-notebook for PAN at CLEF 2017. In: Cappellato, L., Ferro, N., Goeuriot, L., Mandl, T. (eds) CLEF 2017 Evaluation Labs and Workshop - Working Notes Papers, 11–14 September, Dublin, Ireland. CEUR-WS.org (2017)

8. García-Mondeja, Y., Castro-Castro, D., Lavielle-Castro, V., Muñoz, R.: Discovering author groups using a B-compact graph-based clustering-notebook for PAN at CLEF 2017. In: Cappellato, L., Ferro, N., Goeuriot, L., Mandl, T., (eds.) CLEF 2017 Evaluation Labs and Workshop - Working Notes Papers, 11–14 September, Dublin, Ireland, CEUR-WS.org (2017)
9. Mirco Kocher, J.S.: UniNE at CLEF 2017: author clustering-notebook for PAN at CLEF 2017. In: Cappellato, L., Ferro, N., Goeuriot, L., Mandl, T. (eds.) CLEF 2017 Evaluation Labs and Workshop - Working Notes Papers, 11–14 September, Dublin, Ireland, CEUR-WS.org (2017)
10. Duda, R.O., Hart, P.E., Stork, D.G.: Pattern Classification, 2nd edn. Wiley, New York (2000)
11. Sidorov, G., Gelbukh, A., Gómez-Adorno, H., Pinto, D.: Soft similarity and soft cosine measure: similarity of features in vector space model. Computación y Sistemas **18**, 491–504 (2014)
12. Steinbach, M., Ertöz, L., Kumar, V.: The challenges of clustering high dimensional data. In: Wille, L.T. (ed.) New Directions in Statistical Physics, pp. 273–309. Springer, Heidelberg (2004). https://doi.org/10.1007/978-3-662-08968-2_16
13. Gower, J.C.: A general coefficient of similarity and some of its properties. Biometrics **27**, 857–871 (1971)
14. Batyrshin, I., Kubysheva, N., Solovyev, V., Villa-Vargas, L.: Visualization of similarity measures for binary data and 2 x 2 tables. Computación y Sistemas **20**, 345–353 (2016)
15. Tschuggnall, M., et al.: Overview of the author identification task at PAN 2017: style breach detection and author clustering. In: Cappellato, L., Ferro, N., Goeuriot, L., Mandl, T. (eds.) Working Notes Papers of the CLEF 2017 Evaluation Labs, CEUR Workshop Proceedings (2017)
16. Stamatatos, E., et al.: Clustering by authorship within and across documents. In: Working Notes Papers of the CLEF 2016 Evaluation Labs. Volume 1609 of CEUR Workshop Proceedings, CLEF and CEUR-WS.org (2016)
17. Amigó, E., Gonzalo, J., Artiles, J., Verdejo, M.: A comparison of extrinsic clustering evaluation metrics based on formal constraints. **12**, 461–486 (2009)
18. Caliński, T., Harabasz, J.: A dendrite method for cluster analysis. Commun. Stat. **3**, 1–27 (1974)

Exploring the Context of Lexical Functions

Olga Kolesnikova[1]([✉]) and Alexander Gelbukh[2]

[1] Escuela Superior de Cómputo, Instituto Politécnico Nacional,
Mexico City, Mexico
kolesolga@gmail.com
[2] Centro de Investigación en Computación, Instituto Politécnico Nacional,
07738 Mexico City, Mexico
http://www.gelbukh.com/

Abstract. We explore the context of verb-noun collocations using a corpus of the Excelsior newspaper issues in Spanish. Our purpose is to understand to what extent the context is able to distinguish the semantics of collocations represented by lexical functions of the Meaning-Text Theory. For experiments, four lexical functions were chosen: Oper1, Real1, CausFunc0, and CausFunc1. We inspected different parts of the eight-word window context: the left context, the right context, and both the left and right context. These contexts were retrieved from the original corpus as well as from the same corpus after stopwords deletion. For the vector representation of the context, word counts and tf-idf of words were used. To estimate the ability of the context to predict lexical functions, we used various machine-learning techniques. The best F-measure of 0.65 achieved for predicting Real1 by Gaussian Naïve Bayes using the left context without stopwords and word counts as features in vectors.

Keywords: Natural language processing · Lexical functions
Verb-noun collocations · Context representation

1 Introduction

Contemporary research on natural language processing (NLP) issues has achieved high results in various areas; however, there is still a need to improve methods of text understanding and semantic analysis. One of the issues is distinguishing between free word combinations and multiword expressions: free combinations can be interpreted by a compositional analysis, which does not work for multiword expressions. Therefore, another issue is semantic analysis of multiword expressions.

In this work, we focus on verb-noun collocations as a particular type of multiword expressions and use lexical functions of the Meaning-Text Theory proposed by Mel'čuk [7] to interpret their semantic content. Accordingly, first we present a brief overview of the Meaning-Text Theory, and then explain the concept of lexical functions.

1.1 The Meaning-Text Theory

The Meaning-Text Theory (MTT) was proposed by Igor Mel'čuk in the 1960s in Moscow, Russia, as a universal theory powerful enough to describe and model any

© Springer Nature Switzerland AG 2018
I. Batyrshin et al. (Eds.): MICAI 2018, LNAI 11289, pp. 57–69, 2018.
https://doi.org/10.1007/978-3-030-04497-8_5

natural language. Since then, it has been further developed in many research works [2, 4, 5, 8, 13, 19].

The MTT views a natural language as a system of rules that on the one hand enables its speakers to transfer meaning into text in the process of speaking or text construction and on the other hand, transfer text into meaning, that is, to understand or interpret a text. Up to nowadays, the priority in research has been given to the meaning-text transfer because it is supposed that text interpretation can be explained by patterns humans use to generate text.

The MTT sets up a multilevel language model stating that to express meaning, speakers do not produce text directly and immediately but in a series of transformations fulfilled consecutively on various levels.

Thus, beginning from the level of meaning or semantic representation, we first execute some operations to express the intended meaning on the level of deep syntax, then we go to the surface syntactic level, afterwards proceeding to the deep morphological level, then to the surface morphological level, finally arriving at the phonological level where text can be spoken and heard. Another option is a written text which in fact is speech represented by means of an orthographic system created to facilitate human communication. Each transformation level possesses its own units, rules of combining units together, and rules of transfer from a given level to the consecutive one. So at each level we obtain some particular text representation: deep syntactic representation, surface morphological representation, and so on to an observable text as the final stage.

The most significant elements of the Meaning-Text Theory are its syntactic theory, the theory of lexical functions, and the explanatory combinatorial dictionary. The latter is a database of lexical units where a detailed semantic and syntactic information on each unit is given. In this paper, we deal with lexical functions, so the next section presents this concept.

1.2 Lexical Functions

Lexical function (LF) is a concept developed for describing and classifying diverse associations among words in a lexicon. LF is defined as a mapping from a word (LF argument) to a set of other words (LF meaning). Each LF formalizes a specific lexical semantic relation between the LF argument and each word of the LF value set. About 60 lexical functions have been found on the paradigmatic and syntagmatic level, they are described in [7]. In this paper, we will explain and exemplify only four LFs chosen for our experiments, these LFs are defined for relations between a noun and a verb in verb-noun collocations.

So in the case of verb-noun collocations, a lexical function maps a noun to a set of verbs such that the resulting verb-noun collocations share a common meaning and verbs are characterized by a common predicate-argument structure. For example, the lexical function termed Oper1, from Latin *operari* 'do, carry out', is a label for the meaning 'perform the action given by a noun' as in the following examples: *make a decision, make a step, take a shower, take a walk, commit suicide, do an exercise, give a talk, give a smile, have breakfast, pay a visit,* and *lend support.* In all these collocations, different verbs are used, however, all of them communicate the same meaning of 'realizing, performing, carrying out' of what is expressed by the noun.

Integers in the LF notation capture the predicate-argument structure typically used to express the LF meaning in sentences. In Oper1, 1 means that the action is realized by the agent, the first argument of a verb; therefore, Oper1 represents the pattern 'The agent performs what is expressed by the noun'. Consider an example *the professor gave a lecture yesterday*. In this utterance, *the professor* is the agent of the action expressed by *lecture*.

Another lexical function, Real1, from Latin *realis* 'real', represents the concept of fulfilling a requirement imposed by the noun or performing an action typical for the noun: *drive a bus, follow advice, spread a sail, prove an accusation, succumb to illness*, and *turn back an obstacle*.

Oper1 and Real1 discussed in the previous paragraphs represent a single semantic concept, such LFs are called simple. A combination of more than one meaning is denoted by a complex lexical function. In this work we used collocations belonging to the complex LFs named CausFunc0 and CausFunc1.

Caus, from Latin *causare* 'cause', represents the pattern 'do something so that the event denoted by the noun starts occurring'. Func0, from Latin *functionare* 'function', represents the meaning 'happen, occur'. Combining these two meanings, we get CausFunc0 with the semantics 'the agent does something so that the event denoted by the noun occurs'. Such semantics can be observed in the following collocations: *bring about the crisis, create a difficulty, present a difficulty, call elections, establish a system*, and *produce an effect*. Another complex lexical function CausFunc1 represents the pattern 'the non-agentive participant does something such that the event denoted by the noun occurs' as in the following examples: *open a perspective, raise hope, open a way, cause damage*, and *instill a habit into someone*.

In our work, we studied the extent to which the context of collocations in a corpus is able to distinguish among the four chosen LFs using supervised learning methods. Our choice of lexical functions is made due to the data available to us. However, there are more lexical functions defined for verb-noun collocations. The interested reader can find a detailed explanation of all lexical functions, their notation, and meaning illustrated by English examples in [7].

The rest of the paper is organized as follows. Section 2 presents related work. In Sect. 3, we define our objective and describe the experimental setup giving details on our dataset, the corpus, and the methodology. In Sect. 4, we discuss the results. Finally, Sect. 5 presents conclusions and future work.

2 Related Work

A large amount of research has been dedicated to developing methods for analysis of verb-noun constructions since verb and noun are two morphosyntactic categories responsible for transmitting the major part of text contents. Therefore, their adequate identification and semantic interpretation are vital in any semantically oriented language system. In this section, we revise state of the art results on annotation of verb-noun collocations with lexical functions and automatic identification of lexical functions in collocations.

Research papers focused on the topics stated in the previous paragraph can be classified into three categories according to their methodology. In the research of the first category, collocations are manually tagged with LFs by language experts. An example of such work is [12], where a Spanish learner corpus is manually annotated with LFs for second language acquisition purposes.

Research works in the second category use extensive syntactic and lexicon-based information to distinguish LFs in text. Tutin [17] took advantage of a database of French collocations described by Polguère [11], syntactic patterns, and finite-state transducers associated with metagraphs for labeling collocations with LFs in corpora. In the experiments, a precision of 90% and a recall of 86.2% were archived.

In the third category, the authors used hypernymy information from the WordNet semantic database [9] to detect LFs by means of supervised learning methods. Wanner [18, 19] experimented with nine LFs in Spanish verb-noun collocations representing each collocation by a vector whose features were hypernyms of the noun and the verb. In the experiments, an average F-measure of about 70% was obtained. The highest results were shown by the ID3 algorithm with an F-measure of 76% and by the Nearest Neighbor technique with an F-measure of 74%. In our previous work [1], an average F-measure of 74% was achieved on a Spanish verb-noun LF dataset using 68 supervised learning algorithms implemented in WEKA [16, 21] and the same hypernym representation as in [18, 20].

3   Experimental Setup

State-of-the-art research has not yet explored the degree to which the context of verb-noun collocations in corpus can distinguish among lexical functions. Therefore, in this work we intend to detect lexical functions using words of the context as features in vector representation of collocations. An advantage of this approach is that it does not need semantic information from a dictionary or syntactic data generated by a parser, since dictionary compilation requires much human effort and time, while the output of parsers is still far from perfect.

In this section, we describe our data, the corpus, and the methodology employed to identify lexical functions in verb-noun collocations.

3.1   Data and Corpus

In our experiments, we used Spanish verb-noun collocations of the four lexical functions presented in Sect. 1: Oper1, Real1, CausFunc0, and CausFunc1. Our dataset is a part of the list of most frequent Spanish verb-noun collocations [1] manually annotated with lexical functions.[1]

The choice of the four LFs was made due to the limitations of the data available to us. The original list [1] includes 60 samples of Real1 and more than 60 samples of the

---

[1] The complete list of 737 Spanish verb-noun collocations annotated with 36 lexical functions can be accessed at http://148.204.58.221/okolesnikova/index.php?id=lex/ or http://www.gelbukh.com/lexical-functions.

other three LFs: Oper1, CausFunc0, and CausFunc1. To make our dataset fully balanced, we randomly selected 60 samples of each of the three LFs; therefore, the dataset included 60 samples of each LF. Table 1 presents a few examples of our data.

Table 1. Examples of data used in our experiments

Lexical function	Verb-noun collocations	
	Spanish	English translation
Oper1	*realizar un estudio*	do a study
	cometer un error	make an error
	dar un beso	give a kiss
Real1	*alcanzar el nivel*	reach a level
	utilizar recurso	use a resource
	cumplir la función	fulfill the function
CausFunc0	*crear una cuenta*	create an account
	formar un grupo	form a group
	hacer ruido	make noise
CausFunc1	*ofrecer una posibilidad*	offer a possibility
	causar un problema	cause a problem
	crear una condición	create a condition

The corpus used in our experiments is a collection of 1,131 issues of the Excelsior newspaper in Spanish from April 1, 1996 to June 24, 1999. As vector features, we used four words to the left of the verb and four words to the right of the noun without taking into account the words between the verb and the noun.

3.2 Experimental Configurations

In order to explore the context of collocations in more detail, we performed experiments using different parts of the context. Context words were represented using the bag of words model, i.e., the position of each word and the order of words were not taken into account.

First, we lemmatized the corpus using the patern.es package for Python [15]. Secondly, some experiments were performed on the whole corpus and the other experiments were made on the same corpus after removing stopwords. Then, we used three options to capture the context: (1) four words to the left of the verb, (2) four words to the right of the noun, and (3) all eight words, that is, (1) and (2) concatenated. These options produced six experimental configurations:

1. Four words to the left of the verb on the whole corpus.
2. Four words to the left of the verb on the corpus with stopwords removed.
3. Four words to the right of the noun on the whole corpus.
4. Four words to the right of the noun on the corpus with stopwords removed.
5. The left and right contexts concatenated on the whole corpus.
6. The left and right contexts concatenated on the corpus with stopwords removed.

In addition, for each of the six experimental configurations in the above list, we implemented two vector representations of a collocation's context: in the first representation, we used context word counts as vector features, and in the second one, we used tf-idf values for each context word as vector features. Therefore, there were 12 experimental configurations in total.

We experimented with several supervised learning methods using the Scikit-learn package for Python [10]. Here are the methods we used in the experiments; each method has its implementation in the Scikit-learn package with the name given in parenthesis:

- Naïve Bayes (Multinomial NB, Gaussian NB),
- support vector machine (Linear SVC, SVC),
- multi-layer perceptron (MLPClassifier),
- k-nearest neighbors vote (KNeighborsClassifier),
- Gaussian processes for probabilistic classification (GaussianProcessClassifier),
- decision tree multi-class classification (DecisionTreeClassifier),
- random forest algorithm (RandomForestClassifier),
- AdaBoost-SAMME algorithm (AdaBoostClassifier),
- classification via generating a quadratic decision boundary by adjusting class conditional densities according to the data and employing Bayes' rule (QuadraticDiscriminantAnalysis).

All methods were executed using their default parameters in the Scikit-learn package [10]. A subset of 50% of the dataset was used for training, and the other 50% was used for validation.

In the next section, we give the results of our experiments and their discussion.

4   Results and Discussion

Tables 2, 3, 4 and 5 present the results of classifying lexical functions on all 12 experimental configurations explained in Sect. 3.2.

The tables are divided vertically into two parts. Each part includes the six configurations listed in the order indicated in Sect. 3.2. The left part contains the results for six configurations run on the lemmatized corpus using word counts as features in vectors, and the right part includes the results for the same six configurations run on the same lemmatized corpus using tf-idf of context words as features in vectors. The performance of the classifiers was evaluated in terms of precision (P), recall (R), and F-measure (F). For convenience, F-measure values are in bold. The three highest F-measure values for each lexical function are underlined.

For Oper1 (Table 2), the best result was an F-measure value of 0.41 showed by support vector machine (Linear SVC) on the fifth configuration (the left and write contexts concatenated on the whole corpus) with word counts as features in vectors as well as on the first configuration (the left context on the whole corpus) with tf-idf as features in vectors. The next two highest values of 0.40 and 0.39 were achieved by the same technique, however, using other context representations. The value of 0.40 was reached on the first configuration with word counts, and 0.39 was achieved on the fifth

Table 2. Experimental results for Oper1

Classifier	Metrics	1	2	3	4	5	6	1	2	3	4	5	6
		Word counts						tf-idf					
Multinomial NB	P	0.19	0.00	0.19	0.00	0.18	0.00	0.15	0.00	0.15	0.00	0.15	0.00
	R	0.86	0.00	0.86	0.00	0.82	0.00	1.00	0.00	1.00	0.00	1.00	0.00
	F	**0.32**	**0.00**	**0.31**	**0.00**	**0.30**	**0.00**	**0.26**	**0.00**	**0.26**	**0.00**	**0.26**	**0.00**
Gaussian NB	P	0.17	0.08	0.17	0.00	0.15	0.08	0.16	0.08	0.16	0.00	0.15	0.08
	R	0.82	0.25	0.68	0.00	0.73	0.25	0.77	0.25	0.68	0.00	0.68	0.25
	F	**0.28**	**0.12**	**0.27**	**0.00**	**0.25**	**0.12**	**0.27**	**0.12**	**0.27**	**0.00**	**0.24**	**0.12**
Linear SVC	P	0.32	0.00	0.26	0.00	0.34	0.00	0.30	0.00	0.25	0.00	0.29	0.00
	R	0.55	0.00	0.41	0.00	0.50	0.00	0.64	0.00	0.59	0.00	0.59	0.00
	F	**0.40**	**0.00**	**0.32**	**0.00**	**0.41**	**0.00**	**0.41**	**0.00**	**0.35**	**0.00**	**0.39**	**0.00**
SVC	P	0.27	0.00	0.20	0.00	0.18	0.00	0.15	0.00	0.15	0.00	0.15	0.00
	R	0.41	0.00	0.41	0.00	0.32	0.00	1.00	0.00	1.00	0.00	1.00	0.00
	F	**0.33**	**0.00**	**0.27**	**0.00**	**0.23**	**0.00**	**0.26**	**0.00**	**0.26**	**0.00**	**0.26**	**0.00**
MLP classifier	P	0.31	0.00	0.29	0.00	0.22	0.00	0.22	0.00	0.18	0.00	0.17	0.00
	R	0.55	0.00	0.32	0.00	0.09	0.00	0.77	0.00	0.64	0.00	0.64	0.00
	F	**0.39**	**0.00**	**0.30**	**0.00**	**0.13**	**0.00**	**0.34**	**0.00**	**0.28**	**0.00**	**0.27**	**0.00**
KNeighbors classifier	P	0.33	0.00	0.19	0.00	0.23	0.00	0.17	0.00	0.20	0.00	0.23	0.00
	R	0.50	0.00	0.23	0.00	0.32	0.00	0.45	0.00	0.41	0.00	0.73	0.00
	F	**0.40**	**0.00**	**0.21**	**0.00**	**0.26**	**0.00**	**0.25**	**0.00**	**0.27**	**0.00**	**0.35**	**0.00**
Gaussian process classifier	P	0.00	0.00	0.00	0.00	0.00	0.00	0.24	0.00	0.17	0.00	0.20	0.00
	R	0.00	0.00	0.00	0.00	0.00	0.00	0.95	0.00	0.82	0.00	0.91	0.00
	F	**0.00**	**0.00**	**0.00**	**0.00**	**0.00**	**0.00**	**0.38**	**0.00**	**0.29**	**0.00**	**0.33**	**0.00**
Decision tree classifier	P	0.24	0.00	0.21	0.00	0.20	0.00	0.27	0.00	0.24	0.00	0.26	0.00
	R	0.36	0.00	0.32	0.00	0.36	0.00	0.55	0.00	0.45	0.00	0.41	0.00
	F	**0.29**	**0.00**	**0.25**	**0.00**	**0.26**	**0.00**	**0.36**	**0.00**	**0.32**	**0.00**	**0.32**	**0.00**
Random forest classifier	P	0.29	0.00	0.13	0.00	0.25	0.00	0.16	0.00	0.17	0.00	0.14	0.00
	R	0.45	0.00	0.18	0.00	0.36	0.00	0.27	0.00	0.32	0.00	0.27	0.00
	F	**0.36**	**0.00**	**0.15**	**0.00**	**0.30**	**0.00**	**0.20**	**0.00**	**0.22**	**0.00**	**0.18**	**0.00**
AdaBoost classifier	P	0.11	0.00	0.22	0.00	0.16	0.00	0.16	0.00	0.09	0.00	0.13	0.00
	R	0.05	0.00	0.50	0.00	0.27	0.00	0.50	0.00	0.32	0.00	0.32	0.00
	F	**0.06**	**0.00**	**0.31**	**0.00**	**0.20**	**0.00**	**0.24**	**0.00**	**0.14**	**0.00**	**0.19**	**0.00**
Quadratic discriminant analysis	P	0.00	0.08	0.35	0.00	0.07	0.00	0.16	0.08	0.17	0.00	0.12	0.00
	R	0.00	0.25	0.27	0.00	0.09	0.00	0.23	0.75	0.64	0.00	0.14	0.00
	F	**0.00**	**0.12**	**0.31**	**0.00**	**0.08**	**0.00**	**0.19**	**0.15**	**0.26**	**0.00**	**0.12**	**0.00**

configuration with tf-idf. It can be observed that the first configuration with word counts was more successful for Oper1 identification, as the k-nearest-neighbors method (KNeighborsClassifier) showed an F-measure of 0.40, and the multi-layered perceptron (MLPClassifier) reached an F-measure of 0.39 on it.

Now let us discuss how the context of verb-noun collocations distinguishes the Real1 lexical function. The results on Real1 are in Table 3. As one can see, here the values are higher than those for Oper1 in Table 2. The top F-measure value of 0.65 was achieved by Gaussian Naïve Bayes algorithm (GaussianNB) on the second configuration (the left context on the corpus with stopwords removed) with word counts as features. The same method reached 0.64 on the same configuration but with tf-idf

Table 3. Experimental results for Real1

Classifier	Metrics	1	2	3	4	5	6	1	2	3	4	5	6
		Word counts						Word tf-idf					
Multinomial NB	P	0.33	0.48	0.38	0.35	0.39	0.46	1.00	0.00	1.00	0.36	0.00	0.00
	R	0.56	0.73	0.48	0.67	0.63	0.51	0.11	0.00	0.04	0.91	0.00	0.00
	F	**0.42**	**0.50**	**0.43**	**0.46**	**0.48**	**0.49**	**0.20**	**0.00**	**0.07**	**0.51**	**0.00**	**0.00**
Gaussian NB	P	0.38	0.54	0.35	0.38	0.38	0.38	0.36	0.53	0.39	0.37	0.36	0.40
	R	0.44	0.81	0.41	0.88	0.37	0.43	0.48	0.81	0.41	0.85	0.37	0.51
	F	**0.41**	**_0.65_**	**0.38**	**0.53**	**0.38**	**0.41**	**0.41**	**_0.64_**	**0.40**	**0.52**	**0.36**	**0.45**
Linear SVC	P	0.44	0.67	0.30	0.30	0.37	0.57	0.46	0.63	0.41	0.41	0.48	0.46
	R	0.56	0.38	0.26	0.33	0.37	0.22	0.70	0.32	0.56	0.55	0.81	0.16
	F	**0.49**	**0.48**	**0.28**	**0.31**	**0.37**	**0.31**	**0.56**	**0.43**	**0.47**	**0.47**	**0.60**	**0.24**
SVC	P	0.25	0.67	0.25	0.57	0.25	0.80	0.00	0.00	0.00	0.37	0.00	0.00
	R	0.59	0.05	0.52	0.12	0.59	0.11	0.00	0.00	0.00	0.42	0.00	0.00
	F	**0.36**	**0.10**	**0.34**	**0.20**	**0.35**	**0.19**	**0.00**	**0.00**	**0.00**	**0.39**	**0.00**	**0.00**
MLP classifier	P	0.41	1.00	0.25	0.36	0.25	0.00	0.40	0.67	0.40	0.39	0.50	0.37
	R	0.59	0.03	0.56	0.15	0.67	0.00	0.70	0.32	0.22	0.79	0.59	0.19
	F	**0.48**	**0.05**	**0.35**	**0.21**	**0.36**	**0.00**	**0.51**	**0.44**	**0.29**	**0.53**	**0.54**	**0.25**
KNeighbors classifier	P	0.38	0.62	0.26	0.75	0.26	0.83	0.55	0.00	0.36	0.35	0.42	1.00
	R	0.30	0.14	0.19	0.09	0.33	0.14	0.22	0.00	0.37	0.76	0.59	0.05
	F	**0.33**	**0.22**	**0.22**	**0.16**	**0.29**	**0.23**	**0.32**	**0.00**	**0.36**	**0.48**	**0.49**	**0.10**
Gaussian process classifier	P	0.19	0.45	0.19	0.35	0.18	0.45	0.39	1.00	0.24	0.44	0.42	0.00
	R	1.00	0.95	1.00	0.97	1.00	0.97	0.70	0.03	0.30	0.85	0.70	0.00
	F	**0.32**	**_0.61_**	**0.32**	**0.52**	**0.31**	**0.62**	**0.50**	**0.05**	**0.26**	**0.58**	**0.53**	**0.00**
Decision tree classifier	P	0.32	0.58	0.24	0.30	0.28	0.36	0.32	0.55	0.24	0.33	0.22	0.33
	R	0.41	0.30	0.15	0.33	0.30	0.11	0.41	0.30	0.37	0.39	0.26	0.22
	F	**0.36**	**0.39**	**0.18**	**0.31**	**0.29**	**0.17**	**0.36**	**0.39**	**0.29**	**0.36**	**0.24**	**0.26**
Random forest classifier	P	0.35	0.59	0.30	0.28	0.23	0.56	0.28	0.63	0.24	0.45	0.23	0.57
	R	0.33	0.27	0.48	0.33	0.22	0.24	0.30	0.59	0.22	0.61	0.22	0.22
	F	**0.34**	**0.37**	**0.37**	**0.31**	**0.23**	**0.34**	**0.29**	**_0.61_**	**0.23**	**0.52**	**0.23**	**0.31**
AdaBoost classifier	P	0.14	0.42	0.16	0.42	0.28	0.39	0.08	0.80	0.06	0.46	0.21	0.45
	R	0.22	0.38	0.33	0.30	0.19	0.32	0.07	0.22	0.04	0.18	0.30	0.59
	F	**0.17**	**0.40**	**0.22**	**0.35**	**0.22**	**0.35**	**0.08**	**0.34**	**0.05**	**0.26**	**0.24**	**0.51**
Quadratic discriminant analysis	P	0.22	0.44	0.19	0.33	0.18	0.42	0.22	0.50	0.18	0.39	0.29	0.44
	R	0.89	0.95	0.26	0.76	0.70	0.81	0.56	0.49	0.07	0.67	0.70	0.73
	F	**0.35**	**0.60**	**0.22**	**0.46**	**0.29**	**0.55**	**0.32**	**0.49**	**0.11**	**0.49**	**0.41**	**0.55**

values as features. It is interesting to note that on the same configuration with word counts, GaussianProcessClassifier showed an F-measure value of 0.61 and RandomForestClassifier reached an F-measure of 0.61 on the same configuration but with tf-idf as vector features. We can conclude that the left context seems to be a better indicator of Real 1 than the right context.

Table 4 presents the results on CausFunc0 detection. Here, the values are lower than those for Real1 in Table 3, but higher than the numbers for Oper1 in Table 2. All top results were reached with word counts as features in vectors. The highest F-measure value of 0.51 was given by the multi-layer perceptron (MLPClassifier) on the first configuration (the left context on the whole corpus). The second-best value for Real1 of

Table 4. Experimental results for CausFunc0

Classifier	Metrics	1	2	3	4	5	6	1	2	3	4	5	6
		Word counts						tf-idf					
Multinomial NB	P	0.00	0.17	0.00	0.00	0.00	0.12	0.00	0.18	0.00	0.20	0.00	0.18
	R	0.00	0.35	0.00	0.00	0.00	0.29	0.00	1.00	0.00	0.05	0.00	1.00
	F	**0.00**	**0.23**	**0.00**	**0.00**	**0.00**	**0.17**	**0.00**	**0.30**	**0.00**	**0.08**	**0.00**	**0.30**
Gaussian NB	P	0.50	0.30	0.00	0.25	1.00	0.24	0.33	0.32	1.00	0.14	1.00	0.21
	R	0.03	0.41	0.00	0.05	0.03	0.47	0.03	0.41	0.03	0.05	0.03	0.35
	F	**0.05**	**0.35**	**0.00**	**0.08**	**0.06**	**0.31**	**0.05**	**0.36**	**0.06**	**0.07**	**0.06**	**0.26**
Linear SVC	P	0.62	0.20	0.29	0.13	0.44	0.21	0.64	0.26	0.45	0.14	0.53	0.20
	R	0.37	0.18	0.29	0.15	0.43	0.24	0.26	0.29	0.26	0.10	0.26	0.29
	F	**0.46**	**0.19**	**0.29**	**0.14**	**0.43**	**0.22**	**0.37**	**0.28**	**0.33**	**0.12**	**0.35**	**0.24**
SVC	P	0.29	0.19	0.26	0.00	0.26	0.19	0.00	0.18	0.00	0.00	0.00	0.18
	R	0.26	1.00	0.34	0.00	0.29	1.00	0.00	1.00	0.00	0.00	0.00	1.00
	F	**0.27**	**0.32**	**0.29**	**0.00**	**0.27**	**0.32**	**0.00**	**0.30**	**0.00**	**0.00**	**0.00**	**0.30**
MLP classifier	P	0.49	0.15	0.39	0.00	0.39	0.23	0.71	0.22	0.32	0.12	0.44	0.18
	R	0.54	0.18	0.40	0.00	0.37	0.18	0.14	0.24	0.20	0.05	0.11	0.18
	F	**0.51**	**0.16**	**0.39**	**0.00**	**0.38**	**0.20**	**0.24**	**0.23**	**0.25**	**0.07**	**0.18**	**0.18**
KNeighbors classifier	P	0.37	0.12	0.30	0.20	0.27	0.12	0.00	0.00	0.21	0.11	0.67	0.00
	R	0.40	0.18	0.43	0.05	0.34	0.18	0.00	0.00	0.09	0.05	0.06	0.00
	F	**0.38**	**0.14**	**0.35**	**0.08**	**0.30**	**0.14**	**0.00**	**0.00**	**0.12**	**0.07**	**0.11**	**0.00**
Gaussian process classifier	P	0.00	0.00	0.80	0.00	0.00	0.00	0.00	0.15	0.00	1.00	0.00	0.14
	R	0.00	0.00	0.11	0.00	0.00	0.00	0.00	0.29	0.00	0.05	0.00	0.29
	F	**0.00**	**0.00**	**0.20**	**0.00**	**0.00**	**0.00**	**0.00**	**0.20**	**0.00**	**0.10**	**0.00**	**0.19**
Decision tree classifier	P	0.24	0.19	0.21	0.20	0.24	0.21	0.39	0.29	0.38	0.40	0.28	0.19
	R	0.14	0.29	0.17	0.25	0.23	0.29	0.26	0.59	0.17	0.30	0.29	0.24
	F	**0.18**	**0.23**	**0.19**	**0.22**	**0.24**	**0.24**	**0.31**	**0.38**	**0.24**	**0.34**	**0.28**	**0.21**
Random forest classifier	P	0.38	0.18	0.32	0.23	0.42	0.16	0.21	0.10	0.33	0.35	0.32	0.23
	R	0.43	0.35	0.26	0.35	0.46	0.29	0.09	0.12	0.40	0.30	0.29	0.35
	F	**0.41**	**0.24**	**0.29**	**0.28**	**0.44**	**0.21**	**0.12**	**0.11**	**0.36**	**0.32**	**0.30**	**0.28**
AdaBoost classifier	P	0.20	0.26	0.20	0.56	0.34	0.31	0.28	0.24	0.18	0.29	0.29	0.26
	R	0.26	0.53	0.06	0.25	0.34	0.65	0.26	0.76	0.06	0.20	0.14	0.41
	F	**0.22**	**0.35**	**0.09**	**0.34**	**0.34**	**0.42**	**0.27**	**0.36**	**0.09**	**0.24**	**0.19**	**0.32**
Quadratic discriminant analysis	P	0.00	0.33	0.31	0.00	0.00	0.10	0.00	0.00	0.57	0.00	0.29	0.16
	R	0.00	0.06	0.11	0.00	0.00	0.06	0.00	0.00	0.11	0.00	0.14	0.24
	F	**0.00**	**0.10**	**0.17**	**0.00**	**0.00**	**0.07**	**0.00**	**0.00**	**0.19**	**0.00**	**0.19**	**0.19**

0.46 was reached on the same configuration by support vector machine (LinearSVC). At last, an F-measure value of 0.44 was reached by RandomForestClassifier on the fifth configuration (the left and right contexts concatenated on the whole corpus).

Concerning CausFunc1, the results on this lexical function can be seen in Table 5. It is interesting that all three top results were achieved using word tf-idf values as vector features. The best result was an F-measure value of 0.59 reached by the support vector machine (LinearSVC) on the fourth configuration (the right context on the corpus with stopwords removed). The same value was showed by AdaBoostClassifier on the sixth configuration (the left and right context on the corpus with stopwords

Table 5. Experimental results for CausFunc1

Classifier	Metrics	1	2	3	4	5	6	1	2	3	4	5	6
		Word counts						tf-idf					
Multinomial NB	P	0.40	0.00	0.57	0.36	0.75	0.00	0.00	0.00	0.00	0.60	0.00	0.00
	R	0.06	0.00	0.24	0.24	0.09	0.00	0.00	0.00	0.00	0.18	0.00	0.00
	F	**0.10**	**0.00**	**0.33**	**0.29**	**0.16**	**0.00**	**0.00**	**0.00**	**0.00**	**0.27**	**0.00**	**0.00**
Gaussian NB	P	0.29	0.00	0.29	0.50	0.14	0.00	0.40	0.00	0.29	0.50	0.13	0.00
	R	0.06	0.00	0.21	0.24	0.06	0.00	0.06	0.00	0.21	0.24	0.06	0.00
	F	**0.10**	**0.00**	**0.24**	**0.32**	**0.08**	**0.00**	**0.10**	**0.00**	**0.24**	**0.32**	**0.08**	**0.00**
Linear SVC	P	0.43	0.33	0.50	0.21	0.39	0.29	0.56	0.26	0.56	0.48	0.53	0.26
	R	0.35	0.70	0.32	0.24	0.26	0.60	0.56	0.70	0.53	0.76	0.50	0.60
	F	**0.39**	**0.44**	**0.39**	**0.22**	**0.32**	**0.39**	**0.56**	**0.38**	**0.55**	**0.59**	**0.52**	**0.36**
SVC	P	0.43	0.00	0.00	0.20	0.50	0.00	0.00	0.00	0.00	0.27	0.00	0.00
	R	0.18	0.00	0.00	1.00	0.09	0.00	0.00	0.00	0.00	0.88	0.00	0.00
	F	**0.25**	**0.00**	**0.00**	**0.33**	**0.15**	**0.00**	**0.00**	**0.00**	**0.00**	**0.41**	**0.00**	**0.00**
MLP classifier	P	1.00	0.39	0.56	0.20	0.62	0.36	0.60	0.29	0.41	0.45	0.55	0.27
	R	0.03	0.55	0.26	0.88	0.15	0.70	0.18	0.75	0.35	0.53	0.35	0.65
	F	**0.06**	**0.46**	**0.36**	**0.32**	**0.24**	**0.47**	**0.27**	**0.42**	**0.38**	**0.49**	**0.43**	**0.38**
KNeighbors classifier	P	0.42	0.24	0.34	0.20	0.50	0.25	0.38	0.13	0.37	0.17	0.36	0.20
	R	0.29	0.65	0.32	1.00	0.15	0.70	0.74	0.50	0.50	0.06	0.24	0.85
	F	**0.34**	**0.35**	**0.33**	**0.34**	**0.23**	**0.36**	**0.50**	**0.20**	**0.42**	**0.09**	**0.29**	**0.32**
Gaussian process classifier	P	0.00	0.50	0.00	0.33	0.00	0.60	0.50	0.25	0.50	0.45	1.00	0.24
	R	0.00	0.45	0.00	0.06	0.00	0.45	0.15	0.75	0.15	0.76	0.06	0.70
	F	**0.00**	**0.47**	**0.00**	**0.10**	**0.00**	**0.51**	**0.23**	**0.38**	**0.23**	**0.57**	**0.11**	**0.35**
Decision tree classifier	P	0.45	0.34	0.24	0.12	0.33	0.36	0.29	0.46	0.44	0.09	0.33	0.55
	R	0.41	0.50	0.24	0.12	0.26	0.50	0.15	0.60	0.32	0.12	0.18	0.60
	F	**0.43**	**0.41**	**0.24**	**0.12**	**0.30**	**0.42**	**0.20**	**0.52**	**0.37**	**0.10**	**0.23**	**0.57**
Random forest classifier	P	0.47	0.38	0.40	0.15	0.48	0.40	0.19	0.31	0.36	0.27	0.32	0.30
	R	0.24	0.50	0.35	0.18	0.35	0.60	0.15	0.50	0.24	0.47	0.24	0.50
	F	**0.31**	**0.43**	**0.38**	**0.16**	**0.41**	**0.48**	**0.17**	**0.38**	**0.29**	**0.34**	**0.27**	**0.38**
AdaBoost classifier	P	0.48	0.50	0.30	0.28	0.46	0.64	0.25	0.47	0.86	0.22	0.50	0.71
	R	0.32	0.50	0.09	0.65	0.32	0.45	0.03	0.45	0.18	0.88	0.09	0.50
	F	**0.39**	**0.50**	**0.14**	**0.39**	**0.38**	**0.53**	**0.05**	**0.46**	**0.29**	**0.36**	**0.15**	**0.59**
Quadratic discriminant analysis	P	0.31	0.00	0.28	0.12	0.00	0.00	0.60	0.00	0.33	0.11	0.31	0.00
	R	0.29	0.00	0.47	0.06	0.00	0.00	0.18	0.00	0.32	0.06	0.15	0.00
	F	**0.30**	**0.00**	**0.35**	**0.08**	**0.00**	**0.00**	**0.27**	**0.00**	**0.33**	**0.08**	**0.20**	**0.00**

removed). The second best result was an F-measure of 0.57 showed by two methods: GaussianProcessClassifier and DecisionTreeClassifier on the fourth (the right context on the corpus with stopwords removed) and the sixth configuration (the left and right context on the corpus with stopwords removed), respectively.

Now, speaking of all four lexical functions used in the experiments, the best F-measure value was 0.65 reached by Gaussian Naïve Bayes (NB) for Real1 on the second configuration (the left context on the corpus with stopwords removed) with word counts as features in vectors. However, if we look at the performance of the same method on the same configuration in terms of F-measure, we can see that GaussianNB was not able to distinguish CausFunc1 at all (0.00), showed very poor performance for Oper1 (0.12), although slightly better for CausFunc0 (0.35).

The best classifier for Oper1 was the support vector machine (LinearSVC) on the fifth configuration (the left and right context on the whole corpus) with word counts as features in vector representation and on the first configuration (the left context on the whole corpus) with tf-idf values as features. This method showed an F-measure of 0.41. Observing the performance of LinearSVC on the same two configurations, it can be seen that it gave F-measure values of 0.37 and 0.56 for respective configurations on Real1, 0.43 and 0.37 for CausFunc0, and finally, 0.32 and 0.56 for CausFunc1.

CausFunc0 was best identified by the multi-layer perceptron (MLPClassifier) with an F-measure of 0.51 on the first configuration (the left context on the whole corpus) with word counts as features in vectors. On the same configuration, this method showed an F-measure of 0.39 for Oper1, a value of 0.48 for Real1, but it was not able to detect CausFunc1: MLPClassifier gave an F-measure as low as 0.06 for this lexical function.

Now let was see how the best classifier for CausFunc1 performed on other LFs. For CausFunc1, the highest result was showed by the support vector machine (LinearSVC), it gave an F-measure of 0.59 on the forth configuration (the right context on the corpus with stopwords removed) with tf-idf values as features in vectors. Revising the same configurations for the other three LFs, we see that LinearSVC showed an F-measure value of 0.47 on Real1, performed very poorly for CausFunc0 (0.12), and could not detect Oper1 at all (0.00).

Summarizing the observations presented in the previous paragraphs, we can say that there is no single method nor single context representation among those we experimented with good enough to detect all four lexical functions. Therefore, we recommend using an LF-specific classifier and context representation to identify a particular LF automatically.

In addition, we see that the results in general are not as high as in works in which hypernyms were used to distinguish among lexical functions [1, 18, 20], see Sect. 2. On the other hand, our results on using context for LF identification are not too low to state that LFs are not specified by their context at all. Surely, the words in the context of verb-noun collocations are indicators of lexical functions. More data and research are necessary to study the context of LFs in more detail.

5 Conclusions and Future Work

In this work, we explored the context of verb-noun collocations and studied to what extent the context is able to distinguish among four lexical functions (LFs) of the Meaning-Text Theory developed by Mel'čuk [7]: Oper1, Real1, CausFunc0, and CausFunc1. The context was composed of four words to the left of the verb and of the four words to the right of the noun. The words in between the verb and the noun were not taken into account in this research. The bag of words model was used to represent the context.

In our experiments on Spanish texts we used 11 supervised learning methods and 12 context representation configurations to see to what extent LFs can be distinguished by their context. The highest F-measure values reached in our experiments were the following ones: 0.41 for Oper1 showed by the support vector machine using the left and right context with word counts as features in the vector representation as well as the left context with tf-idf as features; 0.65 for Real1 showed by Gaussian Naïve Bayes using the left context without stopwords and word counts as features in vectors; 0.51 for CausFunc0 achieved by the multi-layer perceptron using the left context with word counts as features; and 0.59 for CausFunc1 showed by the support vector machine using the right context without stopwords and tf-idf values as features.

No single method and no single context representation could distinguish all four lexical functions in our experiments; so far, we have seen that methods and context representations are specific for each lexical function.

In our future work, we plan to experiment with other window sizes, context representations and similarity measures [3, 6, 14], as well as methods to see if lexical function detection can be improved. In addition, we will study distribution of context words depending on their positions with respect to the verb and the noun in collocations.

Acknowledgements. The research was done under partial support of Mexican Government: SNI, BEIFI-IPN, and SIP-IPN grants 20182119 and 20181792. The work was done when A. Gelbukh was visiting the Research Institute for Information and Language Processing, University of Wolverhampton, on a grant from the Sabbatical Year Program of the CONACYT, Mexico.

References

1. Gelbukh, A., Kolesnikova, O.: Supervised learning for semantic classification of Spanish collocations. In: Martínez-Trinidad, J.F., Carrasco-Ochoa, J.A., Kittler, J. (eds.) MCPR 2010. LNCS, vol. 6256, pp. 362–371. Springer, Heidelberg (2010). https://doi.org/10.1007/978-3-642-15992-3_38
2. Gerdes, K., Reuther, T., Wanner, L. (eds.): MTT 2007: Meaning-Text Theory 2007: Proceedings of the 3rd International Conference on Meaning-Text Theory, Klagenfurt, Austria (2007)
3. Gómez-Adorno, H., Posadas-Duran, J.-P., Ríos-Toledo, G., Sidorov, G., Sierra, G.: Stylometry-based approach for detecting writing style changes in literary texts. Computación y Sistemas 22(1), 47–53 (2018)
4. Kahane, S.: The meaning-text theory. Dependency and valency. In: An International Handbook of Contemporary Research, vol. 1, pp. 546–570. Walter de Gruyter, Berlin (2003)
5. Machova, S.: Meaning-text theory. Comput. Linguist. 18(1), 108–111 (1992)
6. Majumder, G., Pakray, P., Gelbukh, A., Pinto, D.: Semantic textual similarity methods, tools, and applications: a survey. Computación y Sistemas 20(4), 647–665 (2016)
7. Mel'čuk, I.A.: Lexical functions: a tool for the description of lexical relations in a lexicon. In: Wanner, L. (ed.) Lexical Functions in Lexicography and Natural Language Processing, pp. 37–102. Benjamins Academic Publishers, Amsterdam (1996)
8. Mille, S., Wanner, L., Burga, A.: Treebank annotation in the light of the Meaning-Text Theory. Linguist. Issues Lang. Technol. 7(16), 1–12 (2012)
9. Miller, G.A., Leacock, C., Tengi, R., Bunker, R.T.: A semantic concordance. In: Proceedings of the Workshop on Human Language Technology, pp. 303–308. Association for Computational Linguistics (1993)

10. Pedregosa, F., et al.: Scikit-learn: machine learning in Python. J. Mach. Learn. Res. **12**, 2825–2830 (2011)
11. Polguère, A.: Towards a theoretically-motivated general public dictionary of semantic derivations and collocations for French. In: Proceedings of EURALEX 2000, Stuttgart, Germany (2000)
12. Ramos, A.M., Wanner, L., Veiga, N.V., Vincze, O., Suárez, E.M., González, S.P.: Tagging collocations for learners. In: Granger, S., Paquot, M. (eds.) Proceedings of ELex 2009, pp. 375–380. Presses universitaires de Louvain, Louvain-la-Neuve (2010)
13. Sheremetyeva, S., Babina, O.: Meaning-Text Theory for textual input analysis and proofing in a generation system. In: Apresjan, Y., Iomdin, L. (eds.) Proceedings of the Second International Conference on the Meaning-Text Model, pp. 458–466. Slavic Culture Languages Publishing House, Moscow (2005)
14. Sidorov, S., Gelbukh, A., Gómez-Adorno, H., Pinto, D.: Soft similarity and soft cosine measure: similarity of features in vector space model. Computación y Sistemas **18**(3), 491–504 (2014)
15. Smedt, T.D., Daelemans, W.: Pattern for Python. J. Mach. Learn. Res. **13**, 2063–2067 (2012)
16. The University of Waikato Computer Science Department Machine Learning Group, WEKA download. http://www.cs.waikato.ac.nz/~ml/weka/index_downloading.html
17. Tutin, A.: Annotating lexical functions in corpora: showing collocations in context. In: Apresjan, Y., Iomdin, L. (eds.) Proceedings of the Second International Conference on the Meaning-Text Model, pp. 498–510. Slavic Culture Languages Publishing House, Moscow (2017)
18. Wanner, L.: Towards automatic fine-grained classification of verb-noun collocations. Nat. Lang. Eng. **10**(2), 95–143 (2004)
19. Wanner, L.: Selected Lexical and Grammatical Issues in the Meaning-Text Theory. Honour of Igor Mel'cuk. Benjamins, Amsterdam/Philadelphia (2007)
20. Wanner, L., Bohnet, B., Giereth, M.: What is beyond collocations? Insights from machine learning experiments. In: Proceedings of the 12th EURALEX International Congress, pp. 1071–1084, Turin, Italy (2006)
21. Witten, I.H., Frank, E.: Data Mining: Practical Machine Learning Tools and Techniques, 2nd edn. Morgan Kaufmann, San Francisco (2005)

Towards a Natural Language Compiler

Angel Zúñiga[1]([⊠]) [iD], Gerardo Sierra[1] [iD], Gemma Bel-Enguix[1] [iD],
and Sofía N. Galicia-Haro[2]

[1] Instituto de Ingeniería, Universidad Nacional Autónoma de México,
Mexico City, Mexico
{azunigac,gsierram,gbele}@iingen.unam.mx
[2] Facultad de Ciencias, Universidad Nacional Autónoma de México,
Mexico City, Mexico
sngh@fciencias.unam.mx

Abstract. Being able to create a natural language compiler has been
one of the most sought-after goals to reach since the very beginning of
artificial intelligence. Since then; however, it has been an elusive and difficult task to achieve to the extent of being considered almost impossible
to perform. In this article, we present a promising path by using a grammar formalism which attempts to model natural language; in principle,
by using minimalist grammars as one of the last proposed instances of
formalism of this type. The main idea consists in creating a parser based
on this type of grammars which could recognize and analyze the text
(or input program) written in natural language and use this parser as
a front-end of a compiler. Then, for the rest of the compilation process,
utilize the usual phases of a classic compiler of a programming language.
Moreover, we present a prototype of a natural language compiler whose
specific language is that of arithmetic expressions, in order to show with
evidence that it is indeed possible to implement it, that is to say, to put
the proposed compiler design into practice, showing in this manner that
it is actually possible to create a natural language compiler following this
promising path.

Keywords: Natural language compilers · Natural language parsing
Mildly context sensitive grammars · Minimalist grammars
Natural language processing

1 Introduction

Building a natural language compiler has been regarded as one of the most
important problems in computer science since the beginning of artificial intelligence. However, throughout the years, this task has been elusive and difficult
to achieve, to the extent of being considered impossible. Perhaps because of
this, the current works with this goal are very scarce and very limited (see, for
example, [9,21]). This work has the objective of showing that there is a possible
path of research to follow to build a natural language compiler. For this, we

ⓒ Springer Nature Switzerland AG 2018
I. Batyrshin et al. (Eds.): MICAI 2018, LNAI 11289, pp. 70–82, 2018.
https://doi.org/10.1007/978-3-030-04497-8_6

will present a design and architecture of this type of compilers together with a compiler prototype with minimum functionalities that implements the aforementioned design. The aim is to offer practical evidence that a natural language compiler can be developed and used in real life.

Nowadays, we consider that the programming language compilers field has both theoretical and practical bases that are solid and well-established (see, for example, [1]). So, if it is well known how to build programming language compilers, what stops natural language compilers from being built? In the programming language compiler community, we think a possible answer is that, to date, there is no complete, practical and effective method known for the recognition and analysis of natural language.

This drives us to focus our attention on the problem of natural language parsing. To do that, we have to ask ourselves what type of grammars, if any, model natural language. In regards to the latter, Seki et al. [15] say the following: "Literature on generative grammars shows often a mention of inadequacy of context-free grammars (cfg's) for describing the structures involving discontinuous constituents in natural languages [14]. Context-sensitive grammars (csg's or Type 1 grammars), on the other hand, may not be an adequate model of grammars of natural languages because they are too powerful in generative capacity, and phrase structures which are natural extension of phrase structures in cfg's are not defined in Type 0 and Type 1 grammars". On the other hand, in [12] it is mentioned: "Joshi [11] proposed that the class of grammars that is necessary for describing natural languages might be characterized as *mildly context-sensitive grammars* (MCSG)". Taking the latter into account, in principle, we must choose the type of MCSGs that are the most expressive. For this reason, we chose the minimalist grammars (MGs) introduced by Stabler [17]. MGs are a formalization of Chomsky's [5] minimalist program. Regarding the question whether MGs actually model natural language, Jäger and Rogers [10] in 2012 say the following: "Most experts therefore assume at this time that all natural languages are MG languages". In [10], it is also shown that MGs (and their equivalent grammars) contain any other MCSG.

Now, we must take a look at the MG syntactic analysis. Harkema [8] presents a bottom-up analyzer for this type of grammars that takes time $O(n^{4k+4})$ where k is a constant of the grammar. Stabler [20] offers a MG top-down parser that uses a beam to bound the search space. Although Stabler [20] does not provide an analysis of the complexity of this analyzer, everything indicates that due to the beam's use, its complexity is less than Harkema's analyzer, and it could even be $O(n)$. However, we do not know in detail what the complexity of this analyzer is, and a formal analysis of it, which has not been carried out until now, would be needed.

The syntactic analyzers that are used in programming language compilers (consult, for example, [2]) commonly work with a context-free grammar (CFG) or a subset of them, in particular deterministic CFGs. The main types of analyzers that are used are the following:

- Recursive Descent. This type of analyzers works with CFGs in general and can take $O(c^n)$ time, that is, exponential.
- Top-down Predictive. These take a $LL(k)$ grammar as input and are $O(n)$, that is, they take linear time. $LL(k)$ grammars are a CFG's proper subset.
- Bottom-up Predictive. These analyzers take $LR(k)$ grammars as input, and its complexity is $O(n)$, meaning, they take linear time. Deterministic CFGs are defined to be $LR(k)$ grammars.

The idea that is developed in this article is to use a MG analyzer instead of one of these to try to recognize and analyze natural language. This could be refuted because an analyzer that works with a MG takes too long time to be used in practice. In response, we argue that robust, complex compilers that are widely used in real life, such as the case of GCC and LLVM, use recursive descent analyzers that could take exponential time.

Once the syntactic analysis is done, to conduct the remaining compilation process, we plan to do the typical phases of a classic programming language compiler.

In regards to the rest of the article, we will follow a practical approach and we will show how to build a prototype with basic functionalities that implement the natural language compiler's design that we previously explained. The source language of this compiler will be the arithmetic expressions' language written in natural language. We work with both English and Spanish languages. Nevertheless, to simplify and due to lack of space, we only present the development in Spanish. In regards to the arithmetic expressions grammar presented in this paper, we considered that the reader can obtain the version in English in a simple way by making the pertinent small changes. We chose the arithmetic expressions language because it is a representative classic example in the field of compilers. The target language will be Python. We selected Python because, given that it is simple and clear, it is also one of the most used programming languages nowadays. The complete code of the natural language compiler prototype is available in [22].

Our main contributions are the following:

- The proposal of a design and architecture of a natural language compiler.
- A prototype with basic minimum functionalities that implements this design, specifically a compiler from arithmetic expressions written in natural language to Python.

Note that our intention in this paper is to introduce a general natural language compiler design and architecture that serve as basis for further extensions and refinements and to leave open the possibility of several implementations. Only to show that it is in fact possible to implement this design, we provide a minimum compiler prototype whose source language is that of arithmetic expressions written in natural language. As it might be expected, this prototype can be extended, or other implementations can be developed that follow this design and cover other natural language subsets; of course, the very final goal to reach is to build a compiler that covers all natural language.[1] Next, we sketch the structure of this paper.

First, in Sect. 2, we will briefly show the classic architecture of a programming language compiler, and we will introduce the architecture corresponding to the natural language compiler's design we propose. Then, in Sect. 3, we introduce a CFG that generates the arithmetic expressions language, so after that, introduce a MG that expresses the same language. In Sect. 4, we first show how to use an implementation of Stabler's MG analyzer. Later, we will describe how to encode our MG of arithmetic expressions, so it can be a valid input of this analyzer, for it to be able to recognize arithmetic expressions written in natural language. Afterwards, we will describe how to modify this analyzer to add a small code generator turning it in this way into a compiler. Next, in Sect. 5, we present the evaluation of our compiler prototype. Finally, in Sect. 6, we will discuss the possibility that MGs do not generate natural language in its entirety and how our design takes this case into account. In addition, we present our conclusions, and we will outline some possibilities of future work.

2   Compiler Architecture

The classic programming language compilers have an architecture that takes as basis the one shown in Fig. 1, that is, the vast majority has a phase of: lexical analysis, syntactic analysis, semantic analysis and code generation.

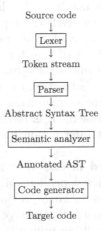

Fig. 1. Classic compiler architecture

The main problem of this architecture for our purposes is that the parser only works with a CFG or a subset of them. Due to this, our main idea is to

---

[1] We believe that it makes sense to start with easy to process and well-defined natural language subsets and subsequently to cover more complex challenging ones.

use a parser that works with a type of grammars that is capable of recognizing and analyzing natural language. The minimum basic architecture of the natural language compiler's design we propose is shown in Fig. 2. Of course, this architecture is designed only for being taken as a starting point to which one or several phases can be added in the future, as appropriate.

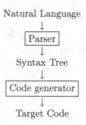

Fig. 2. Natural language compiler architecture

Based on it, the compilation process consists in taking as an input a program written in natural language which the parser will analyze generating a syntax tree as output. From there, the code generator takes as entry the aforementioned tree and generates the corresponding target code as output.

Of course, one can add a lexical analysis phase at the beginning that generates tokens for the parser. We have not done it here simply because the language of arithmetic expressions we will use is so simple that it does not complicate the fact that the syntactic analyzer reads the input directly. For more complex languages, we can add a lexer that could even perform morphological analysis and in this way, have an extra layer of help for the natural language analysis. On the other hand, instead of generating a syntax tree or a derivation tree, we can look for a way to generate an abstract syntax tree (AST) or a variation of it. This is justifiable because, from the computational viewpoint of a compiler (and not from a purely linguistic approach), it is advisable to have an AST, mainly because it only captures relevant information for the subsequent phases of the compilation process.

The semantic analysis is used in compilers mainly to perform two tasks: in general, to verify characteristics dependent of the context (for example, not using a variable that has not been declared previously) and in particular to perform type checking. Regarding natural language, we think that based on the grammar that is used to analyze it, in principle a MG, it is possible to verify the characteristics dependent of the context, reason why the syntactic analyzer would perform this task directly. In this way, it would not be necessary to add a semantic analyzer. On the other hand, it would be interesting to consider the possibility of defining a type system for natural language, particularly based on Curry-Howard's [16] isomorphism approach. This, from a linguistic approach, would be useful as a mean to study and understand natural language phenomena, and from the computational viewpoint, it would lead us to consider the necessity to formally prove that natural language is Turing-complete, or even ask the question of whether natural language has a computational power bigger than that of a Turing machine. From the very point of view of the compiler, it would help to detect (statically, meaning, in compilation time) input programs that make no sense such as: dos más cuatro más Roberto. Given that at the moment there is no semantic analyzer, this kind of errors must be detected in execution time in the style of compilers of languages with dynamic typing (for example, the way the Ikarus Scheme [7] and Chez Scheme [6] compilers do).

In regards to code generation, instead of generating specific target code, it could first generate code for a convenient intermediate representation (for example, we think that for a natural language compiler, a variation of lambda calculus would be a good choice) and from this, generate specific code for one or several target languages, that is to say, our design would become a retargetable compiler design. This type of compilers is common for programming languages, for example, GCC generates code for more than 20 different architectures including the x86.

In Sect. 4, we will show how to build a prototype of a compiler that implements the proposed architecture. As source language, we will use arithmetic expressions written in natural language and Python code as target language.

3   Grammar

This section is dedicated to present the CFG and the MG that respectively generate the language of arithmetic expressions written in natural language that will be the source language of the prototype of our compiler.

First, we present the CFG in Fig. 3a with the purpose that it serves as a point of reference to understand the corresponding MG shown in Fig. 3b. It is worth mentioning that these grammars impose the usual precedence rules.

$$
\begin{aligned}
E &\rightarrow T\ E' & \langle =T\ =E'\ E\rangle(\varepsilon) \\
&\mid T & \langle =T\ E\rangle(\varepsilon) \\
E' &\rightarrow P\ T\ E' & \langle =P\ =T\ =E'\ E'\rangle(\varepsilon) \\
&\mid P\ T & \langle =P = T\ E'\rangle(\varepsilon) \\
T &\rightarrow N\ T' & \langle =N\ =T'\ T\rangle(\varepsilon) \\
&\mid N & \langle =N\ T\rangle(\varepsilon) \\
T' &\rightarrow M\ N\ T' & \langle =M\ =N\ =T'\ T'\rangle(\varepsilon) \\
&\mid M\ N & \langle =M\ =N\ T'\rangle(\varepsilon) \\
P &\rightarrow \textbf{más} & \langle P\rangle(\textbf{más}) \\
M &\rightarrow \textbf{por} & \langle M\rangle(\textbf{por}) \\
N &\rightarrow \textit{num} & \langle N\rangle(\textit{num})
\end{aligned}
$$

(a) CFG   (b) MG

Fig. 3. Arithmetic expressions grammars

At first sight, we can notice that there is a one-to-one correspondence between each of the CFG's rules and the correspondent way of representing them in a

MG. This makes sense if we recall that MGs are more expressive than CFGs, that is, a CFG can be written as a particular case of a MG. An obvious question is, if the language of arithmetic expressions written in natural language can be generated with a CFG, then why to use a MG? The answer is that the language of arithmetic expressions is our minimum basic starting source language. Our strategy is to extend our MG progressively so it generates more and more natural language statements. This would be unachievable with a CFG because in Sect. 1 we saw that the consensus is that CFGs are not capable of completely generating natural language.

Next, we will sketch a more accurate definition of a MG, following Stabler's [17] presentation, with the purpose of gaining intuition of the way these grammars work. Our explanation does not pretend to be deep or detailed, but to take a look at this type of grammars. For a more formal and deep study, please consult [17] directly.

A *minimalist grammar* G is a 4-tuple $G = (V, Cat, Lex, \mathcal{F})$ where
$V = (P \cup I)$,   (non-syntactic features)
$Cat = (base \cup select \cup licensors \cup licensees)$,   (syntactic features)
Lex is a set of expressions built from V and Cat,   (the lexicon)
$F = \{merge, move\}$ is a set of partial functions   (the generating functions)
from tuples of expressions to expressions.

The language defined by such a grammar is the closure of the lexicon under the structure building functions $L(G) = CL(Lex, \mathcal{F})$.

As an exposition, we could explain each of the components of a MG introducing them as a generalization of its corresponding counterpart in a CFG. In this way, V which is also called vocabulary, would correspond to the set of terminal symbols of a CFG. For its part, Cat would correspond to the set of nonterminal symbols; more accurately the subset $base$ would be equivalent to the nonterminals of a CFG whereas the other subsets of Cat have not a CFG counterpart. In rough, the subset $select = \{=x, =X, X= \mid x \in base\}$ indicates the symbols on which the structure building $merge$ operates on, whereas the subsets $licensees = \{-case, -wh, \ldots\}$ and $lincesors = \{+case, +CASE, +wh, +WH, \ldots\}$ indicate the symbols (within an expression) on which $move$ operates. The Lex set would correspond to the set of sentential forms of a CFG; the main difference is that instead of strings, in a MG, trees are used, since Stabler defines the expressions as trees inside Lex. Finally, $\mathcal{F}$ can be regarded as high level functions that work on these trees (expressions) that are absent in a CFG and that allows doing more complex operations, particularly the two that Chomsky proposes in [5], namely $merge$ and $move$ which he argues are part of natural language. It is worth mentioning that in the same way that, from the productions of a CFG one can infer each of a CFG's components; based on the notation used in Fig. 3b, one can also infer each of a MG's components, so it is not necessary to write them explicitly. Stabler uses this notation in [18].

4   Compiler Prototype Implementation

In this section, we will describe how to conduct the implementation of a prototype of the compiler design we propose.

As basis, we took the implementation made in Python of the Stabler's MG top-down beam parser which is available in [19]. Our first task is to encode our grammar from Fig. 3b in Python so it can be a valid input of this parser which is shown as follows:

```
#file arithmeticesp.py
g = [ ([], [('sel', 'T'), ('sel', 'E1'), ('cat', 'E')]),
      ([], [('sel', 'T'), ('cat', 'E')]),
      ([], [('sel', 'P'), ('sel', 'T'), ('sel', 'E1'), ('cat', 'E1')]),
      ([], [('sel', 'P'), ('sel', 'T'), ('cat', 'E1')]),
      ([], [('sel', 'N'), ('sel', 'T1'), ('cat', 'T')]),
      ([], [('sel', 'N'), ('cat', 'T')]),
      ([], [('sel','M'), ('sel','N'), ('sel','T1'), ('cat','T1')]),
      ([], [('sel','M'), ('sel','N'), ('cat','T1')]),
      (['más'],[('cat', 'P')]),
      (['por'],[('cat', 'M')]),
      (['cero'], [('cat', 'N')]),
      (['uno'], [('cat', 'N')]),
      (['dos'], [('cat', 'N')]),
      (['tres'], [('cat', 'N')]),
      (['cuatro'], [('cat', 'N')]),
      (['cinco'], [('cat', 'N')]),
      (['seis'], [('cat', 'N')]),
      (['siete'], [('cat', 'N')]),
      (['ocho'], [('cat', 'N')]),
      (['nueve'], [('cat', 'N')]),]
```

Note that instead of having the token *num*, at the moment that there is no lexical analyzer available, the parser for simplicity directly recognizes only the numbers from zero to nine.

Now, we will see the behavior of this analyzer. To start its execution we write:

```
python mgtdbp.py arithmeticesp E 0.0001
```

If we request it to analyze the input "cinco más dos por cuatro", the following output is obtained:

```
: cinco más dos por cuatro
parse found
0.0116989612579 seconds
(h for help):
```

which indicates that the analysis has been successful. Now, we request through the d option to generate the derivation tree pertinent to this analysis and it generates the tree of Fig. 4 as a result.

Now, we are interested in how from this tree we can generate the Python code corresponding to our input statement. Analyzing the structure and information

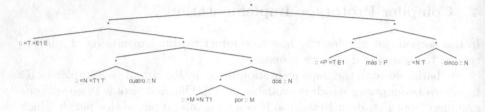

Fig. 4. Derivation tree of "cinco más dos por cuatro"

the tree contains, the answer is: by means of a postorder traversal starting by the right child.

In this manner, we add a small code generator that performs a postorder traversal, visiting in a recursive way: the right child, the left child, and the root of the derivation tree that the parser generates. In addition, we add the option of executing the Python code produced by this generator. Then, the generation and execution of the output code are requested through the option cgae, as follows:

```
(h for help): cgae
Python code generated: 5+2*4
Code evaluation in Python: 13
```

and as we can see, precisely the expected result is obtained and with it, the first basic minimum prototype that implements our proposed design of a natural language compiler.

We end this section showing the output corresponding to a slightly more complex example (Fig. 5):

```
: cinco más dos por cuatro más tres por siete
parse found
0.0524749755859 seconds
(h for help):
```

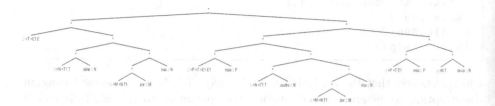

Fig. 5. Derivation tree of "cinco más dos por cuatro más tres por siete"

```
(h for help): cgae
Python code generated: 5+2*4+3*7
Code evaluation in Python: 34
```

If you want to test with more inputs, consult the details of the code generator implementation or of the other compiler's components, please refer to the code of the compiler prototype directly (available in [22]).

5   Prototype Evaluation and Results

Our general strategy to evaluate our compiler prototype is to write a set of tests which consists in a set of arithmetic expressions written in natural language together with its corresponding numeric representation and evaluation. In this way, our plan is to enter the arithmetic expression written in natural language in our compiler prototype and then verify that the generated output is in fact the numeric representation and evaluation written in advance. If both agree, we succeeded, otherwise, there is an error. For example, we write **dos más tres** 2+3 5 in advance; then, we enter **dos más tres** in our compiler, the compiler generates 2+3 5 as output and finally we check that in fact 2+3 5 (the output) is equal to 2+3 5 (written in advance); if for some reason they are not equal, there is an error.

To avoid writing the tests set by hand, we automated this process by developing a test generator:

```
python testg.py itests.txt ovrf.txt 100 20
```

This generates a set of 100 input tests where each arithmetic expression has at most 20 operands. In file **itest.txt** the arithmetic expressions written in natural language were placed (together with the necessary instructions to ask the compiler to translate each of them):

```
cinco más tres mas seis más nueve por tres
cgae
n
uno por cinco por cinco
cgae
n
...
```

whereas in file **ovrf.txt** their corresponding numeric representations and evaluations were written as:

```
5+3+6+9*3  41
1*5*5  25
...
```

now, in a Unix system we can enter the file `itests.txt` as the compiler's input and write the output in `out.txt` as follows:

```
python mgtdbp.py arithmeticesp E
    0.00000000000000000000000001 < itests.txt  > out.txt
```

next, we clean the unnecessary output information generated by the compiler:

```
python clean.py out.txt outc.txt
```

and the cleaned output (in file `outc.txt`) is:

```
5+3+6+9*3 41
1*5*5 25
...
```

Finally, we have to check that each of these numeric representations and evaluations are equal as those written in advance, that is, to check that each of these inside the file `outc.txt` are the same as those in `ovrf.txt`. One simple way to verify this is by checking that files `ovrf.txt` and `outc.txt` are exactly equal, this can be carried out in a Unix system using the `cmp` command:

```
cmp ovrf.txt outc.txt
```

if there is no output (as here) both are equal.

We conducted several experiments with up to 300 arithmetic expressions each one with 50 operands at most and all of them were correct. These experiments were done on a Linux desktop machine with a 1.9 GHz Intel i7 processor. The evaluation tools used in this section are available together with the compiler prototype in [22].

6   Conclusions and Future Work

Building a natural language compiler has been one of the most difficult objectives to achieve since the beginnings of artificial intelligence. We considered that the main problem to build a natural language compiler in the same sense programming language compilers are built today is the natural language syntactic analysis. Due to this, our proposed idea is very simple; to use a natural language syntactic analyzer as front-end and then simply perform the subsequent typical phases of the compilation process of a programming language. In other words, to take the classic design of a programming language compiler as basis, but instead of using a parser that only works with CFGs (like the majority of programming language compilers do) use a parser that works with some type of grammars, in principle MGs, that are capable of expressing natural language.

Throughout this article, we showed that it is possible to build a compiler that implements this idea in practice and that can be used in real life. Of course, this is only a first step that marks the beginning of a long way to walk in the future. Now, we want to support a wider language than simply the language of arithmetic expressions written in natural language. In order to achieve this, our plan is to progressively add more and more natural language statements. Certainly, in this work it might happen that by adding a statement a MG grammar might not be able to express it; however, we think that this would be far from being negative given that this would offer practical evidence that MGs are not able to model all natural language. If this were the case, a possible alternative would be that with the earned feedback, an attempt to strengthen the definition of a MG would be made in order for it to express what is required.

A more radical option would be to use the range concatenation grammars (RCGs) [3,4] which are known for being more expressive grammars than MGs while still, the complexity of its syntactic analysis is polynomial [13], which makes them an ideal candidate to be used in our compiler. At this point, the reader might ask why if RCGs have such ideal properties for our purposes they were not used instead of using MGs. We considered that our main contribution is the natural language compiler's design and architecture that we proposed. To develop a first prototype that would implement this design, we decided to use MGs partly because they are the most expressive of the MCSGs and partly because as Jäger [10] mentions, the majority of experts agree that natural languages are MG languages so, if this is true, there is no need to use more expressive grammars. This is one of the things we wish to confirm in the future. On the other hand, the fact that we have experimented with MGs does not stop us at all from experimenting at par with RCGs; this is another task we plan to do as future work.

All in all, we consider that this work is a first step that opens a path to explore in the future. As a final goal, we think that it is in fact possible to reach the elusive objective of building a natural language compiler.

Acknowledgments. We sincerely thank Edward Stabler for allowing us to freely modify his MG top-down beam parser implementation, for getting acquainted with some of the most recent MG parsing advances and for encouraging us to develop a Spanish MG. This work was supported by the Mexican Council of Science and Technology (CONACYT), fund 2016-01-2225, and DGAPA, fund IN403016.

References

1. Aho, A.V., Lam, M.S., Sethi, R., Ullman, J.D.: Compilers: Principles, Techniques, and Tools, 2nd edn. Addison-Wesley, Boston (2006)
2. Aho, A.V., Ullman, J.D.: The Theory of Parsing, Translation, and Compiling. Prentice-Hall Inc, Upper Saddle River (1972)
3. Boullier, P.: Proposal for a natural language processing syntactic backbone. Research report RR-3342, INRIA (1998). https://hal.inria.fr/inria-00073347
4. Boullier, P.: Range Concatenation Grammars, pp. 269–289. Springer, Dordrecht (2005). https://doi.org/10.1007/1-4020-2295-6_13

5. Chomsky, N.: The Minimalist Program. Current Studies in Linguistics, MIT (1995)
6. Cisco Systems, Inc: Chez scheme. https://github.com/cisco/chezscheme
7. Ghuloum, A.: Ikarus scheme. http://ikarus-scheme.org/
8. Harkema, H.: A recognizer for minimalist grammars, vol. 2000, pp. 111–122, Trento, Italy (2000)
9. Iacob, R., Rebedea, T., Trausan-Matu, S.: NLCP: Towards a compiler for natural language. In: 2017 21st International Conference on Control Systems and Computer Science (CSCS), pp. 252–259, May 2017. https://doi.org/10.1109/CSCS.2017.42
10. Jäger, G., Rogers, J.: Formal language theory: refining the Chomsky hierarchy. Philos. Trans. R. Soc. Lond. B: Biol. Sci. **367**(1598), 1956–1970 (2012). https://doi.org/10.1098/rstb.2012.0077
11. Joshi, A.K.: Tree adjoining grammars: how much context-sensitivity is required to provide reasonable structural descriptions?, pp. 206–250. Studies in Natural Language Processing, Cambridge University Press, Cambridge (1985). https://doi.org/10.1017/CBO9780511597855.007
12. Joshi, A.K., Shanker, K.V., Weir, D.: The convergence of mildly context-sensitive grammar formalisms. Tech. report MS-CIS-09-01, University of Pennsylvania, January 1990. https://repository.upenn.edu/cgi/viewcontent.cgi?article=1571&context=cis_reports
13. Kallmeyer, L.: Parsing Beyond Context-Free Grammars. Cognitive Technologies. Springer, Heidelberg (2010). https://doi.org/10.1007/978-3-642-14846-0
14. Pullum, G.K., Gazdar, G.: Natural languages and context-free languages. Linguist. Philos. **4**(4), 471–504 (1982). https://doi.org/10.1007/BF00360802
15. Seki, H., Matsumura, T., Fujii, M., Kasami, T.: On multiple context-free grammars. Theor. Comput. Sci. **88**(2), 191–229 (1991). https://doi.org/10.1016/0304-3975(91)90374-B
16. Sørensen, M.H., Urzyczyin, P.: Lectures on the Curry-Howard Isomorphism. Studies in Logic and the Foundations of Mathematics, vol. 149. Elsevier, Amsterdam (2006). https://www.sciencedirect.com/bookseries/studies-in-logic-and-the-foundations-of-mathematics/vol/149/suppl/C
17. Stabler, E.: Derivational minimalism. In: Retoré, C. (ed.) LACL 1996. LNCS, vol. 1328, pp. 68–95. Springer, Heidelberg (1997). https://doi.org/10.1007/BFb0052152
18. Stabler, E.: Top-down recognizers for MCFGs and MGs. In: Proceedings of the 2nd Workshop on Cognitive Modeling and Computational Linguistics. pp. 39–48. Association for Computational Linguistics, Portland, June 2011. http://www.aclweb.org/anthology/W11-0605
19. Stabler, E.: MG TD beam parser (2012). https://github.com/epstabler/mgtdb
20. Stabler, E.: Two models of minimalist, incremental syntactic analysis. Top. Cogn. Sci. **5**(3), 611–633 (2013). https://doi.org/10.1111/tops.12031
21. Thomas, J., Antony, P.J., Balapradeep, K.N., Mithun, K.D., Maiya, N.: Natural language compiler for English and Dravidian languages. In: Shetty, N.R., Prasad, N.H., Nalini, N. (eds.) Emerging Research in Computing, Information, Communication and Applications, pp. 313–323. Springer, New Delhi (2015). https://doi.org/10.1007/978-81-322-2550-8_31
22. Zúñiga, A., Sierra, G., Bel-Enguix, G., Galicia-Haro, S.N.: Towards a natural language compiler: the compiler prototype implementation. http://www.pcic.unam.mx/~zuniga.a/nlmgcomp.tgz

Comparative Analysis and Implementation of Semantic-Based Classifiers

Luis Miguel Escobar-Vega[⊠], Víctor Hugo Zaldívar-Carrillo[⊠],
and Ivan Villalon-Turrubiates[⊠]

ITESO (Instituto Tecnológico y de Estudios Superiores de Occidente), 45604
Tlaquepaque, Mexico
{ng700756,victorhugo,villalon}@iteso.mx

Abstract. Text classifiers that extract their features with pure statistical methods are not very useful when there is an extended range of types to classify. They also lack a deeper understanding of the classified data. The use of some semantic methods can improve the efficiency and effectiveness of the purely quantitative approach. This work explores the use of a semantic approach based on a similarity measure to build a vector model containing some semantic evidence. This vector model is used to improve a Maximum Entropy-based text classifier. Experiments show that the F-measures obtained using this approach are competitive. One may conclude that the use of semantic analysis is an excellent complement to statistical approaches and produces better performance and high-grade results.

Keywords: Organizational knowledge · Knowledge management
Semantic technology · Semantic web · Text classification

1 Introduction

The text classification is an integral part of any pattern recognition system. In this way, the classifier is a fundamental piece in the study of the understanding and comprehension of the human knowledge by the computers. The classifier lines up the knowledge according to specific criteria that help to focus and address more efficiently. Mirończuz et al. in [23] make a review of classifiers that extract their classification features with statistical methods. However, these approaches are less effective when there is a poor understanding of the data or when there is an extended range of types of contents to classify. To overcome these statistics issues, this work has turned to use more semantic approaches.

L. M. Escobar-Vega—The authors would like to thank the Instituto Tecnológico y de Estudios Superiores de Occidente (ITESO) of Mexico for the resources provided for this research. Also, the main author would like to thank the National Council of Science and Technology (CONACYT) of Mexico for the sponsoring of this research by the scholarship number 399053.

© Springer Nature Switzerland AG 2018
I. Batyrshin et al. (Eds.): MICAI 2018, LNAI 11289, pp. 83–95, 2018.
https://doi.org/10.1007/978-3-030-04497-8_7

This proposal uses similarity models to select representative characteristics in the text classification; particularly, in extraction task on natural language sources. These models lead to detect if there is semantic evidence, a type of correlation of the meaning between texts and provide a filter of relevant features. The text classification is based on developing a process to extract semantic evidence and build a similarity measure between the n-grams of some texts to be classified. This work explores some classification models based on semantics, semantic distribution models, and similarity measures. Two probability distribution models: Naive Bayes and Maximum Entropy (MaxEnt) are enhanced with this semantic approach. Diverse methods have been proposed to classify the text. However, these methods still have limitations in the way that they do not consider semantic evidence or semantic relations between the characteristics extracted. Therefore, they may overlook some essential elements which could be necessary for text classification. Machine learning-based text classification has recently shown great promise. Further, it has been shown that well-chosen features can improve classification accuracy substantially or reduce the amount of training data needed to obtain the desired level of performance [6].

The fundamental contributions are: (1) A three steps strategy to build a classifier with semantics is presented. (2) Experimental results to illustrate the efficiency and effectiveness of the algorithms are described. (3) It is shown that this classifier can adapt to any number of categories, without the need for re-training or fine-tuning.

2   Related Work

Since computing syntactic tools emerged to analyze information, the text classification has been evolving and has had several different types of models that were proposed for estimating continuous representations of the words. Bag of words (BOW) [11] is a highly efficient model due to its simplicity (economic computing); however, since the order of the words is not kept, there is a significant loss in their meaning or ambiguity.

According to Bayesian's classification, described in [3], this model is of simple implementation, and it has a good computing performance. But, since it is semi-supervised learning, it requires a significant amount of trained data in order to obtain good results. The same as BOW, it does not consider syntactic structures and, even less, semantic structures. The Vector Space Models [28] are highly used in the practice because they are useful in high dimensional spaces, even with very small training sample sizes. They use a subset of training points in the decision function (called support vectors), so they also are memory efficient [29]. Nevertheless, the number of features is much higher than the number of samples and avoiding over-fitting in choosing kernel functions, and regularization terms are crucial. All of them have been acceptable performance solutions but are based on statistical data to set the relation of a word with the class. Therefore, being the aim of this work to create a model of classification based on semantic assumption, it has been initiated with the statistical approaches as a starting point so, in further works, there is a baseline to evaluate works where the formal semantic techniques are used. Although this work does not reach a formal semantic model properly, it follows its bases as semantic fields and traits.

3   Distributional Semantics Approaches

The DS is a branch of study that explores how statistical analysis of large corpora, and in particular word distributions and statistical, can be used to model semantics [16]. In this work, the Latent Semantic Analysis (LSA) and Word2Vec models were chosen because they are two of the most active for word meaning representation, although we do not discard other works such as [25] to mention a few.

The LSA [14] extracts and stands for the meaning of the words in context through statistics treatments applied to a large body of texts. It was developed to capture hidden word patterns in a text document. It is important to highlight that the semantic is taken by understanding terms as references. Mapping of discrete entities in a space and a process simplification with a dimensionality reduction are characteristics that make LSA particularly attractive. However, LSA requires a relatively high computational performance and memory in comparison to other information retrieval techniques and the difficulty in determining the optimal number of dimensions to use [15].

The Word2vec [20,21] consists of two neural network language models [1]. It uses vector-oriented reasoning between words and defining features of Neural Network Language Model, where the words are depicted as high dimensional real-valued vectors and generate a representation model. It aims to identify contextual patterns that get some semantic evidence. Thus, it develops a framework using a model of Recurrent Neural Networks (RNN) that preserves a track of each analyzed sentence hence representing the text with striking syntactic and semantic properties although the model does not have any knowledge of syntax, morphology or semantics. Although it is purely statistical, in [13] the author shows that semantic relationships are often preserved in vector operations on word vectors. Recent optimizations were in [21]. Here, the knowledge encoded in the Word2vec space is integrated with Hyper-parameter free, highly interpretable, and naturally incorporated and it leads to high retrieval accuracy.

The comparative analysis [1,4], shows an intrinsic difference between LSA and Word2vec. Such a difference is that while LSA is a counter-based model, Word2vec is a prediction-based model. Word2vec is used in this proposal since it obtained better results in prediction inference. However, it should be mentioned that Word2vec's performance decrease in small corpora is grounded and it needs several training data to fit its high number of parameters.

4   Semantic Approaches

This work regards the semantic approaches to the ones that are based on the explicit relation between their elements. Usually, their structures are formed by graphs or ontologies. These models are of high interest since they are based on pre-established hierarchies and rules that favor the results of the text classification.

There are a few text classification studies that use isolated semantic analysis to classify. However, approaches such as [19] compared with probabilistic statistical model, the association rule method pays more attention to utilize association relation among features for classification. Other approaches use WordNet [22], a lexical semantic network in which nodes correspond to word senses. Séaghdha [30] uses graph-based kernels [32] on WordNet for a classification and attains very good performance according with SemEval-2012 [31]. Up to now, the previously described approaches achieve good results in specific domains, but they are limited because they use pre-built structures with rules to succeed and supervision is required. Compared with them, this study approach generates semantic structures of texts in the natural language that serve as features in the text classifications and avoids limitations of the approaches previously described. To achieve this, it is proposed to combine the previous DS approaches and the described NLP tasks, together with the ontologies and the rules association methods to have a new classification model. The following section describes a fundamental concept for the generation of semantic structures or proxies.

5   Semantic Evidence

The semantic evidence is linked to the existence of enough connections between the features of two or more elements for sustained evidence and a semantic related to their meaning (meaning of the compared). Such connections are obtained by measuring the space and distance between the elements. Tools of semantic measure estimate the strength of the semantic relationship between units of language through a number according to the comparison of information supporting their meaning. In other words; Semitic evidence is depicted by a value that indicates a grade of the relation between one word and its features, according to with to this competition with another word and its characteristics. Such comparison only can measure with semantic that which we will review later.

The semantic measures compare the relatedness or the similarity of one, or more than one, of the elements. Harispe et al. [10] defines the semantic relatedness as the strength of the semantic interactions between two elements with no restrictions on the types of the semantic links, as well as the semantic similarity as a subset of the notion of semantic relatedness only considering taxonomic relationships in the evaluation of the semantic interaction between two elements. Thus, to extract semantic evidence, there is a great variety of measuring metrics that may be. However, all of them, as a whole, are based on strengths and relations.

Once the evidence is obtained, semantic proxies can be generated, which are the features of a word, originated by the existence of the relations between the semantic elements, e.g., words and concepts; that is, what similar properties are there between the terms based on their meanings. It is important to mention that the quality of the proxies depends, in a high grade, on the techniques that these semantic measures use.

5.1 Similarity Measure

In this work, semantic evidence that is sustained by the similarity of a word was used. The measures between words and features are obtained through Semantic Measures models that are the approaches designed for comparing semantic entities such as units of language, e.g., words, sentences, or concepts and instances defined into knowledge bases. Tversky, in his Studies of Similarity [36], proposed a Feature Model that can be used to analyze the similarity relations between words according to a feature-matching function F which makes use of their common and distinct features. Thus, the similarity is formally described as:

$$sim_F(u, v) = F(U \cap V, U \setminus V, V \setminus U) \tag{1}$$

where U and V are sets of features, F increases when distinct common features are added or removed. Therefore, for the aim of this study, only the featured shared have been used. The differences between U and V are excluded since, in the vector, they represent distant elements. There is a lack of meaning in common, also known as semantic scarcity or non-semantic element. Tversky [36] proposed two models, the contrast model sim_{CM} and the ratio model sim_{RM} to compare two objects u and v represented through sets of features U and V

$$sim_{RM}(u, v) = \frac{\alpha f(u \cap v)}{\alpha f(u \cap v) + \beta f(u - v) + \gamma f(v - u)} \tag{2}$$

The symmetry of the measures produced by the two models can be tuned according to the parameters α and β. This enables the design of asymmetric measures. The major constructs of the feature model is the function f which is used to capture the salience (outstanding feature of a stimulus such as intensity or frequency) of a (set of) features [36]. Therefore, the operators, $\cap$ and $\cup$ are based on feature matching (F) and the function f evaluates the contribution of the common or distinct features to estimate the similarity. In this approach, the ratio model similarity is used as a possible way of normalization. Figuring out the similarity according with the Ratio Feature Model, it happens that if $w_1 = $ beer and $w_2 = $ soda are is compared (where their features were extracted from Wikipedia sources). So, it is obtained that,

$$A = \begin{bmatrix} \text{barley} \\ \text{yeast} \\ \text{alcohol} \\ \text{hop} \\ \text{fermented} \end{bmatrix}, B = \begin{bmatrix} \text{sugar} \\ \text{carbonate} \\ \text{drink} \\ \text{sweetened} \\ \text{flavorings} \end{bmatrix} \quad A \cup B = C$$

A matrix that includes features of each term was generated, and each of the term-feature distances was determined by

$$D = \begin{bmatrix} c_1 & p(w_1, c_1) & p(w_2, c_1) & \cdots & p(w_n, c_1) \\ c_2 & p(w_1, c_2) & p(w_2, c_2) & \cdots & p(w_n, c_2) \\ \vdots & \vdots & \vdots & \ddots & \vdots \\ c_n & p(w_1, c_n) & p(w_2, c_n) & \cdots & p(w_n, c_n) \end{bmatrix} = \begin{bmatrix} c_n & p(w_1, c_n) & p(w_2, c_n) \\ \text{barley} & 0.529 & 0.346 \\ \text{yeast} & 0.517 & 0.386 \\ \text{alcohol} & 0.555 & 0.435 \\ \text{hop} & 0.512 & 0.365 \\ \text{fermented} & 0.600 & 0.493 \\ \text{sugar} & 0.527 & \mathbf{0.444} \\ \text{carbonate} & 0.555 & 0.580 \\ \text{drink} & 0.771 & 0.562 \\ \text{sweetened} & 0.528 & 0.487 \\ \text{flavorings} & 0.443 & 0.451 \end{bmatrix}$$

The first column is formed by characteristics; the second column is the result of the similarity between word 1 and characteristic 1; the third column is the similarity between word 2 and characteristic 1.

Then, just the values with an acceptable similarity degree must be kept. To do so, the longest distance between the features will be fixed as an acceptable limit, e.g. In the matrix C, the closest relation/linear distance that exists between sugar respecting soda, which is 0.444. Any lower value will not be considered (in other words, there is a lack of semantic similarity). At last, the array is sorted according to with the Model of Features; to do so, the current terms features are turned on in order to obtain the following binary arrays:

$$u = \{1,1,1,1,1,1,1,1,1,0\}, \; v = \{0,0,0,0,1,0,1,1,1,1\}$$

where $\alpha = 2$, $\beta = 1$ and $\gamma = 1$, and f the cardinality of sets,

$$u \cap v \text{ is the set of positions } \{i|u_i = v_i = 1\}$$
$$u \setminus v \text{ is the set of positions } \{i|u_i = 1, v_i = 0\}$$
$$v \setminus u \text{ is the set of positions } \{i|u_i = 0, v_i = 1\}$$

substituting, it is found that

$$sim_{RM}(P, Q) = \frac{2|\{5,7,8,9\}|}{2|\{5,7,8,9\}| + |\{1,2,3,4,6\}| + |\{9\}|} = \frac{8}{8+5+1} = 0.571$$

When $\alpha = 0$ instead of a similarity function we get a distance function. Thus, the result is an index of similarity based on features. For this example, features extracted from a vector of Wikipedia are used; however, it is possible to use n-features from different sources. It is necessary to consider that the semantic evidence factor must be homogenized in all the cases.

There are more measures similar to this model, more distinguish are Minkowski L_p Distance metric sim_{Lp} [18], Cosine similarity sim_{COS} [34] and Pearson's product-moment correlation sim_{PER} [33]. However, for this study, there is no much improvement if a more complex variation of the model is used. The model proposal of this work uses the semantic evidence model as binary units to determine the meaning, understanding this meaning as distance and closeness of a word respecting another. These measure metrics do not consider the context nor logic questions as inferences (though, perhaps there would be some more complex models where they are introduced). It has been noted that they are simple mathematical approaches; but, that in a first instance, they could be a starting point for more complex analysis, such as the ones of language processing.

The proposed method relies on to prove the classification methods previously mentioned with the intention of comparing results (particularly, commercial classifiers of an open-closed domain). Especially to know the evidence degree of semantics (SemEval is used to compare the results and to obtain the percentage of effectiveness), and so, to determine if the methods succeeded in a similar way as the one obtained through formal semantic analysis. Regarding the preparation of the components, all the vectors were created by using the WMD algorithm. The corpses came from different sources, especially Spanish corpuses like Ancora lexicons [35], SenSem (Sentence Semantics) [9], Real Academia Española (RAE) [27], and Wikipedia were taken. Besides, Wordnet version three was used as similarity framework and as a validation method in the WMD vectors (recently created); however, WordNet is not finished in Spanish, so a translation English-Spanish was required to enable the language. To cleanse the terms, the algorithms of fuzzy lookup based on Valentin's works were used, for label lookup [38], for the entity matching. For the Natural Language Processing, the Stanford CoreNLP framework was used. It was useful that the 3.9 version included the Spanish universal dependencies.

6   Semantic Classification Process

The earlier tasks of text cleaning were essential for the classification process because they might affect the quality of the final result. The scope of this document is limited only to the classification task. Nevertheless, it is outstanding to mention [12], which proposes a broad reference in spelling correction and noise channel methods.

The semantic classification process consists of three main tasks. The first task determines the existence in the lexicon of the word to be classified. There are two types of reference sources: formals, that means they belong to an organization or group that formalizes the word and assigns one or more definitions, for example, dictionaries or encyclopedias. Also, the second source is the open communities or public domain. The kind of reference source helps to classify the role of the word. The formal sources provide technical terms as classes also called TBox statements, while the other open-resources are taken as an instance, an individual or are called ABox.

The second task aims to find semantic evidence in words to be classified, in another way, that is, extracting relevant features that support the relations with more strength between concepts. The semantic proxies are generated and, later, will be used in the building of a graph that supports the classification. The searching process of semantic evidence is composed of three sub-tasks: 1. The definitions of the words to be classified in reference sources are sought.

2. The named entities are recognized and extracted using Gibbs Sampling [5], these elements are natural to extract and can measure the percentage of entities found. 3. The lexical elements are extracted to give support such as nouns, verbs, and some adjectives, and dependence analysis is used to prioritize the links, discard features and establish relevance in the base of the closeness of the root element of the sentences. The result of this last task is a semantic-graph, where the concepts depict the features extracted from the term to be classified. In this task, it is found that the more iterations, the more characteristics can be extracted. Despite this, the higher the computing process exponentially grows.

In the third task, the terms that were found with a high similitude were selected. The measure is made using vectors. For practical use in this work, it has been established that the lesser scale, the higher the semantic relation, and the opposite happens at a higher distance (the lowest the semantic relation). Thus, features with the highest semantic evidence are selected. A matrix M is created where the classes and distance values from every feature are settled. Using M, a model of probability (MaxEnt) is trained to re-classify the new values. The last step consists of using the trained model with the word to classify. The result is an ordered list with the classes that have a higher percentage.

6.1   An Example of Classification

After trying Decision Trees and Naive Bayes [17], it was determined that the classification for MaxEnt [24] has better performance for this type of text classification. The MaxEnt formula used here is described as:

$$p(c|d) = \frac{1}{Z(d)} exp(\sum_i \lambda_i f_i(d,c)), Z(d) = \sum_c exp(\sum_i \lambda_i f_i(d,c)) \qquad (3)$$

where c is a characteristic and d is a document, with each $f_i(d,c)$ is a feature, λ_i is a parameter to be estimated $Z(d)$ is simply the normalizing factor to ensure a proper probability. A set of weights are parameterized witch combine the joint-features that are generated from a feature set by an encoding. For example, if *Heineken* is to be classified based on the catalog of products of the United Nations (UN), it is necessary to extract from Wikipedia definition, where the information retrieval task returned two proxies, *labels* = $\{barley, beer, alcohol, bottle, pilsner\}$. And, from the UN catalog, five potential classifiers are found: $UN = \{beverage, substance, music, toys, drugs\}$. Now, M is equal to the similarity that was obtained from every classifying-label,

$$M = \begin{pmatrix} label & p_1 & p_2 & p_3 \\ beverage & a:0.549 & b:0.407 & b:\mathbf{0.301} \\ substance & a:0.310 & b:0.298 & c:0.335 \\ music & a:0.211 & b:0.229 & d:\mathbf{0.263} \\ toys & a:0.291 & b:0.244 & e:\mathbf{0.355} \\ drugs & a:0.271 & g:\mathbf{0.239} & h:\mathbf{0.208} \end{pmatrix}$$

Please note that the features p_n may vary according to each classifier; thus, the beverage contains the labels a, b and d while the classifier substance contains the property c that is unique. Classifiers with less similarity, such as drugs, have features g and h. The amount and heterogeneity of classifiers do not affect the model of MaxEnt and allows to use classifiers yet with different feature-labels. Finally, a MaxEnt is applied to the M matrix, and the result is:

$$\begin{array}{cc} label & score \\ \begin{pmatrix} beverage & \mathbf{0.77} \\ substance & 0.43 \\ music & 0.27 \\ toys & 0.25 \end{pmatrix} \end{array}$$

That is, Heineken is 0.77 similar to a beverage. Here, the label "beverage" has a better score, so it is chosen as the class. Lastly, in case of having groups of classes MaxEnt can be used in a series of iterations, where the pre-configured catalogues are organized in groups. That means that every family-level requires a sub-series of new classifications using formula 3.

7 Experimentation and Results

In order to prove this model, A prototype that classifies products and services according to the United Nations Standard Products and Services Catalogue (UNSPSC) [37] was built. The aim was to know the model performance in a real environment where the words are connected with previously defined catalogues. Table 1 shows the taxonomy of the catalog [8], that is a structure with four levels called Segment, Family, Class, and Commodity. The segments are divided into four groups which are ordered in a way that represents how value is added to products in the supply chain.

Because the open datasets do not contain information about categories, a new dataset with 100 terms to be classified was built, starting out of the Goodrelations Dataset [7] (an open data set of services and products) and the labels were included manually. These were divided into four different groups: a are concepts, e.g., beer. b are individuals, such as beer brands, for example, Heineken. c are terms composed by nuns, for example, Economic politics; and finally, d is composed verbs form the most complicated group, for example: eating fast, or adjectives, like European Economic Community.

Table 1. Example of the UNSPSC taxonomy.

Level	Description
Segment	The logical aggregation of families for analytical purposes
Family	A commonly recognized group of inter-related commodity categories
Class	A group of commodities sharing a common use or function
Commodity	A group of substitutable products or services

Semantic evidence was obtained from different sources like Wikipedia, Wiki-Dic, RAE, and WordNet. Two similar classifiers were selected to compare with our proposal: Paralleldots [26] and FastText [2], this approaches are semi-supervised and utilize similar methods.

Table 2 demonstrate that the semi-supervised classifiers are more successful in classifying conceptual words, mainly because the training dataset has a large number of conceptual terms. However, to classify words like individuals or compound data the results were poor. The proposed classifier has more certainty in classifying any words, even with words that were never trained. In general, the classification of concepts is the one that has the best results. The classification of individuals goes as low as 0.79; however, there are some critical ambiguity issues in here, e.g., the term Heineken could be understood as a company, a product, the last name or as a city. Due to this classifier is limited to products and services, heuristics were included to just filter these two groups. The classification of compound concepts, that is, they contain two or more nouns, or they are formed out of verbs, are much more difficult to be classified.

It was found that the complexity of classifying terms is in the task was the semantic evidence is extracted, that is, were the features of a term is searched. The space vector model does not consider the language issues; that means, they do not interpret the meaning of the words. The relation is given statistically and, even though they have a good result, yet semantically speaking, there is not a real comprehension. So, the complex terms do not have good results.

Table 2. Results from the classifier experiments using the F_1 score.

Type	Our proposal	Paralleldots	FastText	Complexity
Concept	0.79	0.92	0.82	Low
Individual	0.68	0.32	0.21	Low
Composed by noun	0.58	0.28	0.19	High
Composed by verb & adj.	0.48	0.10	0.07	High

8   Conclusions

Tversky's is an elementary linear model. Up to this point, it is possible to summarize that there are several techniques to classify text, where SVMs have been the best results given in quantitative scenarios. In SVMs, the distances between found words for the relation in a determined text might create the empiric effect of Semantic-Evidence that is required to have a similarity measure. However, this effect, even though is good, it lacks a formal analysis of both terms. So, the SVM is essential to find more semantic evidence and to generate proxies with more semantic meaning. Moreover, techniques of formal semantic analysis could benefit the tasks of the generation of proxies.

The result of this work proves that it is possible to make a classification of word by using semantic measures and improving statistics models. It was possible to use different resources and compare the results and then to have a better-extracted feature and, consequently, a better classification. However, the classification remained limited, and it worked well with isolated words. It is pending for future works to make some adaptations to improve the process and so identify compound words and the scope of the classification could be higher. Incorporation of formal semantic techniques, like compositional and interpretative, will help in the retrieval module because this technique gives a better understanding of the language than syntactic and statistical approaches.

References

1. Altszyler, E., Sigman, M., Ribeiro, S., Slezak, D.: Comparative study of LSA vs Word2vec embeddings in small corpora: a case study in dreams database. Technical report. arXiv:1610.01520v2 [cs.CL], ArXiV, April 2017
2. Bojanowski, P., Grave, E., Joulin, A., Mikolov, T.: Enriching word vectors with subword information. Trans. Assoc. Comput. Linguist. **5**, 135–146 (2017)
3. Cohen, S.: Bayesian Analysis in Natural Language Processing, 1st edn. Morgan and Claypool, Toronto (2016)
4. Elekes, A., Schäler, M., Boehm, K.: On the various semantics of similarity in word embedding models. In: Proceedings of the ACM/IEEE Joint Conference on Digital Libraries, pp. 1–10. ACM/IEEE, June 2017
5. Finkel, J.: Incorporating non-local information into information extraction systems by gibbs sampling. In: Proceedings of the 43nd Annual Meeting of the Association for Computational Linguistics, vol. 1, no. 1, pp. 363–370, July 2005
6. Forman, G.: An extensive empirical study of feature selection metrics for text classification. J. Mach. Learn. Res. **3**(1), 1289–1305 (2003)
7. GoodRelations: The most powerful Web vocabulary for e-commerce, 2 July 2018. http://wiki.goodrelations-vocabulary.org/Datasets
8. Grenada, R.: Why coding and classifying products is critical to success in electronic commerce, 2 July 2018. https://www.unspsc.org/Portals/3/Documents/Why%20Coding%20and%20Classifying%20Products%20is%20Critical%20to%20Success%20in%20Electronic%20Commerce%20(October%202001).doc
9. GRIAL-Projects: SenSem: Databank of Spanish sentences annotated syntactically and semantically, 17 February 2018. http://grial.uab.es/fproj.php?id=1&idioma=in.
10. Harispe, S., Ranwez, S., Janaqi, S., Montmain, J.: Semantic Similarity from Natural Language and Ontology Analysis, 1st edn. Morgan and Claypool, Toronto (2017)
11. Harris, Z.: Distributional structure. Word **10**(2), 146–162 (1954)
12. Jurafsky, D., James, M.: Speech and language processing, 3rd edn. Prentice-Hall, Upper Saddle River (2017)
13. Kusner, M., Sun, Y., Kolkin, N., Weinberger, K.: From word embeddings to document distances. In: International Conference on Machine Learning, vol. 1, no. 37, pp. 957–966 (2015)
14. Landauer, T., Dumais, S.: A solution to Plato's problem: the latent semantic analysis theory of acquisition, induction, and representation of knowledge. Psychol. Rev. **104**(2), 211–240 (1997)

15. Landauer, T., Laham, D.: An introduction to latent semantic analysis. Discourse Process **25**(1), 259–284 (1998)
16. Lenci, A.: Distributional approaches in linguistic and cognitive research. Ital. J. Linguist. **20**(1), 1–31 (2008)
17. Lewis, D.D.: Naive (Bayes) at forty: the independence assumption in information retrieval. In: Nédellec, C., Rouveirol, C. (eds.) ECML 1998. LNCS, vol. 1398, pp. 4–15. Springer, Heidelberg (1998). https://doi.org/10.1007/BFb0026666
18. Li, Z., Ding, Q., Zhang, W.: A comparative study of different distances for similarity estimation. Intell. Comput. Inf. Sci. **134**(1), 483–488 (2011)
19. Liu, B., Hsu, W., Ma, Y., Ma, B.: Integrating classification and association rule mining. In: Knowledge Discovery and Data Mining, vol. 32, no. 4, pp. 80–86 (1998)
20. Mikolov, T., Corrado, G., Chen, K., Dean, J.: Efficient estimation of word representations in vector space. CoRR **1**(1), 1–2, January 2013
21. Mikolov, T., Sutskever, I., Chen, K., Corrado, G., Dean, J.: Distributed representations of words and phrases and their compositionality. In: NIPS 2013 Proceedings of the 26th International Conference on Neural Information Processing Systems, vol. 2, no. 1, pp. 3111–3119, December 2013
22. Miller, G., Fellbaum, C.: Wordnet then and now. ECML **41**(2), 209–214 (2007)
23. Mirończuk, M., Protasiewicz, J.: A recent overview of the state-of-the-art elements of text classification. Expert Syst. Appl. **106**(1), 36–54 (2018)
24. Nigam, K., Lafferty, J., Mccallum, A.: Using maximum entropy for text classification. In: IJCAI 1999 Workshop on Machine Learning for Information Filtering, vol. 1, no. 1, pp. 61–67, August 1999
25. Pennington, J., Socher, R., Manning, C.: Glove: global vectors for word representation. In: Proceedings of the 2014 Conference on Empirical Methods in Natural Language Processing, pp. 1532–1543. Association for Computational Linguistics, November 2014
26. Pushp, P., Srivastava, M.: Train once, test anywhere: zero-shot learning for text classification. Technical reporty. arXiv:1612.03651 [cs.CL], ArXiV, Dic 2017. https://arxiv.org/abs/1712.05972
27. RAE: Real Academia Española (Jan 9 2018) http://www.rae.es
28. Salton, G., Wong, A., Yang, C.S.: A vector space model for automatic indexing. Mag. Commun. ACM **18**(11), 613–620 (1975)
29. Scikit-learn: Scikit-learn Machine Learning in Python, 29 January 2018. http://scikit-learn.org/stable/
30. Séaghdha, D.: Semantic classification with Wordnet kernels. ECML **37**(1), 237–240 (2015)
31. SemEval: Multilingual and Cross-lingual Semantic Word Similarity, 5 January 2018. http://alt.qcri.org/semeval2017/task2/index.php?id=task-details
32. Shawe-Taylor, J., Cristianini, N.: Kernel Methods for Pattern Analysis, 1st edn. Cambridge, Cambridge (2004)
33. Shevlyakov, G.: Robust Correlation: Theory and Applications, 1st edn. Wiley, West Sussex (2016)
34. Sidorov, G., Gelbukh, A., Gómez-Adorno, H., Pinto, D.: Soft similarity and soft cosine measure: similarity of features in vector space model. Computacion Sistemas **18**(3), 491–504 (2014)
35. Taule, M., Martí, A., Recasens, M.: Ancora: multilingual and multilevel annotated corpora. In: Proceedings of 6th International Conference on Language Resources and Evaluation, vol. 1, no. 1, pp. 96–101, January 2008

36. Tversky, A., Itamar, G.: Studies of similarity. Cogn. Categorization **84**(4), 79–98 (1978)
37. UNSPCP: United Nations Standard Products and Services Code, 25 August 2017. https://www.unspsc.org
38. Yao, X.: Semantic conceptual primitives computing in text classification. In: NAACL Short, vol. 15, no. 3, pp. 66–70 (2015)

Topic–Focus Articulation: A Third Pillar of Automatic Evaluation of Text Coherence

Michal Novák[✉], Jiří Mírovský, Kateřina Rysová, and Magdaléna Rysová

Faculty of Mathematics and Physics, Institute of Formal and Applied Linguistics,
Charles University, Malostranské náměstí 25, 11800 Prague 1, Czech Republic
{mnovak,mirovsky,rysova,magdalena.rysova}@ufal.mff.cuni.cz

Abstract. We present a feature-rich system for automatic evaluation of surface text coherence in Czech essays written by native and non-native speakers. The EVALD system, in addition to basic features covering spelling, vocabulary, morphology and syntax, stands on two main pillars representing the features closely related to the phenomenon of surface coherence: discourse relations and coreference. Newly we add a third pillar, features targeting topic–focus articulation (sentence information structure). Therefore, we propose and implement a procedure for disclosing topic–focus articulation by marking contextual boundness in the text automatically. The experiments show that EVALD enriched with topic–focus articulation features succeeds in outperforming the original system. Further experiments show that the system for essays written by non-native speakers exhibits different signs in terms of importance of individual feature sets and the size of the training data than the system for native speakers.

1 Introduction

This work deals with the automated evaluation of surface coherence in Czech texts. Given a text, the task is to assign to it a grade that reflects to what extent the surface coherence is maintained in the text. Specifically, we target texts in form of essays written by native speakers of Czech as well as by learners of Czech as a foreign language. Such automated evaluator could then be used by teachers to facilitate their work in grading the school essays and also by students who can verify their writing skills in an instant.

In order to address this task, we utilize the EVALD (Evaluator of Discourse) system, which is currently available in two variants: EVALD 2.0 [14] and EVALD 2.0 for Foreigners [15]. EVALD 2.0 targets essays written by native speakers and

---

Supported by the Ministry of Culture of the Czech Republic (project No. DG16P02B016 *Automatic Evaluation of Text Coherence in Czech*). This work has been using language resources developed, stored and distributed by the LINDAT/CLARIN project of the Ministry of Education, Youth and Sports of the Czech Republic (project LM2015071).

© Springer Nature Switzerland AG 2018
I. Batyrshin et al. (Eds.): MICAI 2018, LNAI 11289, pp. 96–108, 2018.
https://doi.org/10.1007/978-3-030-04497-8_8

evaluates their surface coherence in terms of the scale commonly used in Czech high schools: grades 1–5 (excellent–fail). The variant for foreigners adopts the internationally recognized scale established by Common European Framework of Reference for Languages (CEFR), comprising six grades A1–C2 (beginner–mastery).

The EVALD system takes advantage of two types of features: (i) features that use information from lower layers of language description, namely spelling, vocabulary, morphology and syntax, and (ii) text features directly related to surface coherence and reaching also beyond the sentence boundaries. Regarding the latter type of features, the system so far stands on two main pillars, represented by features exploiting (1) coreference, and (2) discourse relations.

In this work, we add a third main pillar – features exploiting topic–focus articulation (sentence information structure). This extension not only makes the system more robust, but it also improves its performance. We then conduct follow-up experiments on the complete system, showing that the variants for native speakers and foreigners considerably differ in two aspects: (1) which types of features play a key role in the evaluation, and (2) how well they could exploit additional training data.

The paper is structured as follows. Section 2 presents the EVALD system and its processing stages. Section 3 describes how we address the topic–focus articulation. In Sect. 4, we describe the datasets on which the system is trained and tested. Then two experimental parts follow. While the first one in Sect. 5 evaluates the topic–focus articulation extension, the second one in Sect. 6 carries out an analysis of the entire system. Finally, we compare our work with related works in Sect. 7 and conclude in Sect. 8.

2   EVALD

EVALD is a pair of applications for automatic evaluation of surface coherence in Czech essays written by native (L1) and non-native (L2) speakers, respectively [23]. The system automatically assigns a grade on an appropriate scale to the given text, i.e. 1–5 (excellent–fail) for L1, and A1–C2 (beginner–mastery) for L2.

The decision making of EVALD is based on standard supervised learning. It thus must be trained on a set of documents that have been manually labeled with grades of surface coherence. The system first preprocesses the documents and augments them with rich linguistic annotation. The annotated text is then described by a set of manually designed features, which are fed along with labels into the machine learning method. We elaborate on these steps in the following text.

Text Preprocessing. First, the system performs an automatic analysis of the raw text data, starting with tokenization and sentence segmentation, and proceeding to morphological analysis/disambiguation, surface-syntax parsing and deep-syntax (tectogrammatical) parsing, following the annotation framework of the Prague Dependency Treebank (PDT).[1] For the analysis from the raw text up to the tectogrammatical layer, the system uses a pre-defined scenario for Czech text analysis in Treex, a modular system for natural language processing [30].

On top of the automatically parsed dependency trees of the tectogrammatical layer, additional Treex modules perform automatic annotation of phenomena closely related to the surface coherence. These phenomena form the three main pillars of the EVALD system. These are:

- explicit discourse relations, i.e. relations expressed by discourse connectives[2] (see [22] for details),
- coreference relations (see [16] for details),
- topic–focus articulation represented by a property of contextual boundness (see Sect. 3 of this paper).

Feature Extraction. From the pre-processed texts, the EVALD system extracts a list of 180 features, divided into the following feature sets: spelling (2 features), vocabulary (4 features), morphology (26 features), syntax (19 features), discourse (24 features), coreference (94 features), and newly topic–focus articulation (TFA; 11 features). Table 1 gives an overview of types of features in the individual feature sets, including the newly added feature set for the topic–focus articulation.

Learning Method. The extracted features are subsequently processed by the Random Forest algorithm implemented in the Waikato Environment for Knowledge Analysis (WEKA) toolkit [7].[3] As a result, the system assigns a mark to the text, evaluating the level of the text coherence: 1–5 for L1, and A1–C2 for L2.

3   Topic–Focus Articulation

Along with discourse relations and coreference (and anaphora generally), topic–focus articulation (TFA; also known as sentence information structure) is

---

[1] The Prague Dependency Treebank (for its last version 3.5 see [6]) is a corpus of Czech newspaper texts (containing almost 50 thousand sentences) with a multi-layer annotation in the theoretical framework of the Functional Generative Description, see [26,27]: it contains manual annotation on morphological, surface syntactic and deep semantico-syntactic (tectogrammatical) layers. On top of the dependency trees of the tectogrammatical layer, the PDT also contains manual annotation of coreference, discourse relations including annotation of discourse connectives, and topic–focus articulation.

[2] As a theoretical background for capturing discourse relations in text, the approach described in [19] is used. It is based on the approach used for the annotation of the Penn Discourse Treebank 2.0 (PDTB; [20]). Both these approaches are lexically based and aim at capturing local discourse relations between clauses, sentences, or short spans of text.

[3] Weka toolkit ver. 3.8.0, downloaded from http://www.cs.waikato.ac.nz/ml/weka/.

Table 1. Overview of feature sets in the EVALD systems. Please note that all absolute numbers are normalized to the length of the text.

Feature set	Overview of features in the set
Spelling	Number of typos, number of punctuation marks
Vocabulary	Richness of vocabulary expressed by several measures, average length of words
Morphology	Percentage of individual cases, parts of speech, degrees of comparison, verb tenses, moods, etc.
Syntax	Average sentence length, percentage of sentences without a predicate, number and types of dependent clauses, structural complexity of the dependency tree (number of levels, numbers of branches at various levels), etc.
Discourse	Quantity and variety of discourse connectives (intra-sentential, inter-sentential, coordinating, subordinating), percentages of four basic classes of types of discourse relations (temporal, contingency, contrast, expansion) and numbers of most frequent connectives, etc.
Coreference	Proportion of 21 different pronoun subtypes, variety of pronouns, percentage of null subjects and several concrete (most commonly used) pronouns, number of coreference chains (intra-sentential, inter-sentential) and distribution of their lengths, etc.
TFA	Variety of rhematizers (focalizers), number of sentences with a predicate on the first or second position, percentage of (contrastive-) contextually bound and non-bound words (more precisely: nodes in the tectogrammatical tree), percentage of SVO and OVS sentences, position of enclitics, percentage of coreference links going from a topic part of one sentence to the focus part of the previous sentence, etc.

considered one of key elements contributing to text coherence (see, e.g., [32]). It deals both with the word order within individual sentences, as well as with the way the text is built from sentence to sentence, i.e. how a sentence connects to the previous one – the most common connections being "continuous topic", see Example 1, and "progression of focus", see Example 2.[4]

(1) *Lucia* has a dog. *She* likes taking it for a walk each day.
(2) *Lucia has* **a dog**. **It** is a yorkshire terrier.

---

[4] A similar phenomenon based on the Centering Theory [9] and its theoretical contribution to evaluation of the text coherence (especially of so-called "rough shifts") was studied in [12].

The theory of topic–focus articulation has been established within the theoretical framework of the Functional Generative Description (see [4]). It describes a bipartition of a sentence into its topic and focus parts, based on a principle of aboutness: the topic is the part of the sentence that is spoken about, while the focus is the part of the sentence that declares something about the topic.[5]

In [3] and [21], the authors show that this bipartition can be automatically derived from contextual boundness of individual nodes in the dependency tree of the tectogrammatical representation of the sentence. Each element (node)[6] of the underlying structure of the sentence (tectogrammatical dependency tree) carries the feature "contextually bound" or "contextually non-bound". In addition, contextually bound elements in the topic of the sentence can be either contrastive, or non-contrastive. Contrastive contextually bound sentence members differ from the non-contrastive ones in the presence of a contrastive stress and in their semantic content – they express contrast to some previous context (e.g. *at home – abroad*).

Annotation of Contextual Boundness in EVALD. The algorithm for automatic annotation of contextual boundness used in the EVALD system is an extension of the algorithm for a pre-annotation of contextual boundness published in [13]. The original rule-based algorithm was now newly re-implemented as a module in Treex and – in order to improve coverage of the annotation – supplemented by many additional rules. The hand-made rules exploit information from all levels of (automatic) text processing in Treex (mostly focusing on information from the deep syntactic – tectogrammatical – layer), see three simplified examples 3, 4 and 5 of such rules (more details in [13]):

(3) Nodes representing the main predicate of the sentence whose lemma does not appear in the previous sentence are marked as contextually non-bound.

(4) Nodes representing an anaphor of a coreference relation are marked as contextually bound.

(5) Nodes representing a free modification specifying the governing noun that is contextually non-bound are marked also as contextually non-bound.

The original system [13] was intended as a first step in a manual annotation of the data and therefore used only highly reliable annotation rules, covering only 41% of TFA-relevant nodes,[7] with precision of 96%. The goal of the new system was to perform a fully automatic annotation of contextual boundness of (most of) the TFA-relevant tectogrammatical nodes, and to this end, it needed to resort also to less reliable rules. The 12 rules (or schemas of rules) from the original system were modified in this manner and further supplemented with 11 additional rules (or schemas of rules). The final version of the annotation

---

[5] The topic/focus categories in general do not need to overlap with given/new information (see [5]).

[6] With some exceptions, see below.

[7] Some types of tectogrammatical nodes, such as coordinating nodes, are not TFA-relevant, i.e. not annotated as contextual bound or non-bound.

procedure covers 94% of all TFA-relevant nodes, with precision of 83%. Table 2 summarizes the results of the automatic annotation of contextual boundness, measured on the evaluation test data of the Prague Dependency Treebank.[8]

Table 2. Results of the automatic annotation of contextual boundness measured on the evaluation test data of the Prague Dependency Treebank; *b* stands for contextually bound, *cb* for contrastive contextually bound, *nb* for contextually non-bound, '-' for not automatically annotated.

PDT\auto	b	cb	nb	-
b	17 596	194	3 018	1 701
cb	741	114	1 110	1 771
nb	6 074	27	36 917	1 063

4 Datasets

In the experiments, we utilize two datasets described in detail in [16]. One dataset consists of texts written by Czech native speakers (L1), the other one by non-native speakers (learners) of Czech (L2). The datasets have been compiled from several language acquisition corpora available for Czech, namely the MERLIN corpus [1], CzeSL-SGT/AKCES 5 [24], and Skript2012/AKCES 1 [25]. Table 3 gives basic statistics about the size of the two datasets and numbers of documents labeled with a particular grade. As the datasets are of a medium size, we do not split them to fixed training and testing portions. We rather exploit all the data for both training and testing in a 10-fold cross-validation fashion.

Table 3. Basic statistics about the datasets for L1 and L2, in total and for individual marks.

L1 dataset	1	2	3	4	5		Total
# documents	484	149	121	239	125		1 118
# sentences	20 986	4 449	2 913	3 382	939		32 669
# tokens	301 238	65 684	40 054	43 797	11 379		462 152
L2 dataset	A1	A2	B1	B2	C1	C2	Total
# documents	174	176	171	157	105	162	945
# sentences	1 802	2 179	2 930	2 302	1 498	10 870	21 581
# tokens	15 555	21 750	27 223	37 717	21 959	143 845	268 049

---

[8] The whole data of the Prague Dependency Treebank have been manually annotated for contextual boundness, see [18].

5 Evaluation of the TFA Extension of EVALD

We have conducted several experiments to test whether the new feature set that captures topic–focus articulation proves useful for the overall performance of the EVALD system. The new system containing the TFA extension is denoted as *EVALD-All+TFA* in the following. First, we describe the evaluation measures and the baseline systems.

Evaluation Measures. A straigthforward measure that can be applied for this task is the standard *accuracy*. We report it, too, but only as a complementary measure.

As the datasets are compiled from multiple sources, their distribution over grades should be considered random. The true distribution of grades would be almost impossible to acquire. It thus seems to be the best strategy to treat the data as if the grades were equally balanced.

The *macro-averaged F-score* satisfies this requirement. For every grade, precision and recall must be first calculated and combined in F-scores. The F-scores are subsequently averaged over all grades, i.e. sum of the F-scores is divided by the number of grades.

Baseline Systems. We compare performance of the TFA extension with two baselines, representing previous versions of the EVALD systems. *EVALD-Coref* is the version of the system proposed and evaluated in [16]. It incorporates vocabulary, discourse and coreference features.[9] *EVALD-All* is the version that involves all the features described in Sect. 2, except for the TFA features.

Evaluation. Table 4 shows results of the evaluation using 10-fold cross-validation. Most importantly, the EVALD system with the TFA extension outperforms both baselines, both for L1 and L2. It shows that the TFA features positively contribute to predicting the surface coherence level of a text, which in general confirms that topic–focus articulation is one of key aspects in maintaining surface coherence.

We also observe a gap of around 15 points between F-scores achieved on the L1 and the L2 datasets. It may indicate that estimating the level of surface coherence is far more difficult in the texts written by native speakers than in the texts created by learners of Czech as a foreign language.

Finally, a disproportion between the accuracy and the macro-averaged F-score on the L1 dataset is a consequence of the skewed distribution of grades in this dataset (see Table 3). Note, e.g., that documents labeled with the mark 1 are four times more frequent than those labeled with the mark 3.

---

[9] In [16], they distinguish pronoun and coreference features. Here, we denote the union of these two sets as coreference features.

Table 4. Performance of the EVALD systems augmented with the TFA features compared with the two previous versions of the system.

System	L1		L2	
	Macro-F	Acc (%)	Macro-F	Acc (%)
EVALD-Coref	46.0	65.7	59.0	64.3
EVALD-All	45.8	66.2	60.7	65.6
EVALD-All+TFA	**47.5**	**67.0**	**62.9**	**68.0**

Table 5. Ablation analysis – performance of the EVALD systems with individual feature sets removed.

System	L1		L2	
	Macro-F	Acc (%)	Macro-F	Acc (%)
EVALD-All+TFA	47.5	67.0	62.9	68.0
– spelling	47.0	66.7	**59.8**	65.0
– vocabulary	**44.9**	65.5	**59.9**	65.3
– morphology	47.5	66.7	60.3	65.8
– syntax	46.3	66.7	62.0	67.0
– discourse	45.6	66.2	60.8	65.3
– coreference	**44.6**	65.8	61.3	65.6
– TFA	45.8	66.2	60.7	65.6

6   Overall Quality of the System

Ablation Analysis. The present results of extending the EVALD system with the TFA features as well as the results from the previous works extending it with discourse-related features [22] and coreference-related features [16] all show that these phenomena are valuable indicators of the level of surface coherence, even if they are modeled automatically. Nevertheless, the comparison among these works is not entirely fair, as the EVALD system has been gradually enriched also with other features of different kinds, namely spelling, vocabulary, morphology and syntax.

We thus carry out an ablation study, which removes each of the feature sets from the full EVALD-All+TFA configuration, one at a time. Results in Table 5 show that the importance of individual feature sets crucially depends on whether the task is being solved on L1 or L2 texts. The only feature set that seems to be among the most valuable ones for both the datasets are the vocabulary features. Richness of the vocabulary thus appears to be the key factor that can signal the level of surface coherence in essays, regardless of whether they were written by native speakers of Czech or learners of Czech as a foreign language.

When it comes to the other feature sets, their impact on performance considerably differs between the datasets. As for the L1 texts, the most valuable features reflect the aspects closely related to surface coherence: coreference, discourse relations and TFA. On the other hand, the power of the morphology, spelling and syntax features to distinguish between the grades is weaker on such texts. It suggests that if a native speaker makes mistakes in surface coherence, it would be signaled by errors in higher levels of language description. Since EVALD is far from modeling these higher levels in their full extent, that is probably also the reason why it performs much worse on the texts written by native speakers.

Performance of EVALD on the L2 dataset paints a different picture. Coreference features have the weakest effect, while spelling and morphology features play an important role here. It suggests that if a learner of Czech as a foreign language makes mistakes in surface coherence, it is probable that they are accompanied by errors in spelling or morphology.

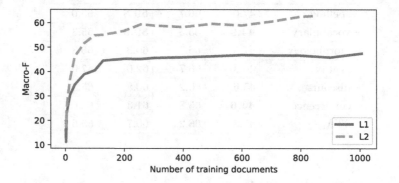

Fig. 1. Learning curves for L1 and L2. (Color figure online)

Learning Curves. Even though performance of the EVALD system has improved with the TFA feature set, it is still far from perfect, especially for the L1 essays. A question arises whether collecting more training data would help. Figure 1 shows learning curves for the EVALD-All+TFA configuration of the system applied on test portions of the L1 (blue solid curve) and the L2 dataset (orange dashed curve). It was conducted by replacing the full training data with their sample of a specified size for each fold of the 10-fold cross-validation, while the test data remained unchanged.

Whereas the learning curve on the L1 dataset seems to have reached its peak with less than 200 training documents, the curve for the L2 dataset appears to be growing further. It suggests that the performance of the system for evaluating surface text coherence in texts written by foreigners has a potential to slightly improve with more data. On the other hand, EVALD for texts written by native speakers is in its current form unlikely to perform better with more training data.

7   Related Work

In the following text, we give an overview of similar and also different approaches in related work and – if possible – compare our results with results presented there. In several cases, we re-evaluated our system to meet evaluation criteria of the related work (a smaller number of predicted classes). Note that in most of the cases, the other systems predict a mark for the overall quality of a text, while our system predicts a mark for a level of local coherence in a text. The comparison of performances of the systems needs to be viewed with this fact in mind, i.e. not only as a comparison of the systems, but possibly also as a comparison of the difficulty of the two different tasks.

Hancke and Meurers [8] use 1 027 German learner texts from the MERLIN corpus divided into a training set (721 texts) and a test set (302 texts), both sets unbalanced, with classes A1–C1. They process the texts up to the surface syntax layer, extract several groups of features (morphological, lexical, a few groups of syntactic features), and use the sequential minimal optimization (SMO) classifier from the WEKA toolkit to predict the overall CEFR level mark of each text (i.e. not a mark for coherence). They test various combinations of the groups of features, with results varying from 57.2% to 62.7% in terms of accuracy. Using all feature groups at once, the system performed worse than any smaller combination of the feature groups (57.2%) and even worse than the lexical group of features alone (60.5%).

Volodina et al. [29] predict the overall CEFR mark on the same scale of A1–C1 in 339 Swedish L2 texts. Their feature set consists of 61 features (lexical, morphological, syntactic and semantic features). Using again the SMO classifier from the WEKA toolkit, they report 67% accuracy (on an unbalanced data set). If similarly restricted only to documents with A1–C1 marks, our EVALD system performs on Czech L2 essays with 63% accuracy on an unbalanced set. Our results are comparable to numbers reported in [8], and slightly worse than those reported in [29].

Vajjala and Lõo [28] perform experiments on 879 Estonian learner texts, predicting the overall CEFR level of the texts. They restrict only to predicting four CEFR levels: A2–C1. Combining lexical and morphological features and using the SMO algorithm from the WEKA toolkit, they reach 79% accuracy and 78.5 macro-averaged F-score on an unbalanced set. If we restrict EVALD for L2 to documents with classes A2–C1, it performs with 65% accuracy and 59 F-score points. Looking at numbers only, their system for Estonian significantly outperforms our system for Czech, however, note that their system targets the overall grade.

Zesch et al. [31] present result of experiments with a set of 96 essays of L2 learners of German, predicting one of three marks (good, fair, poor) as an overall evaluation of the essays. Their system uses a large set of lexical and morphological features (961 features in total) and with the Naive Bayes algorithm and 10-fold cross validation achieves an accuracy of 77%.

For L1, Ostling et al. [17] developed a system for grading high school essays in Swedish with one of four overall marks. It is based on a linear discriminant analysis and incorporates lexical and morphological features. They test the system on a corpus of 1 702 essays with the leave-one-out evaluation method and report the 62.2% accuracy (on an unbalanced data set).

For English, there are also works that focus directly on the evaluation of text coherence (mostly in L1 texts). They, however, evaluate their models on the task of text ordering ranking, comparing an original text with a text created from the original one by a random permutation of its sentences, assuming that the original text is always more coherent than the shuffled one. This allows using raw data with no annotation of grades for training and testing the models, but the performance of such experiment setups are incomparable with the classification task that we focus on. Lin et al. [10] use a discourse parser that automatically annotates a text with discourse relations in the PDTB style, i.e. a framework similar to our approach that captures local coherence of the text. Their system further studies and evaluates a sequence of discourse relations in the text and their transitions, thus assessing global coherence of the text. Feng et al. [2], on the other hand, base their representation of discourse on the Rhetorical Structure Theory [11, RST], which aims from the beginning at a representation of global coherence of a text. They use a parser that annotates the full hierarchical tree-like representation of coherence relations in the text and show that their system slightly outperforms systems based on local coherence models only.[10]

8   Conclusion

In this paper, we introduced an extension to EVALD, a system for automated evaluation of surface coherence in Czech. The extension targets topic–focus articulation and its incorporation to EVALD leads to improvement in terms of macro-averaged F-score of more than 1.5% points for both variants of the system. We also showed that the variants for native speakers (L1) and foreigners (L2) considerably differ in two aspects. First, while the system for L2 texts can benefit from low-level features (e.g. spelling and morphology), it is features more related to the meaning (e.g. coreference, discourse relations, TFA) that play a more important role in the system for L1 texts. The only exception is vocabulary features, which are vital in both variants. Second, with the current approach, evaluation of text coherence in L1 texts seems to be much harder than in L2 texts; it indicates that differences in mastering the language at various levels of learners of Czech as a foreign language (L2) are more pronounced than similar differences for native speakers (L1). Finally, unlike in the case of evaluating L2 texts, additional L1 training data appear not to help anymore.

---

[10] Apart from L1 texts, they also evaluate the system on L2 essays manually annotated with a mark for "text organization", however not as a classification task but again as a pairwise ranking only.

References

1. Boyd, A., et al.: The MERLIN corpus: learner language and the CEFR. In: Proceedings of the 9th International Conference on Language Resources and Evaluation (LREC 2014), pp. 1281–1288. European Language Resources Association, Reykjavík (2014)
2. Feng, V.W., Lin, Z., Hirst, G.: The impact of deep hierarchical discourse structures in the evaluation of text coherence. In: Proceedings of COLING 2014, the 25th International Conference on Computational Linguistics: Technical Papers, pp. 940–949 (2014)
3. Hajičová, E., Havelka, J., Veselá, K.: Corpus evidence of contextual boundness and focus. In: Danielsson, P. (ed.) Proceedings of the Corpus Linguistics Conference Series. vol. 1, pp. 1–9. University of Birmingham, Birmingham (2005)
4. Hajičová, E., Sgall, P., Partee, B.: Topic-Focus Articulation, Tripartite Structures, and Semantic Content. Kluwer, Dordrecht (1998). ISBN 0-7923-5289-0
5. Hajičová, E., Mírovský, J.: Topic/focus vs. given/new: information structure and coreference relations in an annotated corpus. In: 51st Annual Meeting of the Societas Linguistica Europaea, Book of Abstracts, Tallinn, Estonia (in press)
6. Hajič, J., et al.: Prague Dependency Treebank 3.5. LINDAT/CLARIN digital library at the Institute of Formal and Applied Linguistics, Faculty of Mathematics and Physics, Charles University (2018). http://hdl.handle.net/11234/1-2621
7. Hall, M., Frank, E., Holmes, G., Pfahringer, B., Reutemann, P., Witten, I.H.: The Weka data mining software: an update. ACM SIGKDD explor. newslett. **11**(1), 10–18 (2009)
8. Hancke, J., Meurers, D.: Exploring CEFR classification for German based on rich linguistic modeling. Learner Corpus Research, pp. 54–56 (2013)
9. Joshi, A.K., Weinstein, S.: Control of inference: role of some aspects of discourse structure-centering. In: IJCAI, pp. 385–387 (1981)
10. Lin, Z., Ng, H.T., Kan, M.Y.: Automatically evaluating text coherence using discourse relations. In: Proceedings of the 49th Annual Meeting of the Association for Computational Linguistics: Human Language Technologies, vol. 1, pp. 997–1006. Association for Computational Linguistics (2011)
11. Mann, W.C., Thompson, S.A.: Rhetorical structure theory: toward a functional theory of text organization. Text-Interdisc. J. Study Discourse **8**(3), 243–281 (1988)
12. Miltsakaki, E., Kukich, K.: Evaluation of text coherence for electronic essay scoring systems. Nat. Lang. Eng. **10**(1), 25–55 (2004)
13. Mírovský, J., Rysová, K., Rysová, M., Hajičová, E.: (Pre-)annotation of topic-focus articulation in Prague Czech-English Dependency Treebank. In: Proceedings of the 6th International Joint Conference on Natural Language Processing, pp. 55–63. Asian Federation of Natural Language Processing, Nagoya (2013)
14. Novák, M., Rysová, K., Mírovský, J., Rysová, M., Hajičová, E.: EVALD 2.0, data/software. ÚFAL MFF UK, Prague, Czechia (2017)
15. Novák, M., Rysová, K., Mírovský, J., Rysová, M., Hajičová, E.: EVALD 2.0 for Foreigners, data/software. ÚFAL MFF UK, Prague, Czechia (2017)
16. Novák, M., Rysová, K., Rysová, M., Mírovský, J.: Incorporating coreference to automatic evaluation of coherence in essays. In: Camelin, N., Estève, Y., Martín-Vide, C. (eds.) SLSP 2017. LNCS (LNAI), vol. 10583, pp. 58–69. Springer, Cham (2017). https://doi.org/10.1007/978-3-319-68456-7_5
17. Östling, R., Smolentzov, A., Hinnerich, B.T., Höglin, E.: Automated essay scoring for Swedish. In: Proceedings of the Eighth Workshop on Innovative Use of NLP for Building Educational Applications, pp. 42–47 (2013)

18. Panevová, J., Böhmová, A., Hajičová, E., Sgall, P., Ceplová, M., Řezníčková, V.: A manual for tectogrammatical tagging of the Prague Dependency Treebank. Technical report TR-2000-09 (2000)

19. Poláková, L., Mírovský, J., Nedoluzhko, A., Jínová, P., Zikánová, Š., Hajičová, E.: Introducing the Prague Discourse Treebank 1.0. In: Proceedings of the Sixth International Joint Conference on Natural Language Processing, pp. 91–99. Asian Federation of Natural Language Processing, Nagoya (2013)

20. Prasad, R., et al.: The Penn Discourse Treebank 2.0. In: Proceedings of the Sixth International Conference on Language Resources and Evaluation (LREC 2008), pp. 2961–2968. European Language Resources Association, Marrakech (2008)

21. Rysová, K., Mírovský, J., Hajičová, E.: On an apparent freedom of Czech word order. A case study. In: 14th International Workshop on Treebanks and Linguistic Theories (TLT 2015), pp. 93–105. IPIPAN, Warszawa (2015)

22. Rysová, K., Rysová, M., Mírovský, J.: Automatic evaluation of surface coherence in L2 texts in Czech. In: Proceedings of the 28th Conference on Computational Linguistics and Speech Processing ROCLING XXVIII (2016), pp. 214–228. National Cheng Kung University, The Association for Computational Linguistics and Chinese Language Processing (ACLCLP), Taipei (2016)

23. Rysová, K., Rysová, M., Mírovský, J., Novák, M.: Introducing EVALD - software applications for automatic evaluation of discourse in Czech. In: Angelova, G., Boncheva, K., Mitkov, R., Nikolova, I., Temnikova, I. (eds.) Proceedings of the International Conference Recent Advances in Natural Language Processing, pp. 634–641. Bulgarian Academy of Sciences, INCOMA Ltd., Šumen (2017)

24. Šebesta, K., Bedřichová, Z., Šormová, K., et al.: AKCES 5 (CzeSL-SGT), data/software. LINDAT/CLARIN digital library at the Institute of Formal and Applied Linguistics, Charles University, Prague, Czech Republic (2014)

25. Šebesta, K., Goláňová, H., Letafková, J., et al.: AKCES 1, data/software. LINDAT/CLARIN digital library at the Institute of Formal and Applied Linguistics, Charles University, Prague, Czech Republic (2016)

26. Sgall, P.: Generativní systémy v lingvistice [Generative systems in linguistics]. Slovo a slovesnost 25(4), 274–282 (1964)

27. Sgall, P.: Generativní popis jazyka a česká deklinace [Generative Description of Language and Czech Declension]. Academia, Prague (1967)

28. Vajjala, S., Loo, K.: Automatic CEFR level prediction for Estonian learner text. In: Proceedings of the Third Workshop on NLP for Computer-Assisted Language Learning, pp. 113–127 (2014)

29. Volodina, E., Pilán, I., Alfter, D.: Classification of Swedish learner essays by CEFR levels. In: CALL Communities and Culture-Short Papers from EUROCALL 2016, pp. 456–461 (2016)

30. Žabokrtský, Z.: Treex - an open-source framework for natural language processing. In: Information Technologies - Applications and Theory, vol. 788, pp. 7–14. Univerzita Pavla Jozefa Šafárika v Košiciach, Košice (2011)

31. Zesch, T., Wojatzki, M., Scholten-Akoun, D.: Task-independent features for automated essay grading. In: Proceedings of the Tenth Workshop on Innovative Use of NLP for Building Educational Applications, pp. 224–232 (2015)

32. Zikánová, Š., et al.: Discourse and coherence. From the sentence structure to relations in text. Studies in Computational and Theoretical Linguistics, ÚFAL, Praha (2015)

A Multilingual Study of Compressive Cross-Language Text Summarization

Elvys Linhares Pontes[1,2,3]($\boxtimes$), Stéphane Huet[1],
and Juan-Manuel Torres-Moreno[1,2,3]

[1] LIA, Université d'Avignon et des Pays de Vaucluse, 84000 Avignon, France
`elvys.linhares-pontes@alumni.univ-avignon.fr`
[2] Département de GIGL/VERIFORM, École Polytechnique de Montréal,
C.P. 6079, succ. Centre-ville, Montréal, Québec H3C 3A7, Canada
[3] Laboratoire GDAC, Université du Québec à Montréal,
C.P. 8888, succ. Centre-ville, Montrél, Québec H3C 3P8, Canada

Abstract. Cross-Language Text Summarization generates summaries in a language different from the language of the source documents. Recent methods use information from both languages to generate summaries with the most informative sentences. However, these methods have performance that can vary according to languages, which can reduce the quality of summaries. In this paper, we propose a compressive framework to generate cross-language summaries. In order to analyze performance and especially stability, we tested our system and extractive baselines on a dataset available in four languages (English, French, Portuguese, and Spanish) to generate English and French summaries. An automatic evaluation showed that our method outperformed extractive state-of-art CLTS methods with better and more stable ROUGE scores for all languages.

1 Introduction

Cross-Language Text Summarization (CLTS) aims to generate a summary of a document, where the summary language differs from the document language. Many of the state-of-the-art methods for CLTS are of the extractive class. They mainly differ on how they compute sentence similarities and alleviate the risk that translation errors are introduced in the produced summary.

Previous works analyze the CLTS only between two languages for a given dataset, which does not demonstrate the stability of methods for different texts and languages. Recent works carried out compressive approaches based on neural networks, phrase segmentation, and graph theory [1–3]. Among these models, Linhares Pontes et al. introduced the use of chunks and two compression methods at the sentence and multi-sentence levels to improve the informativeness of cross-language summaries [3].

In this paper, we adapt the method presented in [3] to perform CLTS for several languages. More precisely, we modified the creation of chunks and we

© Springer Nature Switzerland AG 2018
I. Batyrshin et al. (Eds.): MICAI 2018, LNAI 11289, pp. 109–118, 2018.
https://doi.org/10.1007/978-3-030-04497-8_9

simplified their multi-sentence compression method to be able to analyze several languages and compress small clusters of similar sentences. To demonstrate the stability of our system, we extend the MultiLing Pilot dataset [4] with two Romance languages (Portuguese and Spanish) to test our system to generate {French, Portuguese, Spanish}-to-English and {English, Portuguese, Spanish}-to-French cross-language summaries. Finally, we carried out an automatic evaluation to make a systematic performance analysis of systems, which details the characteristics of each language and their impacts on the cross-language summaries.

The remainder of this paper is organized as follows. Section 2 details our contributions to the compressive CLTS approach. In Sect. 3, we describe the most recent works about CLTS. Section 4 reports the results achieved on the extended version of the MultiLing 2011 dataset and the analysis of cross-language summaries. Finally, conclusions and future work are set out in Sect. 5.

2   Compressive Cross-Language Text Summarization

Following the approach proposed by Linhares Pontes *et al.* [3], we combined the analysis of documents in the source and the target languages, with Multi-Sentence Compression (MSC) to generate more informative summaries. We expanded this approach in three ways.

In order to simplify and to extend the analysis to several languages, we only use Multi-Word Expressions for the English target language and we replace the analysis of parallel phrases with syntactic patterns to create chunks. Then, we also optimized the MSC method for small clusters by removing the analysis of 3-grams. Unfortunately, we have not found any available dataset for sentence compression in other languages; therefore, we restrict the use of compressive methods to MSC. Finally, differently from [3], compressed versions of sentences were considered in the CoRank method instead of the only original versions, in order to estimate the relevance of sentences for summaries.

The following subsections highlight our contributions to the architecture of the method presented in [3].

2.1   Preprocessing

Initially, source texts are translated into English and French with the Google Translate system[1], which was used in the majority of the state-of-the-art CLTS methods [2,3,5].

Then, a chunk-level tokenization is performed on the target language side [6,7]. We applied two simple syntactic patterns to identify useful structures: $< (ADJ)^*(NP|NC)^+ >$ for English and $< (ADJ)^*(NP|NC)^+(ADJ)^* >$ for French, where ADJ stands for adjective, NP for proper noun and NC for common noun. We also use the Stanford CoreNLP tool [8] for the English translations.

---

[1] https://translate.google.com.

This tool detects phrasal verbs, proper names, idioms and so on. Unfortunately, we did not find a similar tool for French; consequently, the French chunk-level tokenization is limited to the syntactic pattern.

2.2   Multi-Sentence Compression

We aim to generate a single, short, and informative compression from clusters of similar sentences. Therefore, we use the Linhares Pontes *et al.*'s method [3] to create clusters of similar sentences based on their similarity in the source and the target languages.

As the majority of clusters are composed of few similar sentences (normally two or three sentences), 3-grams are not frequent and the associated score is of little interest for the compress process. Therefore, we simplify the Linhares Pontes *et al.*'s method [3,9] to process MSC guided only by the cohesion of words and keywords.

Our MSC method looks for a sentence that has a good cohesion and the maximum of keywords, inside a word graph built for each cluster of similar sentences according to the method devised by Filippova [10]. In this graph, arcs between two vertices representing words (or chunks) are weighted by a cohesion score that is defined by the frequency of these words inside the cluster. Vertices are labeled depending on whether they are or not a keyword identified by the Latent Dirichlet Allocation (LDA) method inside the cluster (see [3] for more details). From these scores and labels, the MSC problem is expressed as the following objective:

$$\text{minimize} \left(\sum_{(i,j) \in A} w(i,j) \cdot x_{i,j} - k \cdot \sum_{l \in L} b_l \right) \tag{1}$$

where x_{ij} indicates the existence of the arc (i,j) in the solution, $w(i,j)$ is the cohesion of the words i and j, L is the set of labels (each representing a keyword), b_l indicates the existence of a chunk with a keyword l in the solution, k is the keyword bonus of the graph[2]. Finally, we generate the 50 best solutions according to the objective (1) and we select the compression with the lowest normalized score (Eq. 2) as the best compression:

$$\text{score}(c) = \frac{e^{\text{opt}(c)}}{||c||}, \tag{2}$$

where opt(c) is the score of the compression c from Eq. 1. We restrict the MSC method to the sentences in the target language in order to avoid errors generated by machine translation, which would be applied in a post-processing step on compressed sentences.

---

[2] The keyword bonus is defined by the geometric average of all weight arcs in the graph and aims at favoring compressions with several keywords.

2.3 CoRank Method

Sentences are scored based on their information in both languages using the CoRank method [5] which analyzes sentences in each language separately, but also between languages (Eqs. 3–7).

$$\mathbf{u} = \alpha \cdot (\tilde{\mathbf{M}}^{\mathbf{sc}})^T \mathbf{u} + \beta \cdot (\tilde{\mathbf{M}}^{\mathbf{tg,sc}})^T \mathbf{v} \tag{3}$$

$$\mathbf{v} = \alpha \cdot (\tilde{\mathbf{M}}^{\mathbf{tg}})^T \mathbf{v} + \beta \cdot (\tilde{\mathbf{M}}^{\mathbf{tg,sc}})^T \mathbf{u} \tag{4}$$

$$M_{ij}^{tg} = \begin{cases} \mathrm{cosine}(s_i^{tg}, s_j^{tg}), & \text{if } i \neq j \\ 0 & \text{otherwise} \end{cases} \tag{5}$$

$$M_{ij}^{sc} = \begin{cases} \mathrm{cosine}(s_i^{sc}, s_j^{sc}), & \text{if } i \neq j \\ 0 & \text{otherwise} \end{cases} \tag{6}$$

$$M_{ij}^{tg,sc} = \sqrt{\mathrm{cosine}(s_i^{sc}, s_j^{sc}) \times \mathrm{cosine}(s_i^{tg}, s_j^{tg})} \tag{7}$$

where $\mathbf{M}^{tg}$ and $\mathbf{M}^{sc}$ are normalized to $\tilde{\mathbf{M}}^{tg}$ and $\tilde{\mathbf{M}}^{sc}$, respectively, to make the sum of each row equal to 1. $\mathbf{u}$ and $\mathbf{v}$ denote the relevance of the source and target language sentences, respectively. α and β specify the relative contributions to the final scores from the information in the source and the target languages, with $\alpha + \beta = 1$.

Finally, summaries are generated with the most relevant sentences in the target language. We add a sentence/compression to the summary only if it is sufficiently different from the sentences/compressions already in the summary.

3 Related Work

Cross-Language Text Summarization schemes can be divided in early and late translations, and joint analysis. The early translation first translates documents to the target language, then it summarizes these translated documents using only information of these translations. The late translation scheme does the reverse. The joint analysis combines the information from both languages to extract the most relevant information.

Regarding the analysis of machine translation quality, Wan et al. [11] and Boudin et al. [12] used sentence features (sentence length, number of punctuation marks, number of phrases in the parse tree) to estimate the translation quality of a sentence. Wan et al. used an annotated dataset made of pairs of English-Chinese sentences with translation quality scores to train their Support Vector Machine (SVM) regression method. Finally, sentences that have a high translation quality and a high informativeness were selected for the summaries. Similarly, Boudin et al. trained an e-SVR using a dataset composed of English and automatic French translation sentences to calculate the translation quality based on the NIST metrics. Then, they used the PageRank algorithm to estimate the relevance of sentences based on their similarities and translation quality. Yao *et al.* devised a phrase-based model to jointly carry out sentence scoring and sentence compression [1]. They developed a scoring scheme for the CLTS task based on a submodular term of compressed sentences and a bounded distortion penalty term.

Wan [5] leverages the information in the source and in the target language for cross-language summarization. He proposed two graph-based summarization methods (SimFusion and CoRank) for the English-to-Chinese CLTS task. The first method linearly fuses the English-side and Chinese-side similarities for measuring Chinese sentence similarity. In a nutshell, this method adapts the PageRank algorithm to calculate the relevance of sentences, where the weight arcs are obtained by the linear combination of the cosine similarity of pairs of sentences for each language. The CoRank method was described in Sect. 2.3.

Recently, Wan *et al.* [2] carried out the cross-language document summarization task by extraction and ranking of multiple summaries in the target language. They analyzed many summaries in order to produce high-quality summaries for every kind of documents. Their method uses a top-K ensemble ranking for candidate summary, based on features that characterize the quality of a candidate summary. They used multiple text summarization and machine translation methods to generate the summaries.

In order to generate abstractive cross-lingual summaries, Zhang *et al.* [13] extended the work of Bing *et al.* [14] that constructs new sentences by exploring noun/verb phrases. Their method first constructs a pool of concepts and facts represented by phrases in English and Chinese translation sentences. Then new sentences are generated by selecting and merging informative phrases in both languages to maximize the salience of phrases and meanwhile satisfy the sentence construction constraints. They employ integer linear optimization for conducting phrase selection and merging simultaneously in order to generate informative cross-lingual summaries with a good translation quality. This method generates abstractive summaries; however, the framework to identify concepts and facts only works for English, which prevents this method from being extended for other languages.

4   Experimental Evaluation

We estimate the performance of our approach in relation to the early and the late translations, SimFusion and CoRank methods[3,4]. All systems generate summaries containing a maximum of 250 words with the best scored sentences without redundant sentences. We regard a similarity score (cosine similarity) with a threshold θ of 0.6 to create clusters of similar sentences for the MSC[5] and a threshold θ of 0.5 to remove redundant sentences in the summary generation.

---

[3] We used the same configuration for SimFusion and CoRank as described in [3].

[4] Unfortunately, the majority of state-of-the-art systems in CLTS are not available. Therefore, we only considered extractive systems in our analysis.

[5] We use the same threshold θ of 0.5 described in [3] for French-to-English cross-language summaries.

4.1 Datasets

We used the English and French language versions of the MultiLing Pilot 2011 dataset [4]. This dataset contains 10 topics which have 10 source texts and 3 reference summaries per topic. These summaries are composed of 250 words. In order to extend the analysis to other languages, English source texts were translated into the Portuguese and Spanish languages by native speakers[6]. Specifically, we use English, French, Portuguese, and Spanish texts to test our system.

4.2 Evaluation

An automatic evaluation with ROUGE [15] was carried out to compare the differences between the distribution of n-grams of the candidate summary and a set of reference summaries. More specifically, we used unigram (ROUGE-1 or R-1), bigram (ROUGE or R-2), and skip-gram (ROUGE-SU4 or R-SU4) analyses.

Table 1 describes the ROUGE f-scores obtained by each system to generate French summaries from English, Portuguese, and Spanish source texts. Despite using the information from both languages, the SimFusion method achieved comparable results with respect to the early and late approaches. On the contrary, CoRank and our approach consistently obtained better results than other baselines, with at least an absolute difference of 0.035 in ROUGE-1 for all languages. The MSC method improved the CoRank method by generating more informative compressions for all languages. The last two lines show that chunks helped our MSC method to generate slightly more informative summaries (better ROUGE-1 scores).

Table 1. ROUGE f-scores for cross-language summaries from English, Portuguese, and Spanish languages to French language.

Methods	English			Portuguese			Spanish		
	R-1	R-2	R-SU4	R-1	R-2	R-SU4	R-1	R-2	R-SU4
Late	0.4190	0.0965	0.1588	0.4403	0.1128	0.1746	0.4371	0.1133	0.1738
Early	0.4223	0.1007	0.1631	0.4386	0.1110	0.1743	0.4363	0.1143	0.1729
SimFusion	0.4240	0.1004	0.1637	0.4368	0.1105	0.1735	0.4350	0.1125	0.1723
CoRank	0.4733	0.1379	0.1963	0.4723	0.1460	0.2006	0.4713	0.1387	0.1942
Our approach	**0.4831**	0.1460	**0.2030**	**0.4784**	0.1511	**0.2045**	**0.4825**	0.1481	0.2050
Our approach w/o chunks	0.4817	**0.1463**	0.2021	**0.4784**	**0.1518**	0.2044	0.4805	**0.1486**	**0.2056**

The Multiling dataset is composed of 10 topics in several languages; however, these topics are expressed in different ways for each language. These dissimilarities implies a variety of vocabulary sizes and sentence lengths, and, consequently,

---

[6] The extension of the MultiLing Pilot 2011 dataset is available at: http://dev.termwatch.es/~fresa/CORPUS/TS/.

of outputs of the MT system from each source language (Table 2). The biggest difference in the statistics is between English source texts and its French translation vocabulary. French translations significantly increased the vocabulary from English source texts and the number of words. These translations also are longer than source texts, except for the Spanish that has similar characteristics. Our simple syntactic pattern created similar numbers of chunks for all languages with the same average length. The addition of these simple chunks did not significantly improve the informativeness of our compressions.

Table 2. Statistics of datasets and their translation to French.

	English		Portuguese		Spanish	
	Source	Fr-Translation	Source	Fr-Translation	Source	Fr-Translation
#words	36,109	39,960	37,339	39,302	40,440	40,269
#vocabulary	8,077	8,770	8,694	8,572	8,808	8,744
#sentences	1,816	1,816	2,002	2,002	1,787	1,787
sentence length	19.9	22.0	18.6	19.6	22.6	22.5
#chunks	–	1,615	–	1,579	–	1,606
Average length of chunks	–	2.1	–	2.1	–	2.1

These differences also act on the clustering process and the MSC method. Table 3 details the number and the average size of clusters with at least two French sentences translated from each source language. French translations from Portuguese produced the shortest compressions (18.6 words) while compressions from Spanish had the highest compression ratio. With respect to other languages, the similarity of the sentences translated from English is lower, which leads to fewer clusters. Summaries from Spanish have a larger proportion of compressions in the summaries than other languages.

Table 3. Statistics about clusters and compressions for texts translated into French.

	English	Portuguese	Spanish
#clusters	50	70	75
Average size of clusters	2.2	2.7	2.8
Average length of clusters	29.1	25.6	35.4
Average length of compressions	21.7	18.6	23.5
Average number of compressions in summaries	0.7	0.9	1.3
Average compression rate of compressions	74.6%	72.6%	66.4%

We apply a similar analysis for the generation of English summaries from French, Portuguese, and Spanish source texts. As observed before for French summaries, the joint analysis still outperformed other baselines (Table 4). While

CoRank obtained a large range of ROUGE scores among different languages (ROUGE-1 between 0.4602 and 0.4715), our approach obtained the best ROUGE scores for all languages with a small difference of ROUGE scores (ROUGE-1 between 0.4725 and 0.4743), which proves that our method generates more stable cross-language summaries for several languages. Chunks spot by the syntactic pattern and the Stanford CoreNLP helped our approach to produce more informative compressions, which results in better ROUGE scores.

Table 4. ROUGE f-scores for cross-language summaries from French, Portuguese, and Spanish languages to English language.

Methods	French			Portuguese			Spanish		
	R-1	R-2	R-SU4	R-1	R-2	R-SU4	R-1	R-2	R-SU4
Late	0.4149	0.1030	0.1594	0.4161	0.1010	0.1576	0.4107	0.1083	0.1603
Early	0.4163	0.1021	0.1602	0.4135	0.1003	0.1580	0.4148	0.1132	0.1644
SimFusion	0.4179	0.1042	0.1607	0.4157	0.0999	0.1582	0.4099	0.1103	0.1616
CoRank	0.4645	0.1326	0.1939	0.4715	0.1415	0.2015	0.4602	0.1414	0.1966
Our approach	**0.4727**	0.1375	**0.1969**	**0.4743**	**0.1466**	**0.2047**	**0.4725**	**0.1458**	**0.2027**
Our approach w/o chunks	0.4704	**0.1391**	0.1963	0.4731	0.1444	0.2037	0.4648	0.1393	0.1975

English translations have fewer words and a smaller vocabulary (difference bigger than 1,000 words) than source texts (Table 5). These translations also have shorter sentences and a more similar vocabulary size than French translations and source texts. The combination of syntactic patterns and the Stanford CoreNLP led to the same characteristics of chunks in terms of numbers and sizes.

Table 5. Statistics of datasets and their translation to English.

	French		Portuguese		Spanish	
	Source	En-Translation	Source	En-Translation	Source	En-Translation
#words	41,071	35,929	37,339	35,244	40,440	37,066
#vocabulary	8,837	7,718	8,694	7,615	8,808	7,703
#sentences	2,000	2,000	2,002	2,002	1,787	1,787
sentence length	20.5	18.0	18.6	17.6	22.6	20.7
#chunks	–	4,302	–	4,327	–	4,324
Average length of chunks	–	2.3	–	2.3	–	2.3

Table 6 details the clustering and the compression processes for the English translations. These translations from French source texts have more clusters because we used a smaller similarity threshold to consider two sentences as similar. English summaries from French have more compressions because of the large number of clusters.

Table 6. Statistics about clusters and compressions for English translated texts.

	French	Portuguese	Spanish
#clusters	128	69	84
Average size of clusters	2.7	2.7	2.8
Average length of clusters	22.1	19.2	27.0
Average length of compressions	16.4	16.3	21.1
Average number of compressions in summaries	2.5	0.9	1.5
Average compression rate of compressions	74.2%	84.9%	78.1%

French and Portuguese source texts have almost the same number of sentences, while English and Spanish source texts have fewer sentences. Comparing the results of English and French translations, English compressions are shorter than French compressions. The use of chunks in MSC improved the results of our cross-language summaries, especially for English translations that have chunks that are more numerous and complex than French translations.

To sum up, our approach has shown to be more stable than extractive methods, thus generating more informative cross-language summaries with consistent ROUGE scores measured in several languages.

5   Conclusion

Cross-Language Text Summarization (CLTS) produces a summary in a target language from documents written in a source language. It implies a combination of the processes of automatic summarization and machine translation. Unfortunately, this combination produces errors, thereby reducing the quality of summaries. A joint analysis allows CLTS systems to extract relevant information from source and target languages, which improves the generation of extractive cross-language summaries. Recent methods have proposed compressive and abstractive approaches for CLTS; however, these methods use frameworks or tools that are available in few languages, limiting the portability of these methods to other languages. Our Multi-Sentence Compression (MSC) approach generates informative compressions from several perspectives (translations from different languages) and achieves stable ROUGE results for all languages. In addition, our method can be easily adapted to other languages.

As future work, we plan to reduce the number of errors generated from the pipeline made of the compression and machine translation processes by developing a Neural Network method to jointly translate and compress sentences. It would also be interesting to include a neural language model to correct possible errors produced during the sentence compression process.

Acknowledgement. This work was granted by the European Project CHISTERA-AMIS ANR-15-CHR2-0001. We also like to acknowledge the support given by the Laboratoire VERIFORM from the École Polytechnique de Montréal and her coordinator Hanifa Boucheneb.

References

1. Yao, J., Wan, X., Xiao, J.: Phrase-based compressive cross-language summarization. In: EMNLP, pp. 118–127 (2015)
2. Wan, X., Luo, F., Sun, X., Huang, S., Yao, J.G.: Cross-language document summarization via extraction and ranking of multiple summaries. Knowl. Inf. Syst. (2018). https://doi.org/10.1007/s10115-018-1152-7
3. Linhares Pontes, E., Huet, S., Torres-Moreno, J.M., Linhares, A.C.: Cross-language text summarization using sentence and multi-sentence compression. In: Silberztein, M., Atigui, F., Kornyshova, E., Métais, E., Meziane, F. (eds.) NLDB 2018. Lecture Notes in Computer Science, vol. 10859. Springer, Cham (2018). https://doi.org/10.1007/978-3-319-91947-8_48
4. Giannakopoulos, G., El-Haj, M., Favre, B., Litvak, M., Steinberger, J., Varma, V.: TAC2011 multiling pilot overview. In: 4th Text Analysis Conference TAC (2011)
5. Wan, X.: Using bilingual information for cross-language document summarization. In: ACL, pp. 1546–1555 (2011)
6. Moirón, B.V., Tiedemann, J.: Identifying idiomatic expressions using automatic word-alignment. In: EACL 2006 Workshop on Multiword Expressions in a Multilingual Context (2006)
7. de Caseli, H.M., Ramisch, C., das Graças Volpe Nunes, M., Villavicencio, A.: Alignment-based extraction of multiword expressions. Lang. Resour. Eval. **44**, 59–77 (2010)
8. Manning, C., Surdeanu, M., Bauer, J., Finkel, J., Bethard, S., McClosky, D.: The Stanford CoreNLP natural language processing toolkit. In: 52nd Annual Meeting of the Association for Computational Linguistics (ACL): System Demonstrations, pp. 55–60 (2014)
9. Linhares Pontes, E., Huet, S., Gouveia da Silva, T., Linhares, A.C., Torres-Moreno, J.M.: Multi-sentence compression with word vertex-labeled graphs and integer linear programming. In: TextGraphs-12: The Workshop on Graph-based Methods for Natural Language Processing, Association for Computational Linguistics (2018)
10. Filippova, K.: Multi-sentence compression: finding shortest paths in word graphs. In: COLING, pp. 322–330 (2010)
11. Wan, X., Li, H., Xiao, J.: Cross-language document summarization based on machine translation quality prediction. In: ACL, pp. 917–926 (2010)
12. Boudin, F., Huet, S., Torres-Moreno, J.M.: A graph-based approach to cross-language multi-document summarization. Polibits **43**, 113–118 (2011)
13. Zhang, J., Zhou, Y., Zong, C.: Abstractive cross-language summarization via translation model enhanced predicate argument structure fusing. IEEE/ACM Trans. Audio Speech Lang. Process. **24**, 1842–1853 (2016)
14. Bing, L., Li, P., Liao, Y., Lam, W., Guo, W., Passonneau, R.J.: Abstractive multidocument summarization via phrase selection and merging. In: ACL, The Association for Computer Linguistics, pp. 1587–1597 (2015)
15. Lin, C.Y.: ROUGE: a package for automatic evaluation of summaries. In: Workshop Text Summarization Branches Out (ACL 2004), pp. 74–81 (2004)

WiSeBE: Window-Based Sentence Boundary Evaluation

Carlos-Emiliano González-Gallardo[1,2]([✉]) and Juan-Manuel Torres-Moreno[1,2]

[1] LIA - Université d'Avignon et des Pays de Vaucluse, 339 chemin des Meinajaries, 84140 Avignon, France
carlos-emiliano.gonzalez-gallardo@alumni.univ-avignon.fr,
juan-manuel.torres@univ-avignon.fr
[2] Département de GIGL, École Polytechnique de Montréal,
C.P. 6079, succ. Centre-ville, Montréal, Québec H3C 3A7, Canada

Abstract. Sentence Boundary Detection (SBD) has been a major research topic since Automatic Speech Recognition transcripts have been used for further Natural Language Processing tasks like Part of Speech Tagging, Question Answering or Automatic Summarization. But what about evaluation? Do standard evaluation metrics like precision, recall, F-score or classification error; and more important, evaluating an automatic system against a unique reference is enough to conclude how well a SBD system is performing given the final application of the transcript? In this paper we propose Window-based Sentence Boundary Evaluation (WiSeBE), a semi-supervised metric for evaluating Sentence Boundary Detection systems based on multi-reference (dis)agreement. We evaluate and compare the performance of different SBD systems over a set of Youtube transcripts using WiSeBE and standard metrics. This double evaluation gives an understanding of how WiSeBE is a more reliable metric for the SBD task.

Keywords: Sentence Boundary Detection · Evaluation
Transcripts · Human judgment

1 Introduction

The goal of Automatic Speech Recognition (ASR) is to transform spoken data into a written representation, thus enabling natural human-machine interaction [33] with further Natural Language Processing (NLP) tasks. Machine translation, question answering, semantic parsing, POS tagging, sentiment analysis and automatic text summarization; originally developed to work with formal written texts, can be applied over the transcripts made by ASR systems [2,25,31]. However, before applying any of these NLP tasks a segmentation process called Sentence Boundary Detection (SBD) should be performed over ASR transcripts to reach a minimal syntactic information in the text.

To measure the performance of a SBD system, the automatically segmented transcript is evaluated against a single reference normally done by a human. But

© Springer Nature Switzerland AG 2018
I. Batyrshin et al. (Eds.): MICAI 2018, LNAI 11289, pp. 119–131, 2018.
https://doi.org/10.1007/978-3-030-04497-8_10

given a transcript, does it exist a unique reference? Or, is it possible that the same transcript could be segmented in five different ways by five different people in the same conditions? If so, which one is correct; and more important, how to fairly evaluate the automatically segmented transcript? These questions are the foundations of Window-based Sentence Boundary Evaluation (WiSeBE), a new semi-supervised metric for evaluating SBD systems based on multi-reference (dis)agreement.

The rest of this article is organized as follows. In Sect. 2 we set the frame of SBD and how it is normally evaluated. WiSeBE is formally described in Sect. 3, followed by a multi-reference evaluation in Sect. 4. Further analysis of WiSeBE and discussion over the method and alternative multi-reference evaluation is presented in Sect. 5. Finally, Sect. 6 concludes the paper.

2 Sentence Boundary Detection

Sentence Boundary Detection (SBD) has been a major research topic science ASR moved to more general domains as conversational speech [17,24,26]. Performance of ASR systems has improved over the years with the inclusion and combination of new Deep Neural Networks methods [5,9,33]. As a general rule, the output of ASR systems lacks of any syntactic information such as capitalization and sentence boundaries, showing the interest of ASR systems to obtain the correct sequence of words with almost no concern of the overall structure of the document [8].

Similar to SBD is the Punctuation Marks Disambiguation (PMD) or Sentence Boundary Disambiguation. This task aims to segment a formal written text into well formed sentences based on the existent punctuation marks [11,19,20,29]. In this context a sentence is defined (for English) by the Cambridge Dictionary[1] as:

> *"a group of words, usually containing a verb, that expresses a thought in the form of a statement, question, instruction, or exclamation and starts with a capital letter when written".*

PMD carries certain complications, some given the ambiguity of punctuation marks within a sentence. A period can denote an acronym, an abbreviation, the end of the sentence or a combination of them as in the following example:

> *The U.S. president, Mr. Donald Trump, is meeting with the F.B.I. director Christopher A. Wray next Thursday at 8 p.m.*

However its difficulties, DPM profits of morphological and lexical information to achieve a correct sentence segmentation. By contrast, segmenting an ASR transcript should be done without any (or almost any) lexical information and a flurry definition of sentence.

---

[1] https://dictionary.cambridge.org/.

The obvious division in spoken language may be considered speaker utterances. However, in a normal conversation or even in a monologue, the way ideas are organized differs largely from written text. This differences, added to disfluencies like revisions, repetitions, restarts, interruptions and hesitations make the definition of a sentence unclear thus complicating the segmentation task [27]. Table 1 exemplifies some of the difficulties that are present when working with spoken language.

Table 1. Sentence Boundary Detection example

Speech transcript	SBD applied to transcript
two two women can look out after a kid so bad as a man and a woman can so you can have a you can have a mother and a father that that still don't do right with the kid and you can have to men that can so as long as the love each other as long as they love each other it doesn't matter	two // two women can look out after a kid so bad as a man and a woman can // so you can have a // you can have a mother and a father that // that still don't do right with the kid and you can have to men that can // so as long as the love each other // as long as they love each other it doesn't matter//

Stolcke and Shriberg [26] considered a set of linguistic structures as segments including the following list:

- Complete sentences
- Stand-alone sentences
- Disfluent sentences aborted in mid-utterance
- Interjections
- Back-channel responses.

In [17], Meteer and Iyer divided speaker utterances into segments, consisting each of a single independent clause. A segment was considered to begin either at the beginning of an utterance, or after the end of the preceding segment. Any dysfluency between the end of the previous segments and the begging of current one was considered part of the current segments.

Rott and Červa [23] aimed to summarize news delivered orally segmenting the transcripts into *"something that is similar to sentences"*. They used a syntactic analyzer to identify the phrases within the text.

A wide study focused in unbalanced data for the SBD task was performed by Liu *et al.* [15]. During this study they followed the segmentation scheme proposed by the Linguistic Data Consortium[2] on the Simple Metadata Annotation Specification V5.0 guideline (SimpleMDE_V5.0) [27], dividing the transcripts in Semantic Units.

---

[2] https://www.ldc.upenn.edu/.

A Semantic Unit (SU) is considered to be an atomic element of the transcript that manages to express a complete thought or idea on the part of the speaker [27]. Sometimes a SU corresponds to the equivalent of a sentence in written text, but other times (the most part of them) a SU corresponds to a phrase or a single word.

SUs seem to be an inclusive conception of a segment, they embrace different previous segment definitions and are flexible enough to deal with the majority of spoken language troubles. For these reasons we will adopt SUs as our segment definition.

2.1   Sentence Boundary Evaluation

SBD research has been focused on two different aspects; features and methods. Regarding the features, some work focused on acoustic elements like pauses duration, fundamental frequencies, energy, rate of speech, volume change and speaker turn [10, 12, 14].

The other kind of features used in SBD are textual or lexical features. They rely on the transcript content to extract features like bag-of-word, POS tags or word embeddings [7, 12, 16, 18, 23, 26, 30]. Mixture of acoustic and lexical features have also been explored [1, 13, 14, 32], which is advantageous when both audio signal and transcript are available.

With respect to the methods used for SBD, they mostly rely on statistical/neural machine translation [12, 22], language models [8, 15, 18, 26], conditional random fields [16, 30] and deep neural networks [3, 7, 29].

Despite their differences in features and/or methodology, almost all previous cited research share a common element; the evaluation methodology. Metrics as Precision, Recall, F1-score, Classification Error Rate and Slot Error Rate (SER) are used to evaluate the proposed system against one reference. As discussed in Sect. 1, further NLP tasks rely on the result of SBD, meaning that is crucial to have a good segmentation. But comparing the output of a system against a unique reference will provide a reliable score to decide if the system is good or bad?

Bohac et al. [1] compared the human ability to punctuate recognized spontaneous speech. They asked 10 people (correctors) to punctuate about 30 min of ASR transcripts in Czech. For an average of 3,962 words, the punctuation marks placed by correctors varied between 557 and 801; this means a difference of 244 segments for the same transcript. Over all correctors, the absolute consensus for period (.) was only 4.6% caused by the replacement of other punctuation marks as semicolons (;) and exclamation marks (!). These results are understandable if we consider the difficulties presented previously in this section.

To our knowledge, the amount of studies that have tried to target the sentence boundary evaluation with a multi-reference approach is very small. In [1], Bohac et al. evaluated the overall punctuation accuracy for Czech in a straightforward multi-reference framework. They considered a period (.) valid if at least five of their 10 correctors agreed on its position.

Kolář and Lamel [13] considered two independent references to evaluate their system and proposed two approaches. The fist one was to calculate the SER for each of one the two available references and then compute their mean. They found this approach to be very strict because for those boundaries where no agreement between references existed, the system was going to be partially wrong even the fact that it has correctly predicted the boundary. Their second app-roach tried to moderate the number of unjust penalizations. For this case, a classification was considered incorrect only if it didn't match either of the two references.

These two examples exemplify the real need and some straightforward solutions for multi-reference evaluation metrics. However, we think that it is possible to consider in a more inclusive approach the similarities and differences that mul-tiple references could provide into a sentence boundary evaluation protocol.

3   Window-Based Sentence Boundary Evaluation

Window-Based Sentence Boundary Evaluation (WiSeBE) is a semi-automatic multi-reference sentence boundary evaluation protocol which considers the per-formance of a candidate segmentation over a set of segmentation references and the agreement between those references.

Let $\mathbf{R} = \{R_1, R_2, ..., R_m\}$ be the set of all available references given a tran-script $T = \{t_1, t_2, ..., t_n\}$, where t_j is the jth word in the transcript; a reference R_i is defined as a binary vector in terms of the existent SU boundaries in T.

$$R_i = \{b_1, b_2, ..., b_n\} \tag{1}$$

where

$$b_j = \begin{cases} 1 \text{ if } t_j \text{ is a boundary} \\ 0 \text{ otherwise} \end{cases}$$

Given a transcript T, the candidate segmentation C_T is defined similar to R_i.

$$C_T = \{b_1, b_2, ..., b_n\} \tag{2}$$

where

$$b_j = \begin{cases} 1 \text{ if } t_j \text{ is a boundary} \\ 0 \text{ otherwise} \end{cases}$$

3.1   General Reference and Agreement Ratio

A General Reference (R_G) is then constructed to calculate the agreement ratio between all references in. It is defined by the boundary frequencies of each ref-erence $R_i \in \mathbf{R}$.

$$R_G = \{d_1, d_2, ..., d_n\} \tag{3}$$

where

$$d_j = \sum_{i=1}^{m} t_{ij} \quad \forall t_j \in T, \quad d_j = [0, m] \tag{4}$$

The Agreement Ratio $(R_{G_{AR}})$ is needed to get a numerical value of the distribution of SU boundaries over $\mathbf{R}$. A value of $R_{G_{AR}}$ close to 0 means a low agreement between references in $\mathbf{R}$, while $R_{G_{AR}} = 1$ means a perfect agreement $(\forall R_i \in \mathbf{R}, R_i = R_{i+1} | i = 1, ..., m-1)$ in $\mathbf{R}$.

$$R_{G_{AR}} = \frac{R_{G_{PB}}}{R_{G_{HA}}} \tag{5}$$

In the equation above, $R_{G_{PB}}$ corresponds to the ponderated common boundaries of R_G and $R_{G_{HA}}$ to its hypothetical maximum agreement.

$$R_{G_{PB}} = \sum_{j=1}^{n} d_j \left[d_j \geq 2 \right] \tag{6}$$

$$R_{G_{HA}} = m \times \sum_{d_j \in R_G} 1 \left[d_j \neq 0 \right] \tag{7}$$

3.2 Window-Boundaries Reference

In Sect. 2 we discussed about how disfluencies complicate SU segmentation. In a multi-reference environment this causes disagreement between references around a same SU boundary. The way WiSeBE handle disagreements produced by disfluencies is with a Window-boundaries Reference (R_W) defined as:

$$R_W = \{w_1, w_2, ..., w_p\} \tag{8}$$

where each window w_k considers one or more boundaries d_j from R_G with a window separation limit equal to R_{W_l}.

$$w_k = \{d_j, d_{j+1}, d_{j+2}, ...\} \tag{9}$$

3.3 *WiSeBE*

WiSeBE is a normalized score dependent of (1) the performance of C_T over R_W and (2) the agreement between all references in $\mathbf{R}$. It is defined as:

$$WiSeBE = F1_{R_W} \times R_{G_{AR}} \quad WiSeBE = [0, 1] \tag{10}$$

where $F1_{R_W}$ corresponds to the harmonic mean of precision and recall of C_T with respect to R_W (Eq. 11), while $R_{G_{AR}}$ is the agreement ratio defined in (5). $R_{G_{AR}}$ can be interpreted as a scaling factor; a low value will penalize the overall *WiSeBE* score given the low agreement between references. By contrast, for a high agreement in $\mathbf{R}$ $(R_{G_{AR}} \approx 1)$, $WiSeBE \approx F1_{R_W}$.

$$F1_{R_W} = 2 \times \frac{precision_{R_W} \times recall_{R_W}}{precision_{R_W} + recall_{R_W}} \tag{11}$$

$$precision_{R_W} = \frac{\sum_{b_j \in C_T} 1 \quad [b_j = 1, b_j \in w \quad \forall w \in R_W]}{\sum_{b_j \in C_T} 1 \quad [b_j = 1]} \tag{12}$$

$$recall_{R_W} = \frac{\sum_{w_k \in R_W} 1 \quad [w_k \ni b \quad \forall b \in C_T]}{p} \tag{13}$$

Equations 12 and 13 describe precision and recall of C_T with respect to R_W. Precision is the number of boundaries b_j inside any window w_k from R_W divided by the total number of boundaries b_j in C_T. Recall corresponds to the number of windows w with at least one boundary b divided by the number of windows w in R_W.

4   Evaluating with *WiSeBE*

To exemplify the *WiSeBE* score we evaluated and compared the performance of two different SBD systems over a set of YouTube videos in a multi-reference environment. The first system (S1) employs a Convolutional Neural Network to determine if the middle word of a sliding window corresponds to a SU boundary or not [6]. The second approach (S2) by contrast, introduces a bidirectional Recurrent Neural Network model with attention mechanism for boundary detection [28].

In a first glance we performed the evaluation of the systems against each one of the references independently. Then, we implemented a multi-reference evaluation with *WiSeBE*.

4.1   Dataset

We focused evaluation over a small but diversified dataset composed by 10 YouTube videos in the English language in the news context. The selected videos cover different topics like technology, human rights, terrorism and politics with a length variation between 2 and 10 min. To encourage the diversity of content format we included newscasts, interviews, reports and round tables.

During the transcription phase we opted for a manual transcription process because we observed that using transcripts from an ASR system will difficult in a large degree the manual segmentation process. The number of words per transcript oscilate between 271 and 1,602 with a total number of 8,080.

We gave clear instructions to three evaluators (ref_1, ref_2, ref_3) of how segmentation was needed to be perform, including the SU concept and how punctuation marks were going to be taken into account. Periods (.), question marks (?), exclamation marks (!) and semicolons (;) were considered SU delimiters (boundaries) while colons (:) and commas (,) were considered as internal SU marks. The number of segments per transcript and reference can be seen in Table 2. An interesting remark is that ref_3 assigns about 43% less boundaries than the mean of the other two references.

Table 2. Manual dataset segmentation

Reference	v_1	v_2	v_3	v_4	v_5	v_6	v_7	v_8	v_9	v_{10}	Total
ref_1	38	42	17	11	55	87	109	72	55	16	502
ref_2	33	42	16	14	54	98	92	65	51	20	485
ref_3	23	20	10	6	39	39	76	30	29	9	281

4.2 Evaluation

We ran both systems (S1 & S2) over the manually transcribed videos obtaining the number of boundaries shown in Table 3. In general, it can be seen that S1 predicts 27% more segments than S2. This difference can affect the performance of S1, increasing its probabilities of false positives.

Table 3. Automatic dataset segmentation

System	v_1	v_2	v_3	v_4	v_5	v_6	v_7	v_8	v_9	v_{10}	Total
S1	53	38	15	13	54	108	106	70	71	11	539
S2	38	37	12	11	36	92	86	46	53	13	424

Table 4 condenses the performance of both systems evaluated against each one of the references independently. If we focus on F1 scores, performance of both systems varies depending of the reference. For ref_1, S1 was better in 5 occasions with respect of S2; S1 was better in 2 occasions only for ref_2; S1 overperformed S2 in 3 occasions concerning ref_3 and in 4 occasions for *mean* (**bold**).

Also from Table 4 we can observe that ref_1 has a bigger similarity to S1 in 5 occasions compared to other two references, while ref_2 is more similar to S2 in 7 transcripts (underline).

After computing the mean F1 scores over the transcripts, it can be concluded that in average S2 had a better performance segmenting the dataset compared to S1, obtaining a F1 score equal to 0.510. But... What about the complexity of the dataset? Regardless all references have been considered, nor agreement or disagreement between them has been taken into account.

All values related to the $WiSeBE$ score are displayed in Table 5. The Agreement Ratio ($R_{G_{AR}}$) between references oscillates between 0.525 for v_8 and 0.767 for v_5. The lower the $R_{G_{AR}}$, the bigger the penalization $WiSeBE$ will give to the final score. A good example is S2 for transcript v_4 where $F1_{R_W}$ reaches a value of 0.800, but after considering $R_{G_{AR}}$ the $WiSeBE$ score falls to 0.462.

It is feasible to think that if all references are taken into account at the same time during evaluation ($F1_{R_W}$), the score will be bigger compared to an average of independent evaluations ($F1_{mean}$); however this is not always true. That is the case of S1 in $v10$, which present a slight decrease for $F1_{R_W}$ compared to $F1_{mean}$.

Table 4. Independent multi-reference evaluation

Transcript	System	ref_1			ref_2			ref_3			$Mean$		
		P	R	F1	P	R	F1	P	R	F1	P	R	F1
v_1	S1	0.396	0.553	0.462	0.377	0.606	<u>0.465</u>	0.264	0.609	0.368	0.346	0.589	0.432
	S2	0.474	0.474	**0.474**	0.474	0.545	**<u>0.507</u>**	0.368	0.6087	**0.459**	0.439	0.543	**0.480**
v_2	S1	0.605	0.548	**0.575**	0.711	0.643	**<u>0.675</u>**	0.368	0.700	**0.483**	0.561	0.630	**0.578**
	S2	0.595	0.524	0.557	0.676	0.595	<u>0.633</u>	0.351	0.650	0.456	0.541	0.590	0.549
v_3	S1	0.333	0.294	<u>0.313</u>	0.267	0.250	0.258	0.200	0.300	0.240	0.267	0.281	0.270
	S2	0.417	0.294	**0.345**	0.417	0.313	**<u>0.357</u>**	0.250	0.300	**0.273**	0.361	0.302	**0.325**
v_4	S1	0.615	0.571	<u>0.593</u>	0.462	0.545	0.500	0.308	0.667	0.421	0.462	0.595	0.505
	S2	0.909	0.714	**0.800**	0.818	0.818	**<u>0.818</u>**	0.455	0.833	**0.588**	0.727	0.789	**0.735**
v_5	S1	0.630	0.618	**<u>0.624</u>**	0.593	0.593	**0.593**	0.481	0.667	**0.560**	0.568	0.626	**0.592**
	S2	0.667	0.436	<u>0.527</u>	0.611	0.407	0.489	0.500	0.462	0.480	0.593	0.435	0.499
v_6	S1	0.491	0.541	**0.515**	0.454	0.563	0.503	0.213	0.590	0.313	0.386	0.565	0.443
	S2	0.500	0.469	0.484	0.522	0.552	**<u>0.536</u>**	0.250	0.590	**0.351**	0.4234	0.537	**0.457**
v_7	S1	0.594	0.578	**0.586**	0.462	0.533	0.495	0.406	0.566	0.473	0.487	0.559	0.518
	S2	0.663	0.523	<u>0.585</u>	0.558	0.522	**<u>0.539</u>**	0.465	0.526	**0.494**	0.562	0.524	**0.539**
v_8	S1	0.443	0.477	0.459	0.514	0.500	<u>0.507</u>	0.229	0.533	0.320	0.395	0.503	0.429
	S2	0.609	0.431	**0.505**	0.652	0.417	**<u>0.508</u>**	0.370	0.567	**0.447**	0.543	0.471	**0.487**
v_9	S1	0.437	0.564	0.492	0.451	0.627	<u>0.525</u>	0.254	0.621	0.360	0.380	0.603	**0.459**
	S2	0.623	0.600	**<u>0.611</u>**	0.585	0.608	**0.596**	0.321	0.586	**0.414**	0.509	0.598	0.541
v_{10}	S1	0.818	0.450	**<u>0.581</u>**	0.818	0.450	**<u>0.581</u>**	0.455	0.556	**0.500**	0.697	0.523	**0.582**
	S2	0.692	0.450	0.545	0.615	0.500	<u>0.552</u>	0.308	0.444	0.364	0.538	0.4645	0.487
Mean scores	S1	—	<u>0.520</u>		—	0.510		—	0.404		—	0.481	
	S2	—	**0.543**		—	**<u>0.554</u>**		—	**0.433**		—	**0.510**	

An important remark is the behavior of S1 and S2 concerning v_6. If evaluated without considering any (dis)agreement between references ($F1_{mean}$), S2 overperforms S1; this is inverted once the systems are evaluated with $WiSeBE$.

5 Discussion

5.1 $R_{G_{AR}}$ and Fleiss' Kappa correlation

In Sect. 3 we described the $WiSeBE$ score and how it relies on the $R_{G_{AR}}$ value to scale the performance of C_T over R_W. $R_{G_{AR}}$ can intuitively be consider an agreement value over all elements of **R**. To test this hypothesis, we computed the Pearson correlation coefficient (PCC) [21] between $R_{G_{AR}}$ and the Fleiss' Kappa [4] of each video in the dataset (κ_R).

A linear correlation between $R_{G_{AR}}$ and κ_R can be observed in Table 6. This is confirmed by a PCC value equal to 0.890, which means a very strong positive linear correlation between them.

5.2 $F1_{mean}$ vs. $WiSeBE$

Results form Table 5 may give an idea that $WiSeBE$ is just an scaled $F1_{mean}$. While it is true that they show a linear correlation, $WiSeBE$ may produce a

Table 5. *WiSeBE* evaluation

Transcript	System	$F1_{mean}$	$F1_{R_W}$	$R_{G_{AR}}$	$WiSeBE$
v_1	S1	0.432	0.495	0.691	0.342
	S2	**0.480**	0.513		**0.354**
v_2	S1	**0.578**	0.659	0.688	**0.453**
	S2	0.549	0.595		0.409
v_3	S1	0.270	0.303	0.684	0.207
	S2	**0.325**	0.400		**0.274**
v_4	S1	0.505	0.593	0.578	0.342
	S2	**0.735**	0.800		**0.462**
v_5	S1	**0.592**	0.614	0.767	**0.471**
	S2	0.499	0.500		0.383
v_6	S1	0.443	0.550	0.541	**0.298**
	S2	**0.457**	0.535		0.289
v_7	S1	0.518	0.592	0.617	0.366
	S2	**0.539**	0.606		**0.374**
v_8	S1	0.429	0.494	0.525	0.259
	S2	**0.487**	0.508		**0.267**
v_9	S1	0.459	0.569	0.604	0.344
	S2	**0.541**	0.667		**0.403**
v_{10}	S1	**0.582**	0.581	0.619	**0.359**
	S2	0.487	0.545		0.338
Mean scores	S1	0.481	0.545	0.631	0.344
	S2	**0.510**	0.567		**0.355**

Table 6. Agreement within dataset

Agreement metric	v_1	v_2	v_3	v_4	v_5	v_6	v_7	v_8	v_9	v_{10}
$R_{G_{AR}}$	0.691	0.688	0.684	0.578	0.767	0.541	0.617	0.525	0.604	0.619
κ_R	0.776	0.697	0.757	0.696	0.839	0.630	0.743	0.655	0.704	0.718

different system ranking than $F1_{mean}$ given the integral multi-reference principle it follows. However, what we consider the most profitable about $WiSeBE$ is the twofold inclusion of all available references it performs. First, the construction of R_W to provide a more inclusive reference against to whom be evaluated and then, the computation of $R_{G_{AR}}$, which scales the result depending of the agreement between references.

6   Conclusions

In this paper we presented WiSeBE, a semi-automatic multi-reference sentence boundary evaluation protocol based on the necessity of having a more reliable way for evaluating the SBD task. We showed how $WiSeBE$ is an inclusive metric which not only evaluates the performance of a system against all references, but also takes into account the agreement between them. According to your point of view, this inclusivity is very important given the difficulties that are present when working with spoken language and the possible disagreements that a task like SBD could provoke.

$WiSeBE$ shows to be correlated with standard SBD metrics, however we want to measure its correlation with extrinsic evaluations techniques like automatic summarization and machine translation.

Acknowledgments. We would like to acknowledge the support of CHIST-ERA for funding this work through the Access Multilingual Information opinionS (AMIS), (France - Europe) project.

We also like to acknowledge the support given by the Prof. Hanifa Boucheneb from VERIFORM Laboratory (École Polytechnique de Montréal).

References

1. Bohac, M., Blavka, K., Kucharova, M., Skodova, S.: Post-processing of the recognized speech for web presentation of large audio archive. In: 2012 35th International Conference on Telecommunications and Signal Processing (TSP), pp. 441–445. IEEE (2012)
2. Brum, H., Araujo, F., Kepler, F.: Sentiment analysis for Brazilian portuguese over a skewed class corpora. In: Silva, J., Ribeiro, R., Quaresma, P., Adami, A., Branco, A. (eds.) PROPOR 2016. LNCS (LNAI), vol. 9727, pp. 134–138. Springer, Cham (2016). https://doi.org/10.1007/978-3-319-41552-9_14
3. Che, X., Wang, C., Yang, H., Meinel, C.: Punctuation prediction for unsegmented transcript based on word vector. In: LREC (2016)
4. Fleiss, J.L.: Measuring nominal scale agreement among many raters. Psychol. Bull. **76**(5), 378 (1971)
5. Fohr, D., Mella, O., Illina, I.: New paradigm in speech recognition: deep neural networks. In: IEEE International Conference on Information Systems and Economic Intelligence (2017)
6. González-Gallardo, C.E., Hajjem, M., SanJuan, E., Torres-Moreno, J.M.: Transcripts informativeness study: an approach based on automatic summarization. In: Conférence en Recherche d'Information et Applications (CORIA), Rennes, France, May (2018)
7. González-Gallardo, C.E., Torres-Moreno, J.M.: Sentence boundary detection for French with subword-level information vectors and convolutional neural networks. arXiv preprint arXiv:1802.04559 (2018)
8. Gotoh, Y., Renals, S.: Sentence boundary detection in broadcast speech transcripts. In: ASR2000-Automatic Speech Recognition: Challenges for the new Millenium ISCA Tutorial and Research Workshop (ITRW) (2000)

9. Hinton, G., et al.: Deep neural networks for acoustic modeling in speech recognition: the shared views of four research groups. IEEE Signal Process. Mag. **29**(6), 82–97 (2012)

10. Jamil, N., Ramli, M.I., Seman, N.: Sentence boundary detection without speech recognition: a case of an under-resourced language. J. Electr. Syst. **11**(3), 308–318 (2015)

11. Kiss, T., Strunk, J.: Unsupervised multilingual sentence boundary detection. Comput. Linguist. **32**(4), 485–525 (2006)

12. Klejch, O., Bell, P., Renals, S.: Punctuated transcription of multi-genre broadcasts using acoustic and lexical approaches. In: 2016 IEEE Spoken Language Technology Workshop (SLT), pp. 433–440. IEEE (2016)

13. Kolář, J., Lamel, L.: Development and evaluation of automatic punctuation for French and english speech-to-text. In: Thirteenth Annual Conference of the International Speech Communication Association (2012)

14. Kolář, J., Švec, J., Psutka, J.: Automatic punctuation annotation in Czech broadcast news speech. In: SPECOM 2004 (2004)

15. Liu, Y., Chawla, N.V., Harper, M.P., Shriberg, E., Stolcke, A.: A study in machine learning from imbalanced data for sentence boundary detection in speech. Comput. Speech Lang. **20**(4), 468–494 (2006)

16. Lu, W., Ng, H.T.: Better punctuation prediction with dynamic conditional random fields. In: Proceedings of the 2010 Conference on Empirical Methods in Natural Language Processing. pp. 177–186. Association for Computational Linguistics (2010)

17. Meteer, M., Iyer, R.: Modeling conversational speech for speech recognition. In: Conference on Empirical Methods in Natural Language Processing (1996)

18. Mrozinski, J., Whittaker, E.W., Chatain, P., Furui, S.: Automatic sentence segmentation of speech for automatic summarization. In: 2006 IEEE International Conference on Acoustics Speech and Signal Processing Proceedings, vol. 1, p. I. IEEE (2006)

19. Palmer, D.D., Hearst, M.A.: Adaptive sentence boundary disambiguation. In: Proceedings of the Fourth Conference on Applied Natural Language Processing, pp. 78–83. ANLC 1994. Association for Computational Linguistics, Stroudsburg, PA, USA (1994)

20. Palmer, D.D., Hearst, M.A.: Adaptive multilingual sentence boundary disambiguation. Comput. Linguist. **23**(2), 241–267 (1997)

21. Pearson, K.: Note on regression and inheritance in the case of two parents. Proc. R. Soc. Lond. **58**, 240–242 (1895)

22. Peitz, S., Freitag, M., Ney, H.: Better punctuation prediction with hierarchical phrase-based translation. In: Proceedings of the International Workshop on Spoken Language Translation (IWSLT), South Lake Tahoe, CA, USA (2014)

23. Rott, M., Červa, P.: Speech-to-text summarization using automatic phrase extraction from recognized text. In: Sojka, P., Horák, A., Kopeček, I., Pala, K. (eds.) TSD 2016. LNCS (LNAI), vol. 9924, pp. 101–108. Springer, Cham (2016). https://doi.org/10.1007/978-3-319-45510-5_12

24. Shriberg, E., Stolcke, A.: Word predictability after hesitations: a corpus-based study. In: Proceedings of the Fourth International Conference on Spoken Language, 1996. ICSLP 1996, vol. 3, pp. 1868–1871. IEEE (1996)

25. Stevenson, M., Gaizauskas, R.: Experiments on sentence boundary detection. In: Proceedings of the sixth conference on Applied natural language processing, pp. 84–89. Association for Computational Linguistics (2000)

26. Stolcke, A., Shriberg, E.: Automatic linguistic segmentation of conversational speech. In: Proceedings of the Fourth International Conference on Spoken Language, 1996. ICSLP 1996, vol. 2, pp. 1005–1008. IEEE (1996)
27. Strassel, S.: Simple metadata annotation specification v5. 0, linguistic data consortium (2003). http://www.ldc.upenn.edu/projects/MDE/Guidelines/SimpleMDE_V5.0.pdf
28. Tilk, O., Alumäe, T.: Bidirectional recurrent neural network with attention mechanism for punctuation restoration. In: Interspeech 2016 (2016)
29. Treviso, M.V., Shulby, C.D., Aluisio, S.M.: Evaluating word embeddings for sentence boundary detection in speech transcripts. arXiv preprint arXiv:1708.04704 (2017)
30. Ueffing, N., Bisani, M., Vozila, P.: Improved models for automatic punctuation prediction for spoken and written text. In: Interspeech, pp. 3097–3101 (2013)
31. Wang, W., Tur, G., Zheng, J., Ayan, N.F.: Automatic disfluency removal for improving spoken language translation. In: 2010 IEEE International Conference on Acoustics Speech and Signal Processing (ICASSP), pp. 5214–5217. IEEE (2010)
32. Xu, C., Xie, L., Huang, G., Xiao, X., Chng, E.S., Li, H.: A deep neural network approach for sentence boundary detection in broadcast news. In: Fifteenth Annual Conference of the International Speech Communication Association (2014)
33. Yu, D., Deng, L.: Automatic Speech Recognition. Springer, London (2015). https://doi.org/10.1007/978-1-4471-5779-3

Readability Formula for Russian Texts: A Modified Version

Marina Solnyshkina[1(✉)], Vladimir Ivanov[2], and Valery Solovyev[1]

[1] Kazan Federal University, 18, Kremlyovskaya Street, Kazan, Russia
maki.solovyev@mail.ru, mesoln@yandex.ru
[2] Innopolis University, 1, Universitetskaya Street, Innopolis, Russia
v.ivanov@innopolis.ru

Abstract. The authors of the article offer new readability formulas for academic texts which provide a comparatively higher degree of accuracy than other Russian readability formulas. The results achieved are due to using original syntactic, lexical and frequency metrics ignored in previous research on Russian readability. The methods applied by the authors include Ridge and linear regression. The new readability formulas were computed on the Corpus of secondary school textbooks on Social Studies and then validated on the Corpus with the total size of 1 mln. tokens. The perspectives of the research lie in further modification of the formula for texts of various genres.

Keywords: Text readability formula · Academic texts
Russian language

1 Introduction

Modern readability studies have lately become interdisciplinary and continue engaging more researchers all over the world. The main reason for this is obvious: the increased number of failures to reach readers with a printed (electronic) text. As target audiences of companies, authorities and organizations become more receptive to audio and visual signals, quality requirements to printed texts grow exponentially and the task to enhance reading outcomes is becoming crucial. As the world itself and human activities are becoming more sophisticated the task to create comprehensible texts has become even more difficult: writers have to use more elaborate words and longer constructions to describe the world. One of the areas where improving the quality of reading materials is especially important and even indispensable is education. It is also true about Russia: after all social and political changes in 1990-s and 2000-s, the country is experiencing a real educational crisis [15,16]. Analysis also proves that the complicated language of school textbooks remains one of the burning issues in Russia today[1]. The quality of educational (printed) material depends largely on the skill of the

---

[1] https://www.hse.ru/news/122263399.html.

© Springer Nature Switzerland AG 2018
I. Batyrshin et al. (Eds.): MICAI 2018, LNAI 11289, pp. 132–145, 2018.
https://doi.org/10.1007/978-3-030-04497-8_11

author, expertise and experience of the editor. At the moment the Ministry of Education is expected to develop new standards of school textbooks expertise[2]. In this regard, the creation of reliable and universally accepted methods of automated verification of text complexity and readability is an urgent task. Another aspect in readability studies is an increasing need for leveled texts, i.e. texts profiled for different readers, in various areas. When performed manually it is time consuming, resource intensive and extremely costly. Therefore an automated tool performing the same functions is very desirable. In this paper we present the research aimed at measuring text complexity of Russian academic texts and offer results of our studies on various metrics of academic texts which successfully allow to profile a text for potential readers' linguistic abilities correlated with a particular grade level.

The current study was conducted to answer two research questions:

(1) How do 'classical' readability metrics work in the corpus of Russian academic texts?
(2) How do the new metrics, offered by the authors of the article, correlate with readability of Russian academic texts?

2   Related Work

The history of Text readability studies are the research aimed at extending the list of text metrics correlating with text complexity. It is also a history of criticism of readability formulas and doubts that the ideal formula profiling the text and the reader may never be derived. In the middle of the last century [5] and [2] proved that the correlation between factors that affect text comprehension is so great that only a few are enough to measure text complexity, but in their search for discovering correlations between text metrics and comprehension levels, researchers only increased the number of metrics: all over the world there have been conducted thousands of experiments with over 200 different text features. At present, there are over two hundred formulas of readability: Gunning fog index, Coleman Liau index, Flesch Kincaid Grade Level etc. for texts in many languages: English, French, German, Dutch, Swedish, Russian and other languages. Below we offer a brief history of views on each of the variable used by the authors in the current research.

2.1   Average Length of Sentences and Words, ASL, ASW

The very first two metrics introduced by Rudolf Flesch and Kincaid in 1948, i.e. ASL (average sentence length (in words)) and ASW (average word length in syllables) [2], are nowadays core components in the majority of readability formulas (see [1]). E.g. English Flesch reading ease, FRE = 206,835 − 1,015 ASL − 84, 6 ASW. It distributes ASL and ASW within the readability range as follows:

---

[2] https://www.kommersant.ru/doc/3614360.

- 100: The text is very easy to read. The average sentence length is 12 or fewer words. There are no words longer than 2 syllables.
- 65: The text is written in plain English. The average sentence length is from 15 to 20 words. An average word consists of 2 syllables.
- 30: The text is rather difficult to read. Sentences contain up to 25 words. Words are disyllabic.
- 0: The text is very difficult to read. An average sentence is 37 words long. An average word has more than 2 syllables.

2.2 Percentage of Long Words in Text, PLW

PLW has been calculated differently in various studies depending on the unit of measurement: either it is a number of characters in a word (letters) or syllables. E.g. one of the variables in Carl-Hugo Björnsson (1968) readability formula for Swedish known as Lix, is the percentage of long words, i.e. words of more than six letters: Lix = WL + SL, where, WL = percentage of words of more than six letters; SL = average number of words per sentence. But as the parameter is different in languages of different morphological types: in analytical languages words are shorter as they have fewer affixes (e.g., in English or in Spanish), while in synthetic languages with their highly developed system of morphemes (e.g. German, Ukranian) words are typically longer. Based on the discriminant analysis of 49 text variables in Russian academic texts [4] arrived at the conclusion that the best correlation (among others) between text metrics and readability are (1) the percentage of words of 11 letters and more and (2) the percentage of words of 13 letters and more.

In many readability formulas for European languages the word length is measured in syllables and researchers use a variable of the so-called 'complex words' which are typically defined as polysyllabic or multisyllabic words. In English, Spanish and French a polysyllabic word is a word made up of three or more syllables. For instance, the SMOG Readability Formula for English computes readability in the following way: SMOG grade = 3 + Square Root of Polysyllable Count (McLaughlin, 1969). Matskovskii [7] proposes to calculate readability of Russian texts based on the percentage of words of 4 or more syllables: Russian text readability = (0.62 × ASL) + (0.123 × percentage of words in the text of 3 or more syllables) + 0.051. Another indirect reason to consider 4, not 3 syllable words as 'complex' in Russian is proposed by I.V. Oborneva [8] who proved that on average an English word (2.97 syllables) is one syllable shorter than a Russian word (3.29 syllables). Thus, in our studies we also observe correlation of 4-syllable words with text readability.

2.3 Type Token Ratio, TTR

For years researchers have been offering different views and developing tools to measure 'lexical diversity' or 'lexical density' of a text. One of the metrics used in many studies is the so-called 'lexical richness of the text', i.e. type-token ratio, the ratio of types of words (unique words) to the total number of words (tokens) in the text [11]. However, later it was proved to be very sensitive to the size of the text. Thus, a number of reformulation of TTR have been offered since that. For example, for Swedish texts a common lexical density measure is OVIX, a word variation index (see [12]), calculated with the help of the natural logarithm. Cvrček and Chlumská [13] introduced two more metrics: (1) standardized (normalized) TTR, the sTTR, which is calculated for every thousand words and (2) zTTR, calculated as the ration of the observed TTR and the reference TTR. Unfortunately none of TTR metrics performs accurately and stable enough in a discourse (see [9,14]).

3   Corpus Description

For the research purposes we compiled a Corpus of two collections of texts (see [17]). The first collection of 7 texts derived as a result of OCR and postprocessing of textbooks on Social Studies by L.N. Bogolubov is marked in the Corpus as "BOG". The textbooks used cover the range of 6 – 11 Grade Levels of secondary and high schools in the Russian Federation. The second collection of 7 texts of textbooks on Social Studies by A.F. Nikitin marked "NIK" in the Corpus comprises the Grade levels of 5 – 11. Both sets of textbooks are from the "Federal List of Textbooks Recommended by the Ministry of Education and Science of the Russian Federation to Use in Secondary and High Schools"[3]. In the study we refer to the joined collection of textbooks as Russian Readability Corpus (RRC). To ensure reproducibility of results, we uploaded the corpus on the website[4], but for copyright purposes we had to shuffle the order of sentences in the uploaded texts of the Corpus. This shuffling, indeed, does not affect the values of features under study as they do not depend on sentence order. Table 1 below provides a numerical description of the RRC.

Table 1. Corpus parameters

Document	Tokens	Sentences	Syllables	Document	Tokens	Sentences	Syllables
1	19,412	1,482	39,964	8	23,019	2,275	42,512
2	26,72	1,907	60,977	9	19,619	1,399	40,739
3	58,391	3,441	138,509	10	28,349	2,009	59,239
4	50,828	2,977	121,407	11	48,844	3,614	108,523
5	90,12	5,051	218,984	12	53,273	3,389	123,358
6	117,251	6,25	287,068	13	47,267	2,711	112,487
7	116,12	6,326	299,019	14	45,943	2,549	111,995

---

[3] http://www.fpu.edu.ru/fpu/.
[4] http://kpfu.ru/slozhnost-tekstov-304364.html.

4   Processing of Texts in the Corpus

For the sake of convenience, we have processed all the texts of the Corpus in the same way. The preprocessing included tokenization, splitting text into sentences (or rather 'sequences of text separated by periods') and part-of-speech tagging (using the TreeTagger for Russian[5]). During the preprocessing stage we categorized two types of outliers and excluded (1) excessively long sequences of words (longer than 120 words) and (2) sequences shorter than five words. The long sequences proved to be either quotations from legal acts, e.g. Constitution of the Russian Federation, or lists generated by the textbook authors to save space in the textbooks. Short sequences of words (separated from nearby sentences with periods) appear to be either names of chapters and sections of books or results of incorrect sentence splitting. The quotations from legal acts were excluded on the presumption that they do not present academic discourse patterns but are typical of legal discourse. The lists and titles with grammatical structures of Nominal Phrases are viewed as outliers and were also excluded from the Corpus as they lack complete grammatical structure of a sentence. The exclusion of those 'sentences' from the Corpus is viewed by the authors as an important preprocessing stage as their metrics may, to a greater extend, influence the average sentence length used in all existing readability formulas. The research shows, that in the Russian discourse, the average sentence length depending on the genre and type of a text varies from ten to 30.

To ensure reproducibility of results, we uploaded the corpus on a website thus providing its availability online[6]. As textbooks we use are protected by copyright rules we shuffled the order of sentences in the Corpus thus limiting the possibility to use the texts in the Corpus for research only.

4.1   Sampling from the Corpus

The RRC contains 14 documents and thus by no means presents a representative sample of the population of all the school textbooks under study. Building a larger corpus is difficult, as it would violate some of the key principles: we either use new texts from different domains, or texts will come from different authors with different writing styles. Both cases of this kind may add noise to the dataset.

In order to overcome the issue of collection size, the following procedure of sampling from the corpus is suggested. The first issue in sampling from the corpus, as Biber (1990) puts it, "concerns the sampling of texts: how linguistic features are distributed across texts and across registers, and how many texts must be collected for the total corpus and for each register to represent those distributions?". Having compared the internal variations of the two texts in the corpus, Biber (1990) concludes that text samples of 1000 words are representative for the text categories under study.

---

[5] http://www.cis.uni-muenchen.de/~schmid/tools/TreeTagger/.
[6] http://kpfu.ru/portal/docs/F1554781210/shuffled.zip.

Both dependent variables (such as readability value) and independent variables (such as ASL) measured for a sample of a text should be close to the complexity value of the whole text. This assumption means, that starting from a certain subset (or sample) of sentences, text complexity of the sample will be almost the same as text complexity of the whole document from corpus. The sample corpus size was set to 5000 tokens. However, preserving order of tokens and sentences is important, otherwise the sampled texts will be less natural, even though they could carry the main features of the documents from the corpus. Thus, we sample 5000 token sequences from each document. We calculate features for readability analysis using the described sampling technique.

5   Text Features for Readability Analysis

In this study we have explored an extended feature set for text readability modeling:

- average number of words per sentence (ASL);
- average number of syllables per word (ASW);
- percentage of long words in text (PLW);
- type-token ratio (TTR), in four variants:
  - type-token ratio for all tokens (TTR);
  - type-token ratio for Nouns only (TTR_N);
  - type-token ratio for Verbs only (TTR_V);
  - type-token ratio for Adjectives only (TTR_A);
- TTR-based ratio of (TTR_A + TTR_N)/TTR_V; at a later stage we denote this feature as(NAV).;
- a relation between number of unique words in text: (number of unique Adjectives + number of unique Nouns)/(number of unique Verbs); at a later stage we denote this feature (UNAV).

The target feature for prediction is the grade level of the text. The feature is represented as real number. The relevance of the set of the above listed features for English text complexity modeling was studied by McNamara et al. in [21]. The features were calculated in the following way.

$$ASL = \frac{\text{total words}}{\text{total sentences}}$$

The ASL metric, i.e. the average sentence length (in words), is a core component in the majority of all readability formulas (see [1]).

$$ASW = \frac{\text{total syllables}}{\text{total words}}$$

The number of syllables was calculated as the number of vowels in a word. In Russian this heuristic gives appropriately good results. If a word does not contain vowels (e.g. some prepositions) it is attached to the adjacent word with vowels.

$$PLW = \frac{\text{total words with 4 and more syllables}}{\text{total words}}$$

$$TTR = \frac{\text{total unique tokens}}{\text{total tokens}}$$

Table 2. Features calculated in Russian readability corpus

Document	ASL	ASW	PLW	TTR	TTR_N	TTR_V	TTR_A	NAV	UNAV	GRADE
1	13.1	2.06	0.17	0.36	0.42	0.52	0.57	1.91	2.19	6
2	14.01	2.28	0.22	0.37	0.38	0.54	0.52	1.67	2.93	7
3	16.97	2.37	0.25	0.35	0.36	0.54	0.45	1.48	2.81	8
4	17.07	2.39	0.26	0.32	0.32	0.52	0.4	1.39	2.81	9
5	17.84	2.43	0.26	0.34	0.35	0.55	0.42	1.39	3.66	10
6	18.76	2.45	0.26	0.34	0.35	0.55	0.45	1.46	3.06	10.5
7	18.36	2.58	0.29	0.35	0.35	0.57	0.4	1.33	3.78	11.5
8	10.12	1.85	0.1	0.37	0.42	0.53	0.54	1.80	2.77	5
9	14.02	2.08	0.18	0.37	0.4	0.56	0.52	1.66	2.55	6
10	14.11	2.09	0.18	0.38	0.4	0.57	0.56	1.68	2.82	7
11	13.52	2.22	0.23	0.38	0.39	0.58	0.47	1.49	2.84	8
12	15.72	2.32	0.25	0.36	0.36	0.56	0.46	1.47	3.30	9
13	17.44	2.38	0.26	0.37	0.38	0.58	0.46	1.45	4.47	10
14	18.02	2.44	0.27	0.32	0.32	0.53	0.39	1.35	3.55	11

6 Model Selection

The problem of readability prediction can be formulated as a regression model. Indeed, most popular readability formulas are simple linear models that use one or several text features. In this section we analyze several regression models for readability prediction. As candidate regression models we consider a simple linear regression, a polynomial regression (i.e. a case when regression is built with the use of polynomial features). Additionally, we measure relative importance of the selected features. To this end, we use a feature selection technique that is based on the F-test. Finally, we apply regularization in order to find a subset of features most useful in prediction. In the end of the section we provide new formulas for readability prediction along with their performance evaluation based on the mean squared error (MSE) measure.

6.1 Linear Models and Feature Selection

Univariate Linear Regression Tests. First, we select features based on the Pearson correlation coefficient between each parameter and the grade level. We use the whole dataset from Table 2 as input data and select top-K (K = 1..8) features. The results of the experiment are presented in form of the ordered list of feature tuples.

- ASW, p-value $= 8.76 \cdot 10^{-7}$
- ASL, p-value $= 2.71 \cdot 10^{-6}$
- PLW, p-value $= 3.79 \cdot 10^{-6}$
- NAV, p-value $= 1.03 \cdot 10^{-5}$
- TTR_A, p-value $= 4.76 \cdot 10^{-5}$
- TTR_N, p-value $= 3.07 \cdot 10^{-4}$
- UNAV, p-value $= 1.95 \cdot 10^{-3}$
- TTR, p-value $= 0.0215$
- TTR_V, p-value $= 0.379$

The second approach to feature selection is the recursive feature elimination. Method starts with full set of features and tries to eliminate features one by one. The result of this method is elimination of two features: TTR and TTR_V. In comparison to the previous technique it confirms that those two features can be eliminated from further investigation.

6.2 Building a Linear Model with Regularization

After elimination of "TTR" and "TTR_V" we can build a simpler and robust linear model for prediction. The common approach to build such formula is to regularize the coefficients in linear regression. This can be done in several ways, including Lasso (L1) regularization, Ridge regression and Elastic-Net regularization [22]. In Table 3 we provide results of building linear regression with the three approaches to regularization. The higher the absolute value of a coefficient, the more important corresponding feature is.

Table 3. Building a linear readability models with regularization

Regularization type	PLW	UNAV	TTR_N	TTR_A	ASL	ASW	NAV
Lasso	0.00	0.78	0.00	0.00	0.32	2.45	−2.10
Ridge	0.35	0.82	−0.35	−0.95	0.36	1.84	−1.69
Elastic-Net	0.00	0.83	0.00	−0.15	0.44	1.23	−1.46

Table 3 shows that ASL and ASW are useful features as well as NAV and UNAV. Corresponding values in a column (weights of a feature) for each feature are close to each other in different regularization techniques. These features could be a basis for a more robust readability formula.

The resulting linear formulas with 2, 3 and 4 features are presented in Table 4. The table contains coefficients (weighs) of corresponding features. In order to measure performance of models, we use well-known measures: mean squared error (MSE) and mean absolute error (MAE).

$$MSE = \frac{1}{N} \sum (Y_{predicted} - Y_{observed})^2$$

Table 4. Coefficients of linear models.

Model name	ASL	ASW	UNAV	NAV	Intercept
M_0	0.28	6.2	-	-	−10.12
M_1	0.24	3.48	0.75	−2.38	−1.87
M_2	0.26	3.55	-	−3.74	2.07
M_3	0.25	4.98	0.89	-	−9.53
M_4	-	-	0.89	−8.42	18.6

$$MAE = \frac{1}{N} \sum |Y_{predicted} - Y_{observed}|$$

Finally, we have run a brute-force search for a formula that can give lowest MSE in training set. The following formulas with 4 and 5 regressors are provided below:

$$F_4 = 0.83UNAV - 6.73TTR\_A + 0.24ASL + 3.36ASW - 2.41$$

$$F_5 = 0.81UNAV - 5.47TTR\_A + 0.24ASL + 3.28ASW - 0.6NAV - 1.79$$

Additionally, we experimented with a polynomial regression (of degree 2 and 3). The selected features were squared before fitting a linear model and the Ridge regression was then applied to select better features.

6.3   Building a Quadratic Model with Regularization

An alternative way to build a readability formula is making multiplication of features and hence producing more complex quadratic model. For instance, given a list of 3 features: ASL, ASW and NAV, one could generate the following list of 6 feature products: ASL ASL, ASW ASL, ASW ASW, ASL NAV, ASW NAV and NAV NAV. These new features are used to fit a linear regression model, making it possible to explore combinations of existing features as terms in the readability formula. Note, that initial three features are added to the final set of features. Thus, the resulting formula will have 10 parameters overall (9 for features and 1 for the intercept). In the experiment with a quadratic model, the initial set of features is limited to the following: ASW, ASL, UNAV and NAV. After generation of feature products, the set of features contains 15 features (1 for the intercept, 4 initial features, and 10 features generated as pair products).

We also tried three different regularization techniques, but Lasso and Elastic-Net performed outrageously bad. In contrast, Ridge regression performed better than the existing linear models. In fact, during validation the absolute error exceeded 1.0 only three times. The formula for the quadratic model (Q) is the following (the intercept was fitted to zero):

$Q = - 0.124ASL + 0.018$ ASW $- 0.007$ UNAV $+ 0.007$ NAV -0.003 $ASL^2 +$
$+ 0.184$ ASL ASW $+ 0.097$ ASL UNAV $- 0.158$ ASL NAV $+ 0.09$ $ASW^2 +$

$+\ 0.091$ ASW UNAV $+\ 0.023$ ASW NAV $-\ 0.157$ UNAV^2 $-$
$-\ 0.079$ UNAV NAV $+\ 0.058$ NAV^2 .

In the rest of this section we provide evaluation of the derived formulas and compare them with the existing formulas for readability of Russian texts.

6.4   Evaluation of Models Performance

Given the small size of the corpus, to evaluate formulas we use Leave-One-Out Cross-validation (LOOCV). In this setting test set contains a single text and the training set contains all remaining documents. Thus, in corpus of 14 documents it is possible to generate 14 splits. For each such split we build a model, evaluate MSE and then calculate the average. The result of LOOCV is provided in Table 5.

Table 5. Performance of models measured with LOOCV.

	Linear							Quadratic
	M_0	M_1	M_2	M_3	M_4	F_4	F_5	Q
LOOCV MSE	0.76	0.74	0.68	0.67	1.13	**0.62**	0.83	0.54
MSE on training set	0.42	0.25	0.34	0.28	0.58	0.24	0.24	0.18
LOOCV MAE	0.73	0.76	**0.72**	0.68	0.85	**0.72**	0.84	0.68
MAE on training set	0.55	0.44	0.49	0.45	0.62	0.44	0.44	0.37

6.5   Comparison to Existing Readability Formulas

The Flesh Reading Ease formula was adopted for the Russian language only in the late 1970-s: first by M.S. Matskovskiy in 1976. Later, I.V. Oborneva has proposed a readability formula for Russian. In 1976, M.S. Matskovskiy computed the first readability formula for the Russian language:

$$Z_1 = 0.62ASL + 0.123X_3 + 0.051,$$

where Z_1 is text readability (or difficulty), ASL is the average sentence length (in words); X_3 is the percentage of words of more than 3 syllable in the text. Another formula which became quite popular in Russian readability studies is the one developed by I.V. Oborneva (2005):

$$Z_2 = 0.5ASL + 8.4ASW - 15.59,$$

To compute the coefficents in the formula, the researcher compared:

- the average length of syllables in English and Russian words in 100 parallel English-Russian literary texts and;
- the percentage of multi-syllable words in dictionaries for Russian and English.

I.V. Oborneva concluded that an average English word is formed of 2.97 syllables, while an average Russian word consists of 3.29 syllables. We evaluate the formulas Z_1, Z_2 on the corpus compiled for the study and compare the results of readability prediction for each text separately. Results of the comparison are provided in Figs. 1 and 2.

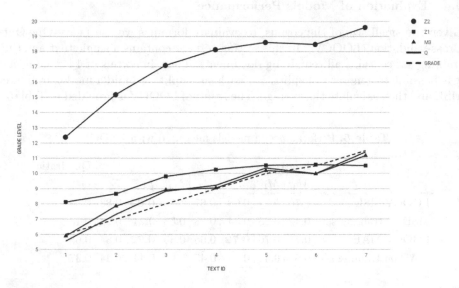

Fig. 1. Comparison of the formulas Z_1, Z_2, M_3 and Q on the "BOG" subcorpus of the Russian academic corpus. Dashed line represents the ground truth.

7   Analysis of Results

By now there have been formed two approaches to automatic assessment of text complexity. The classical approach, which we pursue in this paper, implies selecting a limited number of most relevant parameters for estimating text complexity and developing a text complexity formula based on the linear regression method. Another approach presupposing selecting the largest possible number of parameters - 100 or more - and applying a classifier such as Random Forest. The second approach is applied, in particular, in works of Reynolds [18], Laposhina [19] and Sadov [20]. The drawback of this approach is lack of transparency for the end-user. As for the first approach it proved to be useful for testing and applying TTR metrics measured on the Corpus of Russian academic texts. For the first time in Russian readability studies we applied a two-step method including assessment of correlation of coefficients, using Ridge regression and other methods of selecting the most informative parameters and finally we applied modified TTR parameters. As a result, in the new linear formula designed to measure Russian texts readability we use three parameters: ASW, ASL, UNAV ((number of unique Adjectives + number of unique Nouns)/(number of unique Verbs)).

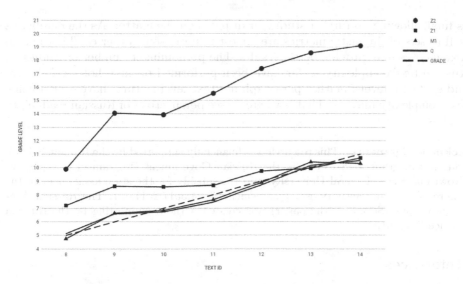

Fig. 2. Comparison of the formulas Z_1, Z_2, M_3 and Q on the "NIK" subcorpus of the Russian academic corpus. Dashed line represents the ground truth.

The curves in Figs. 1 and 2 show that the new formula provides a much higher accuracy measuring text complexity than the formula of I.V. Oborneva and a better accuracy than the formula of M.S. Matskovskiy. Another formula offered by the authors is also innovative in terms of its being not linear but quadratic. It also provides a higher accuracy in comparison with linear ones. However, it is not intuitively perceivable, and the achieved improvement in accuracy is not comparatively much higher. Taking into account that in this and previous works of the authors a rather large number of basic parameters (at the sentence level) of a text were explored and obtained improvements in accuracy is relatively not high, we can make a preliminary conclusion that as a result of the studies conducted we managed to develop models close to optimal.

8   Conclusion

The article offers new formulas to measure the level of complexity of Russian texts. This study is carried out on a text corpus of secondary and high school textbooks in Social Studies that we compiled earlier. We show that the previously proposed formulas do not correctly determine complexity level of academic texts in Russian. Solving the problem we studied and applied a number of parameters which were never used in Russian text complexity assessment, though successfully applied for assessing English and other languages text complexity. We offer original metrics in two innovative readability formulas, i.e. a quadratic (introduced for the first time in Russian readability studies) and linear. The accuracy of both exceeds the accuracy of all previously computed readability formulas for Russian texts. The latter does not imply the research conducted is to be viewed as final, allowing no further studies or disputes on the matter. As the number as well as sets of linguistic metrics are almost infinite, some other combinations of text metrics may provide better results. The predominant number of studies on Russian text complexity so far have been performed on morphological, lexical and syntactic levels: neither paragraph nor text level features have ever become text complexity metrics. That is where we see perspectives of Russian readability studies.

Acknowledgements. This research was financially supported by the Russian Science Foundation, grant #18-18-00436, the Russian Government Program of Competitive Growth of Kazan Federal University, and the subsidy for the state assignment in the sphere of scientific activity, grant agreement # 34.5517.2017/6.7. The Russian Academic Corpus (Sect. 3 in the paper) was created without supporting by the Russian Science Foundation.

References

1. Solnyshkina, M.I., Harkova, E.V., Kiselnikov, A.S.: Comparative Coh-Metrix analysis of reading comprehension texts: Unified (Russian) State Exam in English vs Cambridge First Certificate In English. English Lang. Teach. **7**(12), 65–76 (2014)
2. Flesch, R.: A new readability yardstick. J. Appl. Psychol. **32**, 221–233 (1968)
3. McLaughlin, G.: SMOG grading: a new readability formula. J. Reading **12**(8), 639–646 (1969)
4. Nevdakh, M.M.: Research of information characteristics of educational text using methods of multidimensional statistical analysis. Appl. Inform. **4**(16), 117–130 (2008)
5. Lorge, I.: Predicting readability. Teacher's Coll. Rec. **45**, 404–419 (1944)
6. Flesch, R.: Estimating the comprehension difficulty of magazine articles. J. Gen. Psychol. **28**, 63–80 (1943)
7. Matskovskii, M.S.: Problems of Readability of Printed Material. Semantic Perception of a Speech Message in Mass Communication, pp. 126–142. Nauka, Moscow (1976)
8. Oboroneva, I.V.: The automated estimation of complexity of educational texts on statistical parameters. Diss. Ped. n. M., 2006. 165 p
9. Falkenjack, J., Jonsson, A.: Classifying easy-to-read texts without parsing. In: Proceedings of the 3rd Workshop on Predicting and Improving Text Readability for Target Reader Populations (PITR) (2014)
10. Falkenjack, J., Heimann, M., Jönsson, A.: Features indicating readability in Swedish text. In: Proceedings of the 19th Nordic Conference of Computational Linguistics (NODALIDA 2013), pp. 27–40 (2013)
11. Piotrovsky, R.G. and others: Mathematical linguisticstick. Textbook. manual for ped. in-tov. M.: Higher School, 383 p. (1977)
12. Hultman, T.G., Westman, M.: Gymnasistsvenska. Liber, Lund (1977)
13. Cvrček, V., Chlumská, L.: Simplification in Translated Czech: A New Approach to Type-Token Ratio-Russian Linguistics, pp. 309–325. Springer, Dordrecht (2015). https://doi.org/10.1007/s11185-015-9151-8
14. Romanishin, G.V.: The study of the lexical wealth of scientific texts in New information technologies in automated systems: materials of the nineteenth scientific and practical seminar. M.: IPM them. M.V. Keldysh. - 352 p. (2016)

15. Karmanova, D.: Crisis of Russian higher education: towards the issue of aspectization labyrinth. J. Soc. Hum. Res. **1**, 78–84 (2012)

16. Stepanov, V.I., Stepanova, O.T.: The crisis of education in Russia: the ways and causes of the exit. In: Non-State-Walled Education in Russia, Novosibirsk (1996)

17. Ivanov, V.V., Solnyshkina, M.I., Solovyev, V.D.: Efficiency of text readability features in Russian academic texts. Comput. Linguist. Intellect. Technol. **17**, 277–287 (2018)

18. Reynolds, R.: Insights from Russian second language readability classification: complexity-dependent training requirements, and feature evaluation of multiple categories. In: Proceedings of the 11th Workshop on Innovative Use of NLP for Building Educational Applications, pp. 289–300 (2016)

19. Laposhina, A.N.: Analysis of relevant characteristics for automatic assessment of complexity of Russian texts used in courses for Russian as a foreign language [Electronic resource]: URL: http://www.dialog-21.ru/media/3993/laposhina.pdf. Accessed 10 July 2018

20. Sadov, M.A.: Development of an approach for measuring Russian text readability. Master course thesis. NRU HSE. 2018

21. Crossley, S., Allen, D., McNamara, D.: Text readability and intuitive simplification: a comparison of readability formulas. Read. Foreign Lang. **23**(1), 84–101 (2011)

22. Press, W.H., Teukolsky, S.A., Vetterling, W.T., Flannery, B.P.: Numerical Recipes: The Art of Scientific Computing. Cambridge University Press, Cambridge (2007)

Timed Automaton RVT-Grammar
for Workflow Translating

Alexander Afanasyev, Nikolay Voit$^{(\boxtimes)}$ ⓘ, and Sergey Kirillov

Ulyanovsk State Technical University, Ulyanovsk, Russia
{a.afanasev,n.voit}@ulstu.ru, kirillovsyu@gmail.com

Abstract. The paper studies grammar for workflow translating including semantic analysis. The main purpose of the translation is to expand the methods of semantic analysis of the grammatical model of distributed workflows due to the capabilities of the translation language. The article describes grammar, algorithm of its construction, differences from usual RV-grammar and author's modifications. At the end of the work the result of the experiment of translating the BPMN language diagrams into a temporary Petri net is presented.

Keywords: Automated systems · Diagrammatic models · Workflows
Automaton grammars · Timed Petri nets · Diagram translation

1 Introduction

Large manufacturing enterprises have complex business processes. Management of business processes while providing customers with services and products has become key for such enterprises [9]. As a rule, the analysis of a large amount of information is necessary for decision-making by the manager. It uses data mining, data warehousing, on-line analytical processing to obtain unbiased useful information [5]. In most cases, the presentation format of business processes (for example, BPMN, IDEF3, eEPC) should be translated into a modeling format that is understandable for modeling tools [11]. When analyzing business processes for errors, designers are forced to translate the internal representation of the system work flows into a view that is suitable for modeling [8]. Diagrammatic model of visual modeling languages (e.g. UML [1], IDEF [2], eEPC [3], BPMN [4], SharePoint, ER, DFD) are widely used in the practice of designing complex automated systems (as) especially at the conceptual stage [5]. Such languages are flexible and allow you to build diagrams that can be applied to different subject areas. The flexibility of languages is due to the incompleteness or informality of their description, as a result of which the resulting diagrams can be interpreted ambiguously. Machine processing of such graphic diagrams is difficult [11]. Language flexibility can lead to a family of languages, i.e. many languages that conceptually have a common basis but different interpretation specific to the subject area of their application. Most of the existing approaches consider such languages in isolation, although it is sufficient to determine the generalizing semantics for the language family (perhaps for some elements it will be abstract) and to specialize the semantic component of the individual elements of the language and (or) the diagram before its interpretation.

© Springer Nature Switzerland AG 2018
I. Batyrshin et al. (Eds.): MICAI 2018, LNAI 11289, pp. 146–155, 2018.
https://doi.org/10.1007/978-3-030-04497-8_12

The paper has the following structure. Section 2 has related work that short describes major relevant research studies. Section 3 provides a brief overview of the semantics of graphical languages. Section 4 contains RVTITg-grammar. Section 5 presents the example RVTITg-grammar for basic BPMN. Conclusions and further research directions are presented in Conclusion.

2 Related Work

The authors investigate some works that consider the specification of document flow, verification and translation. Several papers focused on the definition of formal semantics and validation methods for workflows using Petri nets, process algebra, abstract state machine, see for example [12–22]. In [18, 19], Decker and Weske propose a Petri net-based formalism for determining choreographies, properties as realizability and local applicability, and a method for verifying these two properties. However, they consider only synchronous communication and does not explore the association with languages modeling of interaction of a high-level BPMN. Bultan and Fu [23] determine a sufficient condition for analyzing the feasibility of choreographies defined using UML collaboration diagrams (CD). In [24], Salaün and Bultan modify and extend this work with the feasibility analysis method by adding a synchronization message among peers. This method controls the realizability of CDs for bounded asynchronous communication. The feasibility problem for Message sequence diagrams (MSCs) has also been studied (e.g. [25, 26]). In [26], the authors offer bounded MSCS graphs which are bounded by BPMN 2.0 because branching and looping behavior are not supported by CDs and MSCs (there is no selection in CDs, there are no some looping behaviors in MSCs, and only Self-loops in CDs). In [27] BPMN behavior is studied from the semantic point of view and several BPMN patterns are proposed. This work is not theoretically justified and is not complete, it discusses only some of the laws. Lohmann and Wolf [28] propose to analyze existing patterns and control them with compatible patterns. In [29], the authors focused on the translation of BPMN into the algebra of processes for the analysis of choreography using model checking and equivalence. The main limitation of these methods is that they do not work when there are different types of diagrams at the same time, which means that in some cases the input diagrams cannot be analyzed.

3 The Semantics of Graphical Languages

All existing graphic languages can be divided into the following types according to the language formality

1. Formal. The syntax and semantics of such languages are formally defined.
2. Semi-formal. The syntax of the language is formal, and semantics can have different interpretations.
3. Informal. The syntax and semantics of the language are informal [6].

The vast majority of popular graphic languages are semi-formal. For them, it is worth investigating the methods of formalization. In [5, 8, 9, 11–29] mainly offer to give a semantic technicality language in the following ways:

1. To specialize the language. To give it some capabilities of the language to simplify it, to give new opportunities that will positively affect formalization.
2. To determine the semantics dynamically. Dynamic semantics involves the transformation of the diagrams of the basic graphic language into a target language.

The second method is more promising, because it makes it possible to implement diagrams of different graphic languages on the basis of one universal tool. The method develops the ideas of primitive libraries. In this case, libraries store the interpretation of the graphical image in terms of the target graphic or text language. And there can be several interpretations for the same image, each of them is assigned a unique name to avoid ambiguous interpretation and incorrect use. The target language is a more formal language relative to the base language.

4 RVTITg-Grammar

RVTITg-grammar (Timed RV-grammar for graphic translation) is an RVTI-grammar (Timed RV-grammar [7]) development in which the grammar scheme products are expanded to store the correspondence in terms of the target formal description, and the internal memory stores the information necessary for the translation process [7, 10]. Temporal RVT-grammar of a language L (G) is an ordered n-tuple of eleventh non-empty sets

$$G = \left(V, U, \Sigma, \tilde{\Sigma}, M, F, C, E, R, T, r_0\right) \tag{1}$$

where $V = \{v_e, e = \overline{1, L}\}$ is auxiliary alphabet; $U = \{u_e, e = \overline{1, K}\}$ is auxiliary alphabet of the target language; $\Sigma = \{a_t, t = \overline{1, T}\}$ is terminal alphabet graphic language; $\tilde{\Sigma} = \{\tilde{a}_t, t = \overline{1, T}\}$ is quasi terminal alphabet; $M = TT \cup TN$ is combining terminal (TT) and non-terminal (TN) characters of the target language; $F = \{generate\_input(), generate\_output(), select\_output(), stick\_connection\_points()\}$ is many translation functions for working with elements of the set; C is set of clock identifiers; E is the set of temporal relations "Before", "During", "After" (initialization of the clock $\{c := 0\}$, relations of the form $\{c \sim x\}$, where x the variable (the identifier of the clock), c is a constant, $\sim \epsilon \{=, <, \leq, >, \geq \}$); $R = \{r_i, i = \overline{0, I}\}$ is grammar G schema (set of names of complexes of products, each complex r_i consists of a subset P_{ij} of products $r_i = \{P_{ij}, j = \overline{1, J}\}$); $T \in \{t_1, t_2, t_3 \ldots, t_n\}$ is a set of time stamps; $r_0 \in R$ is RV-axiom grammar.

Graphical objects containing more than one input or output are loaded with translation functions from the set to ensure that the inputs and outputs of such objects match the base and target languages. Assignment of translation functions and operations performed by them.

1. generate_input() is a formation of a set of input connection points, except for the one on which this graphic object was reached. It is performed in the primary analysis of graphical objects containing more than one input.
2. generate_output() is a formation of a set of outgoing connection points. It is performed in the primary analysis of graphical objects containing more than one output, or when the only output is supposed to be used as a link – label, i.e. it is necessary to change the direction of analysis.
3. select_output() is a select the outgoing connection point of the element as the continuer of the target language chain. An event function that is executed for graphical circuits with a dynamically changing number of outgoing connection points after the formation of a set of connection points for outgoing connections, it is selected from such points. In General, the selection algorithm is not regulated, i.e. the choice is random.
4. stick_connection_points() is a link connection points and object. An event function that is performed when secondary graphical objects that contain more than one input are analyzed. Binds the incoming link to the connection point of the object, information about which is stored in the internal memory.

The presence of these functions allows you to create an algorithm for building the output chain, the main operations in which are the selection of the point – continuer analysis for objects containing more than one output, and the layout of a complete sequence of already analyzed objects containing more than one input, and associated with them analyzed objects. The order of application of functions is shown in Fig. 1.

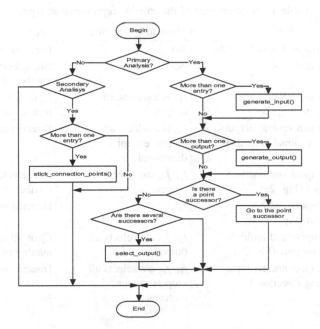

Fig. 1. The main algorithm of translation.

150   A. Afanasyev et al.

For Fig. 2 all possible display options are presented. Table 1 describes the corresponding types of the choice of the successor and the application of f function. On the maps (see Fig. 2) hollow points show the guide points, shaded – the points through which the analyzer reaches the graphical object, and filled with points – all other incoming and outgoing points. In the following tables, the encoding of translation functions is accepted. The value f_{gi} in the table cells corresponds to the function call generate_input(), f_{go} is a generate_output(), f_{so} is a select_output(), f_{scp} is a stick_connection_point().

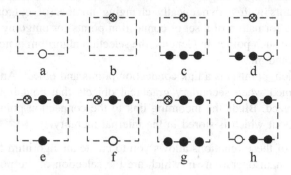

Fig. 2. Possible display types.

Table 1. A description of the possible representation types.

Representation description	Translating functions	Succesors
With one directional output (Fig. 2a)	No	Directed output
With one input (Fig. 2b)	No	No, selected from the number of label links
With one input and several non-directional outputs (Fig. 2c)	f_{go} is a selects all outputs	f_{so} or selected from the number of label links
With one input and several outputs, including the directional (Fig. 2d)	f_{go} is a selects all outputs except directional	Directed output
With multiple inputs and one non-directional output (Fig. 2e)	f_{gi}, f_{go} are selects all outputs	No, selected from the number of label links
With multiple inputs and one directional output (Fig. 2f)	f_{gi}	Directed output
With multiple inputs and multiple non-directional outputs (Fig. 2g)	f_{gi}, f_{go} are selects all outputs	f_{so} or selected from the number of label links
With multiple inputs and multiple outputs, including directional (Fig. 2h)	f_{gi}, f_{go} are selects all outputs except directional	Directed output

5   Example RVTITg-Grammar for Basic BPMN

The base language from which the broadcast will be produced is a well-known BPMN. A standard Business Process Model and Notation (BPMN) will provide businesses with the capability of understanding their internal business procedures in a graphical notation and will give organizations the ability to communicate these procedures in a standard manner [4]. The BPMN specification also provides a mapping between the graphics of the notation and the underlying constructs of execution languages, particularly Business Process Execution Language (BPEL) [4]. As a translation language, a timed Petri net was chosen. Representation in terms of the timed Petri network for the language elements of the BPMN presented in Table 2.

Table 2. Representation in terms of the timed Petri network for the language elements of the BPMN.

	Element name	BPMN representation	Timed Petri nets representation
A	Start event		
B	End event		
C	Action		
D	Exclusive gateway		
E	Parallel gateway		
F	Intermediate event "Timer"		
H	Link flow		

Table 3. Translation grammar example.

N	State	Quasi term	Next state	Operation with memory	
				Base language	Target language
1	r_0	A_0	r_1	$\varnothing$	$\varnothing$
2	r_1	rel	r_3	$\varnothing$	$\varnothing$
3	r_2	$label_{EG}$	r_3	$W_2(b^{1m}, b^{t(6)})$	$W_2(b^{3m})$
4		$label_{PG}$	r_3	$W_2(b^{2m}, b^{t(6)})$	$W_2(b^{4m})$
5	r_3	A_i	r_1	$\varnothing$	$\varnothing$
6		A_{im}	r_1	$\varnothing$	$\varnothing$
7		A_{it}	r_1	$W_1(t_s^{t(6)})$	$\varnothing$
8		A_{kl}	r_2	$\varnothing/W_3(!e^{1m}, !e^{2m})$	$\varnothing/W_3(!e^{3m}, !e^{4m})$
9		A_k	r_4	$\varnothing$	$\varnothing$
10		A	r_1	$W_1(t_s^{t(6)})$	$\varnothing$
11		A_{it}	r_3	$W_1(t_s^{t(6)})$	$\varnothing$
12		EG_c	r_1	$W_1(\ t^{1m^{(n-1)}})/W_3(k=1)$	$W_1(\ t^{3m^{(n-1)}})/W_3(k=1)$
13		EG	r_2	$W_1(1^{t(1)}, k^{t(2)})/W_3(e^{t(2)}, k!=1)$	$W_1(1^{t(7)}, k^{t(8)})/W_3(e^{t(8)}, k!=1)$
14		$\_EG$	r_2	$W_1(inc(m^{t(1)})/W_3(m^t_{(1)} < k^{t(2)})$	$W_1(inc(m^{t(7)})/W_3(m^t_{(7)} < k^{t(8)})$
15		$\_EGe$	r_1	$W_1(\ t^{1m^{(n-1)}})/W_3(m^{t(1)} = k^t_{(2)}, p!=1)$	$W_1(\ t^{3m^{(n-1)}})/W_3(m^{t(7)} = k^t_{(8)}, p!=1)$
16		$\_EG_{me}$	r_1	$o/W_3(m^{t(1)} = k^{t(2)}, p=1)$	$o/W_3(m^{t(7)} = k^{t(8)}, p=1)$
17		PG_f	r_1	$W_1(\ t^{2m^{(n-1)}})/W_3(k=1)$	$W_1(\ t^{4m^{(n-1)}})/W_3(k=1)$
18		PG	r_2	$W_1(1^{t(3)}, k^{t(4)})/W_3(e^{t(3)}, k!=1)$	$W_1(1^{t(9)}, k^{t(10)})/W_3(e^{t(9)}, k!=1)$
19		$\_PG$	r_2	$W_1(inc(m^{t(3)})/W_3(m^t_{(3)} < k^{t(4)})$	$W_1(inc(m^{t(9)})/W_3(m^t_{(9)} < k^{t(10)})$
20		$\_PG_e$	r_1	$W_1(\ t^{2m^{(n-1)}})/W_3(m^{t(3)} = k^t_{(4)}, p!=1)$	$W_1(\ t^{4m^{(n-1)}})/W_3(m^{t(9)} = k^t_{(10)}, p!=1)$
21		$\_PG_{je}$	r_1	$W_1(\ t^{2m^{(n-1)}})/W_3(m^{t(3)} = k^t_{(4)}, p=1)$	$W_1(\ t^{4m^{(n-1)}})/W_3(m^{t(9)} = k^t_{(10)}, p=1)$
22	r_4	no_label	r_5	*	*
23	r_5				

Final tabular form RVTg-grammar for the languages BPMN and timed Petri nets are presented in Table 3.

When the analysis is complete, the stores in the internal memory of the target language must be empty and all tape cells must contain the "0" character. Checking the memory status is described by the operation indicated by the "" symbol. Take an example of an abstract diagram in BPMN. It is depicted in the Fig. 3.

Representation of temporal BPMN diagram example in timed Petri nets presented in Fig. 4.

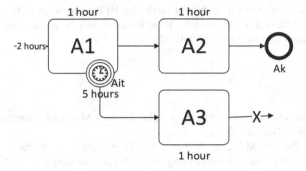

Fig. 3. Temporal BPMN diagram example.

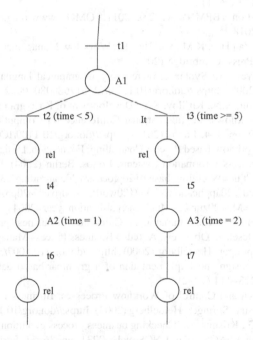

Fig. 4. Representation of temporal BPMN diagram example in timed Petri nets.

6 Conclusions

Presented RVTITg-grammar based on RVTI-grammar, which takes into account the temporal characteristics and broadcast diagrammatically models in different graphical languages. An example of the translation of BPMN diagrams into a timed Petri net. Further directions of work are the expansion of the possibilities of semantic analysis of diagrammatic models from the point of view of coordinating text attributes of diagrams with project documentation.

Acknowledgment. The reported study was funded by RFBR according to the research project № 17-07-01417 and Russian Foundation for Basic Research and the government of the region of the Russian Federation, grant № 18-47-730032.

References

1. Booch, G., Jacobson, I., Rumbaugh, J.: The Unified Modeling Language User Guide. Addison-Wesley, Boston (1998)
2. Mayer, R.J., Painter, M.K., de Witte, P.S.: IDEF Family of Methods for Concurrent Engineering and Business Re-engineering Applications. Knowledge Based Systems, College Station (1994)
3. Santos, P.S., Almeida, J.P.A., Pianissolla, T.L.: Uncovering the organisational modelling and business process modelling languages in the ARIS method. Int. J. Bus. Process Integr. Manag. **5**(2), 130–143 (2011)
4. Model, B.P.: Notation (BPMN), v. 2.0, 2011. OMG. www.omg.org/spec/BPMN/2.0. Accessed 01 Sept 2018
5. Van Der Aalst, W., Van Hee, K.M., van Hee, K.: Workflow Management: Models, Methods, and Systems. MIT Press, Cambridge (2004)
6. Sharov, O., Afanasyev, A.: Syntax error recovery in graphical languages. Prog. Comput. Softw. **34**, 44–48 (2008). https://doi.org/10.1134/S0361768808010052
7. Afanasyev, A.N., Voit, N.N., Kirillov, S.Y.: Development of RYT-grammar for analysis and control dynamic workflows. In: International Conference on Computing Networking and Informatics (ICCNI), pp. 1–4. Lagos (2017). https://doi.org/10.1109/ICCNI.2017.8123797
8. Zur Muehlen, M.: Workflow-based Process Controlling: Foundation, Design and Application of Workflow-Driven Process Information Systems. Logos, Berlin (2004). http://scholar.google. com/scholar_lookup?title=Workflow-based%20process%20controlling%3A%20foundation% 2C%20design%20and%20application%20of%20workflow-driven%20process%20information %20systems&author=M.%20zur%20Muehlen&publication_year=2004
9. Becker, J., Rosemann, M., von Uthmann, C.: Guidelines of business process modeling. In: van der Aalst, W., Desel, J., Oberweis, A. (eds.) Business Process Management. LNCS, vol. 1806, pp. 30–49. Springer, Heidelberg (2000). https://doi.org/10.1007/3-540-45594-9_3
10. Maurer, P.M.: The design and implementation of a grammar-based data generator. Softw. Pract. Exp. **22**(3), 223–244 (1992)
11. Reijers, H.A.: Design and Control of Workflow Processes: Business Process Management for the Service Industry. Springer, Heidelberg (2003). https://doi.org/10.1007/3-540-36615-6
12. Poizat, P., Salaün, G., Krishna, A.: Checking business process evolution. In: Kouchnarenko, O., Khosravi, R. (eds.) FACS 2016. LNCS, vol. 10231, pp. 36–53. Springer, Cham (2017). https://doi.org/10.1007/978-3-319-57666-4_4. https://hal.inria.fr/hal-01366641
13. Martens, A.: Analyzing web service based business processes. In: Cerioli, M. (ed.) FASE 2005. LNCS, vol. 3442, pp. 19–33. Springer, Heidelberg (2005). https://doi.org/10.1007/ 978-3-540-31984-9_3
14. Raedts, I., Petkovic, M., Usenko, Y.S., van der Werf, J.M.E., Groote, J.F., Somers, L.J.: Transformation of BPMN models for behaviour analysis. In: MSVVEIS 2007, pp. 126–137 (2007)
15. Dijkman, R.M., Dumas, M., Ouyang, C.: Semantics and analysis of business process models in BPMN. Inf. Softw. Technol. **50**(12), 1281–1294 (2008). https://doi.org/10.1016/j.infsof. 2008.02.006

16. Wong, P.Y.H., Gibbons, J.: A process semantics for BPMN. In: Liu, S., Maibaum, T., Araki, K. (eds.) ICFEM 2008. LNCS, vol. 5256, pp. 355–374. Springer, Heidelberg (2008). https://doi.org/10.1007/978-3-540-88194-0_22

17. Wong, P.Y., Gibbons, J.: Verifying business process compatibility (short paper). In: The Eighth International Conference on Quality Software, 2008. QSIC 2008, pp. 126–131. IEEE, August 2008. https://doi.org/10.1109/QSIC.2008.6

18. Decker, G., Weske, M.: Interaction-centric modeling of process choreographies. Inf. Syst. **36** (2), 292–312 (2011). https://doi.org/10.1016/j.is.2010.06.005

19. Decker, G., Weske, M.: Local enforceability in interaction petri nets. In: Alonso, G., Dadam, P., Rosemann, M. (eds.) BPM 2007. LNCS, vol. 4714, pp. 305–319. Springer, Heidelberg (2007). https://doi.org/10.1007/978-3-540-75183-0_22

20. Güdemann, M., Poizat, P., Salaün, G., Dumont, A.: VerChor: a framework for verifying choreographies. In: Cortellessa, V., Varró, D. (eds.) FASE 2013. LNCS, vol. 7793, pp. 226–230. Springer, Heidelberg (2013). https://doi.org/10.1007/978-3-642-37057-1_16

21. Mateescu, R., Salaün, G., Ye, L.: Quantifying the parallelism in BPMN processes using model checking. In: Proceedings of the 17th International ACM Sigsoft Symposium on Component-Based Software Engineering, pp. 159–168. ACM, June 2014. https://doi.org/10.1145/2602458.2602473

22. Kossak, F., et al.: A rigorous semantics for BPMN 2.0 process diagrams. In: Kossak, F., et al. (eds.) A rigorous semantics for BPMN 2.0 process diagrams, pp. 29–152. Springer, Cham (2014). https://doi.org/10.1007/978-3-319-09931-6_4

23. Bultan, T., Fu, X.: Specification of realizable service conversations using collaboration diagrams. SOCA **2**(1), 27–39 (2008). https://doi.org/10.1109/SOCA.2007.41

24. Salaün, G., Bultan, T.: Realizability of choreographies using process algebra encodings. In: Leuschel, M., Wehrheim, H. (eds.) IFM 2009. LNCS, vol. 5423, pp. 167–182. Springer, Heidelberg (2009). https://doi.org/10.1007/978-3-642-00255-7_12

25. VBPMN Framework. https://pascalpoizat.github.io/vbpmn/. Accessed 01 Sept 2018

26. Alur, R., Etessami, K., Yannakakis, M.: Realizability and verification of MSC graphs. Theor. Comput. Sci. Autom. Lang. Program. **331**(1), 97 (2005). https://doi.org/10.1016/j.tcs.2004.09.034

27. Lotos, I.S.O.: A formal description technique based on the temporal ordering of observational behaviour. ISO8807, 1XS989 (1989)

28. Lohmann, N., Wolf, K.: Realizability is controllability. In: Laneve, C., Su, J. (eds.) WS-FM 2009. LNCS, vol. 6194, pp. 110–127. Springer, Heidelberg (2010). https://doi.org/10.1007/978-3-642-14458-5_7

29. Poizat, P., Salaün, G.: Checking the realizability of BPMN 2.0 choreographies. In: Proceedings of the 27th Annual ACM Symposium on Applied Computing, pp. 1927–1934. ACM, March 2012. https://doi.org/10.1145/2245276.2232095

Extraction of Typical Client Requests from Bank Chat Logs

Ekaterina Pronoza[1(✉)], Anton Pronoza[2], and Elena Yagunova[1]

[1] Saint Petersburg State University, Saint Petersburg, Russian Federation
katpronoza@gmail.com, iagounova.elena@gmail.com
[2] Saint Petersburg Institute for Informatics and Automation of the Russian
Academy of Sciences, 14-Th Liniya, 39, Saint Petersburg 199178, Russia
antpro@list.ru

Abstract. In this paper we propose a simple but powerful method of extracting key client requests from bank chat logs. Many companies nowadays are interested in building a chat bot to optimize their business, and are ready to provide chat bot developers with large amounts of data, but such data often need special preparation to be successfully used for a chat bot system. We propose a method of data preparation which includes not only data cleaning but also data mining: we extract key notions from chat logs and retrieve typical client requests in generalized form. The method uses simple metrics as well as word embeddings and additional semantic resources to extract typical client requests from client-manager chat logs.

Keywords: Chat logs · Chat bot · Keywords · Collocations · Client requests

1 Introduction

Building a successful chat bot system is an important NLP task nowadays, especially in banking industry. Most companies are undoubtedly willing to introduce a chat bot system into their business (if they have not done it already for some reasons). They collect large amounts of data (for example, logs of manager-client dialogues) which can be used by chat bot developers to train chat bot systems, however, such data often need cleaning and special preparation before they could be used as input for a dialogue system. For example, such chat logs may (and do) contain numerous misprints, repeated clarifying questions from clients and information totally irrelevant to the subject of a client's request.

In this paper we propose our method of filling this niche. In other words, we propose a method of not only cleaning raw data, but of extracting key notions from the text and, what is more important, retrieving typical client requests in generalized form. As a result of such parsing of chat logs, we obtain not only request patterns (which could be used in a chat bot system) but also key topics of the chat logs (which is a useful side effect). Our request patterns are represented by noun and verb phrases and consist of 2–3 words. Although most of them are not grammatically correct word combinations, these phrases (according to the results of the experiments) clearly reflect the purpose of the underlying clients' requests in 73% cases. In the other 27% cases, although the request might be unclear, the general topic of the request is obvious.

© Springer Nature Switzerland AG 2018
I. Batyrshin et al. (Eds.): MICAI 2018, LNAI 11289, pp. 156–164, 2018.
https://doi.org/10.1007/978-3-030-04497-8_13

2 Related Work

In a few recent papers conversation logs between customers and managers are dealt with, and similar (to our problem) tasks are being solved.

For example, in [4], over 170 thousand chat logs are analyzed to determine the factors which affect customer's satisfaction. The authors use an existing state-of-the-art sentiment analysis tool and conduct surface statistical analysis of the chat logs corpus (e.g., extract discriminative n-grams indicating customers' dissatisfaction).

N-gram-based approach (in combination with self-customized hyperlink-induced topic search (HITS) algorithm) is used in [1] to solve the problem of finding over-lapping users' interests and their social ties based on users' chat logs.

In [5] healthcare chat logs mining task is solved: the authors present a new prob-abilistic model for the extraction of topics in chat logs.

In terms of NLP tasks, the task closest to ours is topic modelling. In topic modelling community, the most widely used approach relies on the use of generative probabilistic models and includes techniques like singular value decomposition (see LDA, for example). However, such method is rather costly when it comes to implementation. Another group of methods coming from the neural networks community is represented by word embedding techniques (e.g., word2vec by Google, fastText by Facebook). The latter techniques are convenient to use and not as costly as LSA/LDA-based ones, therefore, in our research, we follow such an approach. Our topic modelling algorithm is based on the combination of shallow statistical metrics, word embeddings and specific semantic resources (YARN – Yet Another Russian WordNet [2] and the dictionary of Russian word formation [6]).

3 Data

Our data consist of 58267 chat logs of dialogues (668200 written messages) between bank managers and clients in Russian. The data are anonymized and annotated with tags referring to the authors of messages.

4 Method

Our client requests extraction method consists of several stages. Firstly, chat logs are lemmatized using MyStem[1] and then cleaned. As our raw data are full of misprints and mistakes made by bank clients in their questions to managers, we apply the following procedure: all the words (in their lemma form) are checked against a universal word embedding Russian model from RusVectores [3] (with vocabulary size of about 400 thousand words). Out-of-vocabulary words (not found in the RusVectores vector model) are compared to the in-the-vocabulary words, and if Levenshtein distance between an out-of-vocabulary word and in-the vocabulary word is 1 and word length is

---

[1] Morphological analyzer for Russian by Yandex (https://tech.yandex.ru/mystem/).

at least 4 letters, they are considered equal, and the appropriate replacements are conducted in the whole text.

Secondly, after the data are cleaned, we calculate frequencies of 1- and 2-grams (on the lemma level) occurring in the data. Then we adjust the obtained scores: the scores of words in title case, geographical locations, person names, foreign words and quotations are multiplied by special coefficients (values of these coefficients are selected empirically as a result of preliminary experiments). We further extract top n most frequent lemma collocations using Normalized Pointwise Mutual Information (NPMI), where n is also selected empirically as 1000 with threshold value set to 3. Then these collocations are checked against the large word embeddings model from RusVectores (this model contains collocations as well as single words). Finally, only the collocations present in the RusVectores model are selected.

Thirdly, having obtained a list of top frequent words (with frequency value above the threshold value) and collocations together with their adjusted frequency scores, we merge synonymous words and phrases into single notions (to be able to work on the notions level further). Words or phrases are merged into single notions if they are either synonymous in terms of words embedding model or in terms of YARN-based dataset, or if they share the same root. Since Euclidean product is a costly procedure for a large chat logs corpus, instead of calculating semantic similarity scores for each pair of words from the lexicon, we expand words or phrase by checking up their top n (where n is empirically set to 20) similar words/phrases in terms of (a) the large Russian words embedding model from RusVectores, (b) the YARN-based dataset where similarity scores of each pair of words are calculated as Jaccard scores between their synsets, (c) the words formation dictionary. Synonymous words and phrases from (a), (b) and (c), together with the original words and phrases, constitute generalized notions. They are added to these notions if their similarity scores (the scores between the new and the original words) are not less than the empirically set threshold values and if these words or phrases already occur in the chat logs corpus. If they are not present in the corpus, but their similarity scores are high enough (higher than 0.9 – another empirically set coefficient), they are also added to our generalized notions.

Thus, some of the words or phrases from the lexicon (top frequent words and collocations) are expanded so that they become a set of words/phrases denoting some notions relevant to the bank domain. After this expansion procedure, the original weights of top frequent words and collocations are adjusted: if a word/phrase is a part of a notion (i.e., has synonyms in the lexicon), its weight becomes higher than that of a single word/phrase without any synonyms (it is multiplied by another empirically set coefficient). The final adjustment to the scores of key words and phrases is made to diminish the importance of common words which are not so relevant for bank domain: the scores are multiplied by the inverse frequency of top 20% most frequent words from the Russian National Corpus (the respective frequency values are freely available at the RNC website[2]) multiplied by the empirically set coefficient. And finally, for each notion (represented by a group of synonymous words or phrases) we select a title for this notion using the following scheme: for each word/phrase inside the groups we

---

[2] http://ruscorpora.ru/corpora-freq.html.

search for common hypernyms (if present, one of such hypernyms becomes the title for the notion), if some of the hypernyms are among the words in the group, then one of such hypernyms becomes the title for the notion. If the title could not be selected using hypernyms, then it is selected as the centroid (of all the words in the group) in terms of vector space model.

Finally, we select top n nouns or noun phrases out of the collected list of notions, and they constitute our key notions set.

Having collected the key notions from the corpus, we expand these notions to obtain short phrases. These procedure is conducted in two stages.

Firstly, for each key notion found in the clients' questions (here and further in the paper we only consider the client part of the chat logs corpus) we collect adjectives, adverbs and nouns occurring in the context of this notion (taking into account that notions can be represented by several nouns or noun phrases we repeat the search procedure for each of the words/phrases) with window size equal to 3. Then we calculate Dice coefficient for the collected pairs of the original key notions and surrounding adjectives/adverbs/verbs and select the collocations with Dice coefficient values above the threshold.

Secondly, we repeat the procedure described in the previous paragraph, but this time we only search for verbs in the context of the obtained collocations. If the collocations already include verbs, they are skipped.

Finally, we obtain a list of triples which contain a key notion and a verb (obtained at the second stage) and pairs of key notions and verbs (obtained at the first stage). Such pairs and triples are supposed to represent typical client requests made to the bank managers.

5 Results

Key notions (top 30) obtained as described in Sect. 4 are shown in Table 1

Top 30 requests are shown in Table 2. The requests where the purpose of the client's request is absolutely clear from the request pattern, and which are valid according to the corpus of chat logs, are marked with "+" in the "Valid" column.

Thus, in 73% phrases from top-30 requests (22 phrases out of 30), the underlying client's request is clear and valid. In the 8 leftover phrases (27%) the topic of the request is clear, but its purpose is not. For example, in "карта счет ип"/"card account individual entrepreneur" the topic of the request has to do with the card linked to the account of an individual entrepreneur, but it is not clear whether the client wants the card to be linked to the account or closed, etc. It might be caused by the absence of a verb in this particular request pattern. However, all the rest 7 invalid requests do contain a verb and are still unclear as to the purpose of the client's requests. If we get a closer look at those patterns, we can see that in most of them the verb is too common (while the general meaning of the phrase is conveyed by the noun phrase), and it requires some context for a human to understand the meaning of the whole phrase (e.g., with verbs like "указывать"/specify, "оставаться"/remain, "происходить"/occur) – i.e., such patterns evidently require some additional details to be included into them.

Table 1. Top 30 key notions

Title	Words/phrases
счет/account	счет/account
выписка/extract	выписка/extract
платеж/payment	платеж/payment
статус/status	статус/status
платежный поручение/payment order	платежка, платежный поручение, платежок, поручение/payment order, order
сертификат/certificate	сертификат, сертификация/certificate, certification
расчетный счет/checking account	расчетный счет/checking account
поле/field	поле/field
контрагент/contractor	контрагент/contractor
документ/document	документ/document
ограничение/limit	ограничение/limit
запрос/request	запрос/request
подпись/signature	подпись/signature
файл/file	плагин, скрипт, утилита, файл/plugin, script, tool, file
информация/information	данные, информация/data, information
реквизит/requisite	реквизит/requisite
новый дизайн/new design	новый версия, новый дизайн/new version, new design
дбо сбербанк/dbo sberbank	дбо сбербанк/dbo sberbank
смс/sms	смс/sms
пароль/password	пароль/password
код/code	код/code
режим работа/business hours	режим работа/business hours
деньги/money	денежки, денежный средство, деньги, средство/money, cash, fund
справочник/catalogue	справочник/catalogue
карта/card	карта/card
консультация/consultation	консультация/consultation
последний цифра/last digit	последний цифра/last digit
абс/asb	абс/asb
заявление/claim	заявление/claim

To visualize the collected top key notions and phrases (including request patterns), we built the respective graphs using Gephi[3]. In these graphs, the nodes correspond to the notions (or phrases/requests), while the thickness of connections between them indicates the frequency of co-occurrence of two notions (or requests) in the same chat logs. Node communities (clusters) were identified using Gephi.

---

[3] https://gephi.org/.

Table 2. Top 30 users' requests

Request	Valid
деньги карта переводить/money card transfer	+
деньги счет переводить/money account transfer	+
ограничение счет снимать/limit withdraw money	+
действие подпись истекать/expiration date signature expire	+
подпись срок истекать/signature expire	+
корреспондент справочник отсутствовать/correspondent catalogue absent	+
запись корреспондент подтверждать/record correspondent confirm	
код ошибка происходить/code error occur	
абс статус отказывать/asb status deny	+
неизвестный карта писать/unknown card say	+
деньги счет списывать/money account withdraw	+
выписка счет распечатывать/extract account print	+
карта счет ип/card account individual entrepreneur	
деньги счет поступать/money account transfer	+
ограничение счет налагать/limit account impose	+
ошибка сертификат выдавать/error certificate show	+
документ подпись подписывать/document signature sign	+
выписка период распечатывать/extract period print	+
подразделение услуга не предоставляться/department service not provide	+
карта счет привязывать/card account link	+
выписка период сформировывать/extract period generate	+
ограничение счет открывать/limit account open	
ошибка сертификат происходить/error certificate occur	
запрос сертификат создавать/request certificate generate	
карта номер оставаться/card number remain	
действие сертификат заканчивать/certificate expire	+
номер счет указывать/number account specify	
ошибка сертификат деактивировать/error certificate deactivate	+
код смс приходить/code sms come	+

It can be seen from the graph in Fig. 1 that there are four clusters of key notions which correspond to four different topics:

1. Bank card/account operations: "счет"/"account", "карта"/"card", "операция"/ "operation", "выписка"/"extract".
2. Payment issues: "платеж"/"payment", "поручение"/"order", "статус"/"status".
3. Online banking system issues (errors, etc.): "сообщение"/"message", "ошибка"/ "error", "вход"/"login".
4. SMS/password issues: "смс"/"sms", "пароль"/"password", "телефон"/"phone", "подтверждение"/"confirmation".

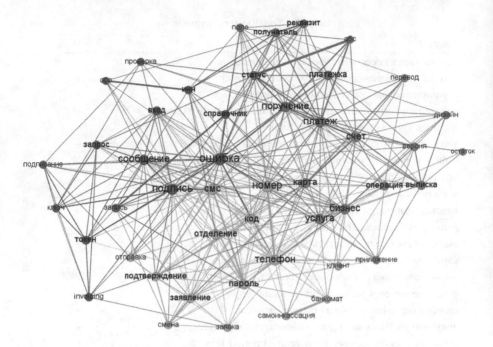

Fig. 1. Graph of key notions extracted from bank chat logs

The graph in Fig. 2 includes 5 clusters: obviously, one of the clusters from the graph in Fig. 1 is split into two node communities. The graph in Fig. 2 is actually a graph from Fig. 1 where key notion nodes are expanded with nodes corresponding to key phrases (those obtained during stages 2 and 3 of the request patterns extraction algorithm – see Sect. 4). In the second graph, there are five clusters: cluster #3 from the first graph (online banking system issues) is split into two subclusters:

– catalogue of the banking system: "справочник"/"catalogue", "запись"/"record", "корреспондент"/"correspondent", "запись заблокировать"/"record block", "справочник добавлять"/"catalogue add",
– operations in the banking system: "подпись"/"signature", "документ подпись подписывать"/"document signature sign", "подписание"/"signing", "запрос"/ "request", "вход"/"login".

In fact, the graph in Fig. 2 is incomplete due to space limits of the paper, but if all the extracted request patterns were included in the graph, other clusters (i.e., #1, #2 or #4) would probably also be split into subclusters describing key topics of the chat logs in detail.

In both graphs key notions and phrases extracted from chat logs represent the main topics mentioned in the corpus of chat logs, and each topic can easily be inferred from the nodes (key notions and phrases) of the corresponding cluster in the graph. The topics themselves meet one's expectations about the content of bank chat logs.

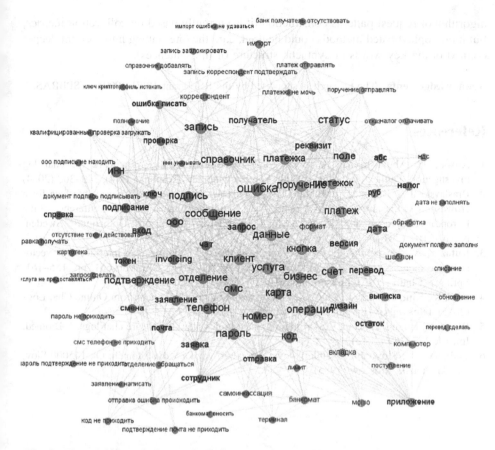

Fig. 2. Key notions and phrases (including request patterns) extracted from bank chat logs

6   Conclusion

In this paper we described a method of cleaning and mining bank chat logs. This method can be used against logs of client-manager dialogues, not necessarily in banking sphere, but in any industry where a chat bot system needs to be built.

The method proposed in this paper is quite simple and relies on the combined use of simple metrics, word embeddings and semantic resources. We work with the Russian language, but the algorithms described in the paper are actually language independent.

As a result, key notions referring to the key topics of chat logs, are obtained, but the main outcome of the proposed algorithm is the list of request patterns: typical requests of the clients represented by short 2–3 word phrases, and such request patterns can undoubtedly be of great help for a chat bot system.

According to the results of the experiments, 73% of the extracted request patterns clearly indicate the purpose of the corresponding request, while others only indicate the general topic of the request. Our future work directions include the improvement of the algorithm of request patterns construction (it is currently based on collocation metrics, but more sophisticated methods could be used, e.g., the ones taking into account deeper context of the key words or syntactic structure of the sentences).

Acknowledgements. The research is supported by the RSF grant 18-71-10094 in SPIIRAS.

References

1. Anwar, T., Abulaish, M.: A social graph based text mining framework for chat log investigation. Digit. Invest.: Int. J. Digit. Forensics Incident Response **11**(4), 349–362 (2014)
2. Braslavski, P., Ustalov, D., Mukhin, M.: A spinning wheel for YARN: user interface for a crowdsourced thesaurus. In: Proceedings of the Demonstrations at the 14th Conference of the European Chapter of the Association for Computational Linguistics, Gothenburg, Sweden, pp. 101–104 (2014)
3. Kutuzov, A., Kuzmenko, E.: WebVectors: a toolkit for building web interfaces for vector semantic models. In: Ignatov, D.I., et al. (eds.) AIST 2016. CCIS, vol. 661, pp. 155–161. Springer, Cham (2017). https://doi.org/10.1007/978-3-319-52920-2_15
4. Park, K., Kim, J., Park, J., Cha, M.: Mining the Minds of Customers from Online Chat Logs (2015). https://arxiv.org/pdf/1510.01801.pdf
5. Wang, T., Huang, Z., Gan, C.: On mining latent topics from healthcare chat logs. J. Biomed. Inf. **61**, 247–259 (2016)
6. Tikhonov, A.: Slovoobrazovatelnij slovar' russkogo yazika v dvuh tomah: Ok 145000 Slov. Russkiy Yazik, Moscow (1985)

A Knowledge-Based Methodology for Building a Conversational Chatbot as an Intelligent Tutor

Xavier Sánchez-Díaz[✉], Gilberto Ayala-Bastidas, Pedro Fonseca-Ortiz, and Leonardo Garrido

Departamento de Computación Regional Norte, Tecnológico de Monterrey, Eugenio Garza Sada 2501 Sur, Col. Tecnológico, 64849 Monterrey, Mexico
{A01170065,A00819406,A00805772,leonardo.garrido}@itesm.mx

Abstract. Chatbots are intelligent agents with which users can hold conversations, usually via text or voice. In recent years, chatbots have become popular in businesses focused on client service. Despite an increasing interest for chatbots in education, clear information on how to design them as intelligent tutors has been scarce. This paper presents a formal methodology for designing and implementing a chatbot as an intelligent tutor for a university level course. The methodology is built upon first-order logic predicates which can be used in different commercially available tools, and focuses on two phases: knowledge abstraction and modeling, and conversation flow. As main result of this research, we propose mathematical definitions to model conversation elements, reasoning processes and conflict resolution to formalize the methodology and make it framework-independent.

Keywords: Chatbots · Knowledge modeling · Methodology
Conversation design · Intelligent tutoring

1 Introduction

Chatbots are computer programs designed to hold conversations with users using natural language [15]. Some of them have human identities and personalities to make the conversation more natural. Ranging from Twitter bots with random responses to more complex counseling service agents, chatbots have become increasingly common in recent years. According to Tsvetkova et al. [16], nearly half of the online interactions between 2007 and 2015 involved a chatbot. They have been documented for use in a variety of contexts, including education [7] and commerce [3].

Chatbots have proven to be useful tools in academic courses, too. One example is the development of Jill Watson, an intelligent tutor developed by Goel et al. [5] at Georgia Tech for an artificial intelligence MOOC. Anderson et al. [2] classify intelligent tutors as intelligent computer-aided instruction software which

© Springer Nature Switzerland AG 2018
I. Batyrshin et al. (Eds.): MICAI 2018, LNAI 11289, pp. 165–175, 2018.
https://doi.org/10.1007/978-3-030-04497-8_14

can respond to a student's specific problem-solving strategies. On the same note, Reyes-González et al. [9] emphasize the importance of an individualized interactive process between the tutor and the student. The effectiveness of Jill Watson showed the potential of chatbots in a massive online class and is a fine example of the new educational era where artificial intelligence may play a major roll. In Goel's words: "it represents an educational technology for supporting learning at a big scale" [5]; sometimes so big that personalized attention from the instructor may result impossible. Using a conversational bot as an intelligent tutor has some other advantages like the fact that it can be available 24/7, giving the student the freedom of learning at their own pace, at any moment and from anywhere with Internet access. Having a chatbot also lightens the work load of the course instructor, as shown in [4].

With so many tools available to develop conversational agents nowadays, the building and deployment of a chatbot may look fairly simple. However, providing the chatbot with suitable information to be able to work as an educational tutor could be difficult. At the time of writing, information on how to design the tutor is scarce and scattered across blog entries and articles which are focused on the chatbot implementation rather than the design and modeling of knowledge. Many of the current development tools (such as Dialogflow and Chatfuel) only handle the implementation phase. The task of abstracting and organizing knowledge is up to the instructor designers. Thus, a methodology for knowledge abstraction and organization is essential, since having a conversation with an instructor is rather different than chatting with a sales agent.

This work proposes a methodology that formally defines and models the chatbot structure as an intelligent tutor. Its aim is to help multidisciplinary teams looking to design and implement chatbots in university courses. The methodology focuses on describing a first-order logic framework which can be implemented on different commercially available tools, and sheds some light on how to represent, expand and maintain the knowledge base.

The rest of the document is formatted as follows. Section 2 reviews human-computer conversation methodologies and available technologies. Section 3 formalizes the chatbots concepts and describes the proposed framework. Implementation and a case study are then discussed in Sect. 4. Finally, Sect. 5 presents concluding remarks.

2   A General Approach to Human-Computer Conversation

When talking about human-computer conversation, the very first technologies like ELIZA (1966) and ALICE (1995) come to mind due to their importance. ELIZA was created to demonstrate that natural conversation with a computer was possible, but failed to pass the Turing test since its implementation was based on string matching and no context was taken into account for the written responses. ALICE introduced the famous AIML (Artificial Intelligence Markup

Language), which uses pattern matching rules to give 'meaning' to the actual words: topic, themes and categories were now taken into account [10].

Since then, chatbots have come a long way. Mobile devices now include simple chatbots which (or perhaps we should say whom) handle simple requests like making phone calls and setting alarms. Furthermore, automation tools have been developed for the creation of more complex chatbots and are now available for commercial use. The ever-increasing presence of machine learning in daily life has made the entire process of 'thinking' a bit more powerful than simple pattern matching, and perhaps the most common examples are those available directly from tech giants like Google, Facebook and Amazon. Despite each framework having its own implementation details and limitations, most tools derive from a general idea—receive raw data, give it meaning, and then act appropriately according to a knowledge base. Figure 1 illustrates this process for a text-based chatbot.

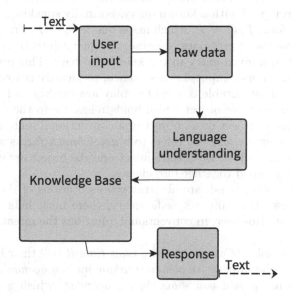

Fig. 1. General methodology for a text-based conversational agent

To process natural language, chatbots rely on pattern recognition and bigram identification [14], a procedure which is usually handled by conversation frameworks. However, it is up to the chatbot designers to generate the knowledge base and provide the learning engine with appropriate examples.

3 Proposed Methodology

Our proposal considers two main phases:

Knowledge Modeling. This phase determines how knowledge is represented and stored in the knowledge base.

Conversation Flow. Both the lexicon used by the tutor and the order in which ideas are presented should be defined in this phase.

This section presents the formal definitions and foundations of the proposed methodology first. Later, each phase is described and contrasted with real-life queries. Finally, since implementation details vary across different conversation frameworks, they are not covered per se in the methodology. Nevertheless, conflict resolution and good implementation practices are broadly presented along a case study.

3.1 Formal Definitions and Foundations

A chatbot can be described as a conversational agent which gives an appropriate response when prompted with a known query. Formally speaking, a chatbot is a function f of the form $f: Q \rightarrow R$, which maps queries $q \in Q$ to responses $r \in R$.

In order to give the appropriate response, the query must be converted from natural language to a given entry in the knowledge base. This process consists on breaking down the user input, i.e. a sentence, to identify key concepts of the conversation. The most notable concepts at play are *entities* and *intents*.

An entity is an abstract object which holds relevance to the user. It can be thought of both as a subject or an object in a conventional sentence: *The quick brown fox jumps over the lazy dog* refers to a *quick brown fox* as a subject, and to the *lazy dog* as the object. Both the subject and the object are entities in this sense and can be grouped together into classes.

Intents, on the other hand, are abstract representations of the user's intentions. Since the user is asking a specific query, there must be something they want to do or know. However, in conventional questions the intent is not always present.

When a person asks, '*Can you tell me what time it is?*' their intention is to *find out the time*. The imperative sentence '*Show me my agenda*', for example, can be rewritten as '*Would you show me my agenda?*' which is a question in which the user wants *their agenda* (the entity) to be *shown* (the intent). Nevertheless, there are some other queries where the intention cannot be extracted from a rearrangement of their words. A user asking, '*Why do we snore?*' wants to know the *reason* of why we *snore* when we sleep. The intention of finding out the *reason* is not in the input text.

Queries as Functions of First-Order Logic. First-order logic is a branch of the study of reasoning, dealing with inference and 'belief' management using formulas in the form of predicates. It uses truth-functional connectives like $\neg$ (not), $\wedge$ (and), $\vee$ (or) and many others; along the use of functions to describe the state of a variable which is either true or false [8]. In the context of chatbots, this truth value can be seen as the presence or the absence of a condition to trigger a response if it exists in the knowledge base. At first glance, intents may resemble verbs in a common sentence, but in fact they are relations of variables to a truth state: first-order logic functions. A query, then, has the form:

$$g^n(t_1, t_2, \ldots, t_n) \tag{1}$$

where $g^n \in G$ is an n-ary function symbol in the set of functions G, and t_i is a term in the set of terms T. The set of terms T is the set of known entities and G is the set of known intents in the knowledge base.

For example, in the statement `show(agenda)`, `show()` is a function which receives a single parameter to generate a response. The entity `agenda` is then shown to the user. Under this assumption, `show(flight)` could also be a possible query, when the user wants to know the status of a recently booked flight. If the user wanted to know the `time` right now, the function would look similar: `tell_time(now)`. Another example could be `tell_time(here)`, in case the bot is programmed to handle different time zones or countries. The query *'What's the time in New Zealand?'* could be translated to `tell_time(nz)`, or something along those lines. In the question *'What's the difference between x86 and x64 architectures?'* the intention is to know the `difference` between two entities, the *x86* and the *x64* processor instruction sets. The relation here is associating a pair of entities to a response, which is the difference between these two concepts. Following the function notation used throughout this work, it can be written as `difference(x86, x64)`. The `difference` function is binary. However, n-ary functions may be defined, as stated in Eq. 1.

It is important to consider that questions triggers are not stored in the knowledge base as text. Instead, a first-order predicate with the form defined in Eq. 1 is built for each query. First-order logic has limited expressivity [13] since some concepts which cannot be expressed using this formal system exist. Nevertheless, an adequate delimitation of the agent's reach can guide the knowledge modeling process, speeding up the implementation phase and reducing the training complexity.

3.2 Phase I: Knowledge Modeling, Extraction and Storing

As described in Sect. 3.1, the sets of queries and responses are to be defined first, and there are several ways to perform that task. To deal with knowledge extraction and representation, for example, Huang et al. [6] use tuples of the form `<input, response>`, which are constructed by ranking the replies of a web forum thread as either 'fascinating', 'acceptable' or 'unsuitable'. Ales et al. [1], on the other hand, focus on automatic emotion detection of news headlines to give meaning to the input using self-organizing maps. Both ideas revolve around a medium-sized knowledge base, with a couple of concepts and queries.

However, a tutor designed for a college courses may deal with hundreds of different concepts and definitions. For this, the input of an expert is recommended, especially if aiming for accurate pedagogical explanations using an adequate lexicon. The tutor language may be either casual and relaxed, or a bit more abstract and rigorous depending on the context. For example, in the field of mathematics—using abstract constructs and precise notations—the input from an expert is advised, since technology alone does not guarantee that students learn mathematics better than using only a regular textbook [12]. Moreover, problem situations can be represented in several ways, even in natural language. The use of an appropriate lexicon is important for students to incite them to translate everyday situations to mathematical models [11].

In order to comply with the trigger-response approach, knowledge should be separated into 'units', extracted from the expert and then laid down on a knowledge base with certain queries in mind for each knowledge unit. Each of these units represent a single query, a unique combination of functions and parameters that yields a certain response.

Although many data structures exist to store the knowledge base, most chatbot conversation structures are based on trees [15]. Each node in the tree represents a unique response, from a simple greeting to detailed information about previous queries. It is also important to note that in order for the conversation service to determine which response the user is looking for, the similarity between the user input and all known queries must be calculated. This process is usually done by machine learning algorithms using similarity measures between sentences in which each word or character may represent a single dimension, and its accuracy is refined by providing thousands of correctly labeled examples of user inputs. Therefore, it is advised to group knowledge units by similarity of the user input that will trigger them, rather than clustering by topic.

For instance, grouping the examples provided in Sect. 3.1 by intent, one can see that all three queries using the function `tell_time` have a similar input:

- *What's the time now?*
- *What time is it here?*
- *What is the time in New Zealand?*

The `tell_time` function input is somewhat different from that of the `show` function, in which the phrase *Show me* is predominant. Creating branches according to intents therefore reduces the complexity of the search.

3.3 Phase II: Conversation Flow

Once knowledge is separated into small atomic units, designing how to present them is the next step. An efficient way to do that is the creation of a glossary and a naming convention to keep track of the available queries and manage their trigger order. For instance, for the creation of the intelligent tutor for the introductory mathematics course in our institution, each knowledge unit in the tree was given a unique ID. The ID was generated automatically from abbreviations of the names of the intents and the entities, with a hyphen separating the intent from the entities, and the entities separated by a plus sign: `def-N` was used to represent the definition of the natural numbers, corresponding to the question *'What is a natural number?'* or *'What is the definition of natural numbers?'*.

Some conversation frameworks allow entities to be grouped into categories, as it is the case of IBM Watson. The glossary, then, may contain entities clustered like the following:

– **Numbers.** Named sets of numbers, like *natural numbers*, *integers*, *rationals*, *reals*, etc.
– **Terms.** Mathematical terms related to the course material, e.g. *infinite decimal expansion* and *numerical representation*.
– **Equations.** Terms and notation specifically related to equations: *linear equations system* and *solution of a linear equation*.
– **Proofs.** Mathematical proofs that require a rigorous and a logical explanation, for example *proof that $\sqrt{2}$ is irrational* or *proof that the cardinalities of the naturals and the integers are equal*.
– **Modifiers.** Terms that modify the behavior of the bot. Usually used in conjunction of already defined entities, like *no solution [of something else]* and *positive [something]* or *negative [something]*.
– **Algebraic components.** Terms related directly with algebraic procedures, e.g. *substitute* or *x of t*, referring to the notation $x(t)$.
– **Wrong terms.** This category could be used to encapsulate frequently asked terms that are either mistaken or nonexistent. A fine example is the term *unreal numbers*: since the set of rational numbers and irrational numbers complement each other, unreal numbers *should* be a complement of the reals. This is obviously a wrong assumption.

Entities are then combined with intents to formulate a unique set of conditions that are needed to trigger their response, which were written by the expert considering the pedagogical aspect of the language employed in the answer. In this way, instructors can easily work along knowledge engineers to effectively model the queries and generate the knowledge base.

An example of an abstract representation of the knowledge base is presented Fig. 2, where queries are grouped according to their intents. Starting from the top of the tree (usually a greeting or a welcome message), the conversation framework will determine which branch contains the desired query and start iterating over the contents until it finds it or reaches the end. Chatbot designers are encouraged to include an error notice at the end of a branch in case the query is not found.

This approach works if the tutor is receiving queries in any order. However, more complex conversations can be modeled by branching each of the different intents into smaller pieces of conversation as needed.

4   Implementation: A Case Study

Following the proposed methodology, two chatbots were designed in our institution: F-1001 and MA-1001, which are the introductory Physics and Mathematics courses respectively. The tutors were implemented in IBM Watson Conversation

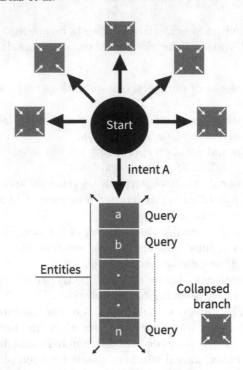

Fig. 2. An abstract representation of the knowledge base built by grouping queries by intent.

Service, using JSON structures for the tree, and providing examples and synonyms via comma separated value (CSV) files.

A small sample of the intents and entities of the intelligent tutor of the introductory math course is presented in Table 1.

Despite the fact that branching by intents guides the search process, the conversation flow may require adjustments due to technical reasons of the framework. It is important to consider how the knowledge base is traversed so that the chatbot answers are sound and complete.

In IBM Watson, the knowledge base is stored in a tree. This tree is usually traversed from top to bottom: taking as a starting point the first node (greeting), and then reviewing a branch if its condition is triggered. Then, each of the nodes in that branch is visited sequentially until one triggers: each branch of the tree is represented by an array. If no answer is found, then the default response is returned, which in the case of the tutors in our institution was a failure notice in the form of "*Sorry, I could not understand your request*". In reality, the knowledge base is a tree of arrays, in which each index of the array is a query.

This behavior may be prone to giving incorrect answers if the query nodes are not in the right order. Consider Examples 1 and 2. The former presents a formal definition of the problem and the latter rewords it into an specific use case. A visual representation of the conflict is also shown in Fig. 3

Table 1. Glossary for the MA-1001 intelligent tutor.

ID	Intent	Entities
def-N	Definition	Natural number
ex-Q	Example	Rational number
notation-Z	Notation	Integers
expl-change+uniform	Explanation	Uniform, change
def-obj_rest	Definition	Object at rest
ex-fin_dec_exp	Example	Finite decimal expansion
subset-Q+R	Subset	Rational number, real number
card-Qc	Cardinality	Irrational number
comp_card-N+Q	Compare cardinality	Natural number, rational number

Example 1. Let the sequence of conditions $B = \langle a, a \wedge b \rangle$ be the array where the desired branch is stored, and the queries $q_1 = a \wedge b$ and $q_2 = a$ be the conditions detected by the chatbot. Query q_1 contains both condition a and condition b. Since the first element of B, B_1 is triggered if all its conditions are met, then q_1 will trigger B_1, when it is highly probable that the user wanted to get the response in B_2. However, q_2 will not trigger B_2 since condition b is missing and not all conditions are met. On the contrary, q_2 will trigger B_1 as the requirements for its activation are all met.

Example 2. Assume a student wants to know the *definition* of *constant velocity*. Since the knowledge base also contains other topics on *velocity*, it would be better to model the query as a binary function, def(velocity, constant), instead of using *constant velocity* as a single entity. However, since there is already a response for the *definition of velocity*, def(velocity), there could be trouble handling the request if def(velocity) is found first.

To prevent unwanted responses to trigger if a condition is partially met, it is recommended to order the queries in the branch from more specific to less specific, so that a branch B ends up looking as $B = \langle a \wedge b, a \rangle$ and not the other way around. In Example 2, the problem can be solved if the *definition of constant velocity* is found first, and the more general query for the *definition of velocity* is presented last.

Another alternative is to add an additional level of branching and ask for extra information when dealing with similar situations. For instance, the tutor could explicitly ask the user *which of the following definitions on velocity are you interested in?*. However, this approach for conflict resolution is highly discouraged, as one may find different *topics* on velocity in the definition branch as well as in the examples branch.

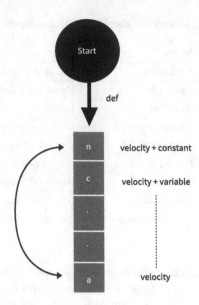

Fig. 3. A visual representation of the conflict presented in Example 2, which is solved by placing more specific queries first.

5   Conclusions

Chatbots have become incredibly common nowadays, however there is few information on how to implement them as intelligent tutors for university courses. This work provides a formal methodology for organizing knowledge and arranging it for an appropriate flow to be implemented in commercially available conversation frameworks. The methodology presented focuses on using first-order logic predicates to represent knowledge units extracted from an expert as n-ary functions, which can be later grouped to simplify the search process. The paper also reviewed a case study of the creation of two intelligent tutors using the proposed methodology, which served as an example of conflict resolution when two queries share a trigger condition and are in the same level of the branch. This is just a first step towards formalizing conversational agents in education. Future work should consider modeling of more complex conversation scenarios, for example those requiring additional branching in the knowledge base. Incorporating automated methods for training and response refinement should also be considered. Is also important to develop a methodology to analyze student–chatbot interaction to ensure the chatbot qualifies as an intelligent tutor for higher education.

Acknowledgements. The authors would like to acknowledge the financial support of Writing Lab, TecLabs and Tecnológico de Monterrey, Mexico, for the production of this work.

References

1. Alès, Z., Duplessis, G.D., Şerban, O., Pauchet, A.: A methodology to design human-like embodied conversational agents. In: International Workshop on Human-Agent Interaction Design and Models (HAIDM 2012), Valencia, Spain (2012). https://hal.archives-ouvertes.fr/hal-00927488
2. Anderson, J.R., Boyle, C.F., Reiser, B.J.: Intelligent tutoring systems. Science **228**(4698), 456–462 (1985). https://doi.org/10.1126/science.228.4698.456. http://science.sciencemag.org/content/228/4698/456
3. Angeli, A.D., Johnson, G.I., Coventry, L.: The unfriendly user: exploring social reactions to chatterbots. In: Proceedings of International Conference on Affective Human Factor Design, pp. 467–474. Asean Academic Press (2001)
4. Dutta, D.: Developing an intelligent chat-bot tool to assist high school students for learning general knowledge subjects. Technical report, Georgia Institute of Technology (2017)
5. Goel, A.K., Polepeddi, L.: Jill Watson: a virtual teaching assistant for online education. Technical report, Georgia Institute of Technology (2016)
6. Huang, J., Zhou, M., Yang, D.: Extracting chatbot knowledge from online discussion forums. In: Proceedings of the 20th International Joint Conference on Artifical Intelligence, pp. 423–428. IJCAI 2007, Morgan Kaufmann Publishers Inc., San Francisco (2007)
7. Jia, J.: CSIEC (computer simulator in educational communication): an intelligent web-based teaching system for foreign language learning. CoRR (2003). http://arxiv.org/abs/cs.CY/0312030
8. Makinson, D.: Sets. Logic and Maths for Computing. Springer, Heidelberg (2012)
9. Reyes-González, Y., Martínez-Sánchez, N., Díaz-Sardiñas, A., Patterson-Peña, M.: Conceptual clustering: a new approach to student modeling in intelligent tutoring systems. Revista Facultad de Ingeniería **0**(87), 70–76 (2018). https://doi.org/10.17533/udea.redin.n87a09, https://aprendeenlinea.udea.edu.co/revistas/index.php/ingenieria/article/view/327565
10. Russell, S., Norvig, P.: Artificial Intelligence: A Modern Approach, 3rd edn. Prentice Hall, Upper Saddle River (2010)
11. Salinas, P., Alanís, J.A., Pulido, R., Santos, F., Escobedo, J.C., Garza, J.L.: Cálculo aplicado: competencias matemáticas a través de contextos (Tomo 1). Cengage Learning, México (2012)
12. Salinas, P., Quintero, E., Sánchez, X.: Math and motion: a (coursera) MOOC to rethink math assessment. In: Zaphiris, P., Ioannou, A. (eds.) LCT 2015. LNCS, vol. 9192, pp. 313–324. Springer, Cham (2015). https://doi.org/10.1007/978-3-319-20609-7_30
13. Serafini, L.: Expressive power of logical languages. Internet, July 2012. http://iaoa.org/isc2012/docs/expressivity.pdf
14. Setiaji, B., Wibowo, F.W.: Chatbot using a knowledge in database: human-to-machine conversation modeling. In: 2016 7th International Conference on Intelligent Systems, Modelling and Simulation (ISMS), pp. 72–77. Thailand, January 2016. https://doi.org/10.1109/ISMS.2016.53
15. Shawar, B.A., Atwell, E.: Chatbots: are they really useful? LDV Forum **22**, 29–49 (2007)
16. Tsvetkova, M., García-Gavilanes, R., Floridi, L., Yasseri, T.: Even good bots fight. PLos One **12**(2) (2017). https://arxiv.org/abs/1609.04285

Top-k Context-Aware Tour Recommendations for Groups

Frederick Ayala-Gómez[1,2](✉), Barış Keniş[2], Pınar Karagöz[3],
and András Benczúr[4]

[1] Faculty of Informatics, Eötvös Loránd University, Pázmóny P. sny. 1/C.,
Budapest 1117, Hungary
fayala@caesar.elte.hu
[2] Computer Science Department, Aalto University, 02150 Espoo, Finland
baris.kenis@aalto.fi
[3] Computer Engineering Department, METU, A-404, 06800 Ankara, Turkey
karagoz@ceng.metu.edu.tr
[4] Institute for Computer Science and Control, Hungarian Academy of Sciences
(MTA SZTAKI), Budapest 1111, Hungary
benczur@sztaki.mta.hu

Abstract. Cities offer a large variety of Points of Interest (POI) for leisure, tourism, culture, and entertainment. This offering is exciting and challenging, as it requires people to search for POIs that satisfy their preferences and needs. Finding such places gets tricky as people gather in groups to visit the POIs (e.g., friends, family). Moreover, a group might be interested in visiting more than one place during their gathering (e.g., restaurant, historical site, coffee shop). This task is known to be the orienteering under several constraints (e.g., time, distance, type ordering). Intuitively, the POI preference depends on the group, and on the context (e.g., time of arrival, previously visited POIs in the itinerary). Recent solutions to the problem focus on recommending a single itinerary, aggregating individual preferences to build the group preference, and contextual information does not affect the scheduling process. In this paper, we present a novel approach to the following setting: Given a history of previous group check-ins, a starting POI, and a time budget, find top-k sequences of POIs relevant to the group and context that satisfy the constraints. Our proposed solution consists of two primary steps: training a POI recommender system for groups, and solving the orienteering problem on a candidate set of POIs using Monte Carlo Tree Search. We collected a ground-truth dataset from Foursquare, and show that the proposed approach improves the performance in comparison to a Greedy baseline technique.

Keywords: Recommender systems for groups
Tour recommendation · Orienteering

© Springer Nature Switzerland AG 2018
I. Batyrshin et al. (Eds.): MICAI 2018, LNAI 11289, pp. 176–193, 2018.
https://doi.org/10.1007/978-3-030-04497-8_15

1   Introduction

The web changed the way people interact with each other and with organizations [13]. As a part of the web, Location-Based Social Networks (LBSNs) help people to track their leisure and tourism activities. In populous cities, LBSNs help users navigate the large variety of Points of Interest (POI) such as parks, museums, restaurants, etc. Due to the large amount of POIs, finding venues that are relevant to the user preferences is time-consuming. To help with this task, POI recommendation is a popular research problem. Most of the research in POI recommendation focus on recommending to one user [25]. However, as people gather in groups – which is a frequent situation in social life (e.g., holidays, weekend, birthdays), the preferences of the individuals might change and conflict. Research on POI recommendations for groups is scarce but a promising and emerging area. In this paper, we focus on building recommendations for groups (i.e., gatherings of more than one user). More specifically, groups have at least two members. A user might be a member of many groups. And groups are considered to be different if any of their members is added or removed.

When groups are planning to visit more than one POI in their gathering, to a sequence of POIs, the task is more challenging, as additional constraints (e.g., time) must be satisfied. This task is known as the group orienteering problem. In the group orienteering problem, the preferences of a group of people are considered. Intuitively, group's preferences vary according to the members. For example, the group might prefer different venues when it is a *best friends gathering* compared to a family gathering or meeting with colleagues after work. Moreover, the group may have specific constraints (e.g., category sequence, duration, time). In this work, we investigate the group orienteering problem. More specifically, we study the problem *recommending top-k sequences of POIs for a known group, given a starting POI, time constraint, and contextual information (e.g., popularity of the venue at time of arrival)*. Our study focuses on LBSNs as the data source, as they enable groups to share their activities by doing *check-ins* on the visited venues. Therefore, it is possible to find group traces of visits as well as individual visits. These traces provide important insights of group preferences.

The proposed solution differs from related studies in the literature in several ways. Recent work on group orienteering (i.e. itinerary planning for group ([2,9])) modeled groups preference in terms of individual preferences. In our work, we model group preference on the basis of observed group behavior. The motivation behind this choice is that, in [3], it is reported that recommending POIs for groups based on the group preferences performs better than the aggregation of individual preferences. Thus, we model group preferences directly from the group behavior observations and use. Additionally, building the itinerary depends on contextual information such as the venue's popularity at the time of arrival, previous visited POIs, and POI category transitions. For example, assume that the group is visiting a restaurant. It is likely that the next POI will not be a restaurant, but rather a coffee shop, or a plaza. Moreover, if at the time of expected arrival to the next POI, the venue is unpopular, selecting the venue is less likely, such as arriving to a night-club in the afternoon. Recent work does not incorporate contextual information during the itinerary planning. In our work, we incorporate such features in our solution. Finally, recent work focus on recommending one itinerary for the group, and our work focus on recommending top-k itineraries.

Our proposed approach for building top-k tour recommendations for groups consists of two main steps. In the first step, we model group preferences using group check-ins and train a recommender systems to learn the group preferences. In the second step, we generate a candidate set of POIs near the starting POI, rank the POIs using the recommender system, and solve the orienteering problem using Monte Carlo Tree Search (MCTS), where the expected reward of the POIs depend on the POI recommendation score and on contextual features.

We conducted the experiments on a real dataset of check-ins collected from Swarm by Foursquare, a popular LBSN, and use it as a ground-truth. The collected dataset contains group check-ins as well as individual check-ins. The proposed solution differs from related studies in the literature in several ways. First, we analyze the coverage of the generated POI candidates near a given venue. Second, we compare the performance of various recommender systems and different individual aggregation strategies. Implicit matrix factorization for groups performed better than content-based, item-to-item. Moreover, we show that training a recommender systems using group profiles perform better than aggregating individual recommendations. Finally, our results show that MCTS constructs better itineraries in terms of recall and precision, than the greedy baseline.

The paper is organized as follows. Section 2, summarizes previous studies on group recommendation and itinerary planning. Section 3 presents our proposed approach for building top-k itineraries for groups. Section 4 explains the data used as ground truth, and a data analysis. Section 5 presents the evaluation methodology, and experimental results. Finally, Sect. 6 presents findings and conclusions.

2   Related Work

We categorize related studies in the literature into two: studies on item recommendations for groups, and studies on itinerary recommendations. In this section, we summarize the literature for both of these problems.

2.1   Item Recommendation for Groups

In [1], as one of the earliest works on group recommendation, the problem is defined as a consensus function aiming to maximize relevance and minimize disagreement. Their algorithm uses a disagreement list, and several variations of the algorithm are presented. In [19], personal impact of groups members on group decisions are modeled as a latent variable. The model is further refined by exploring the relationship between personal impact and other features (e.g., social network). In [26], the proposed method for group recommendation is based on aggregating preferences of the group member with different weights. In [12], rather than aggregating member preferences, group preference is modeled as a distribution of member's ratings. On the basis of this distribution, item selection is solved through a multi-criteria decision making process. The study in [6] focuses on diversity and novelty aspects in group recommendation. The different preferences of the group members are considered as group hesitation, and group preference is modeled through Hesitant Fuzzy Sets (HFSs). In accordance with the focus of the paper, evaluation involves diversity and novelty measurement in addition to accuracy. The study in [24] models the group recommendation problem as a multi-criteria optimization problem to maximize the satisfaction of each group member and to minimize the unfairness within the group. Fairness metric employed aim to minimize the individual utility differences in the group. In the study, optimality is defined as *Pareto-optimality*, and to solutions are presented in order to achieve Pareto-efficient solution. In [21], in addition to location context, they focus on passive users who do not explicitly specify preferences. Such users are represented through topic modeling. In order to improve time efficiency of the proposed method, the authors additionally propose an index structure, namely *Topic-aware R-tree*. The work in [3] proposes a class of hybrid recommendation algorithms that combine a set of features including geographical preferences, venue category and group check-in history in order to recommend a list of POIs to a group. Empirical analysis reveal that the individual preferences deviate from preference of groups. Hence recommendation under group preference model performs better than aggregation of individual preferences.

2.2 Itinerary Recommendation

Research on itinerary recommendation mostly focus on recommendation for individuals. In [8], it is aimed to generate inter-city travel itineraries by using geo-temporal traces (i.e., breadcrumbs) left by travelers in social media platforms. They use Flickr[1], by mapping the photos taken by users to POIs used to construct paths. The generated paths are mined under several constraints to construct travel itineraries. The work given in [10] also uses geo-temporal traces of travelers on LBSNs. Their aim is to construct tour itinerary within a city. The authors propose two algorithms to construct itinerary under several constraints, including those on venue types and distance. In [23], authors consider itinerary planning as an iterative process. At each step, the user gives feedback on the POIs selected by the system. This feedback is used to pick the next steps in the itinerary. The problem is modeled as a rooted orienteering problem, and the proposed optimization solution is based on finding a Hamiltonian path in a hypercube. For time-efficiency, heap-based data structures are used. The work presented in [7] aim to construct more general itineraries involving several days. A two-step solution is proposed. In the first step, all possible single-day itineraries

---

[1] https://www.flickr.com/.

are constructed. In the second step, the single-day itineraries are combined selectively for optimal multi-day itinerary. Hence, the itinerary planning problem is modeled as a set-packing problem with good approximate algorithms, instead of NP-complete tour orienteering problem. In [17], itinerary planning problem is considered as orienteering problem as in the previous studies. However, the major difference in the paper is that the user preference/interest for a POI is weighted though visit duration rather than visit frequency. This weighting approach is further enhanced by considering recency of past visits. In [16], the same authors studied the itinerary planning problem under queuing time consideration. The objective function of this new version of the problem aims to maximize user interest and POI popularity whereas to minimize queuing time.

Recently in the literature, itinerary recommendation for groups has attracted attention as a research problem. In [9], a multi-user itinerary planning solution is proposed. The proposed algorithm is based on aggregating individual preferences while considering agreement and disagreement among the group members. The authors propose two algorithms: the first constructs the optimal itinerary under limitations, and the second one generates an approximate solution with bounded approximation ratio. In [2], the authors consider the task to be the *orienteering* problem, as well, where the goal is to find a path in a graph within a time budget, that has the best value for the group. The group value is an objective function that aggregates individual preferences of the POIs in the path. Their solution for *TourGroup* has two steps. In the first step, they compute the individual preference for the POIs using a content-based recommender system. In the second step, they build the itineraries for groups by solving *TourGroup* for different objective functions: *(i) TourGroupSUM*, where they added the individual values; *(ii) TourGroupMIN*, where the goal is to find the maximum of the least liked individual path; *(iii) TourGroupFAIR*, where the score is the average value of the individual paths minus a weighted standard deviation. For these problems, the authors applied various search methods including, greedy heuristics, dynamic programming, and ant colony optimization. They found the best results for itinerary construction using Ant Colony, although it did not vary much from the greedy approach.

In our work, one of the basic differences from the last two summarized work is that rather than aggregating group members' preferences, we use the known group preferences. In that sense, we have a common approach with the study in [3]. Moreover, we focus on recommending a list of itineraries instead of a single itinerary. In terms of the methods used to solve the orienteering task, we use Monte Carlo Tree Search (MCTS), and added contextual features during the itinerary planning such as time and transition probabilities.

3   Proposed Approach

At a glance, our proposed approach consists of two steps. In the first step, we train a recommender system using the known check-ins from a ground-truth data set. In the second step, from a starting POI, we generate a set of K neighbor

POIs used as candidate set, and score them with the recommender system (from step one). These ranked candidates are used as input to solve the orienteering problem with Monte Carlo Tree Search. The overview of the process is presented in Fig. 1.

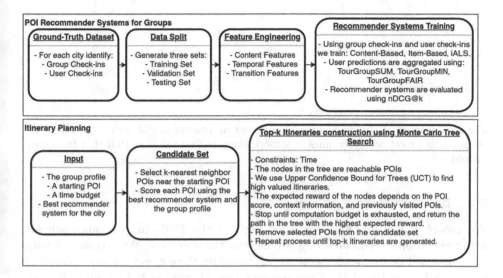

Fig. 1. Overview of the proposed approach.

We define the problem of top-k tour recommendations for groups as follows. Given a group G of LBSN users, a set of POIs P, and a POI p_0 as a starting point, the aim is to construct K paths $I = <p_0, ..p_n>$, where $p_i \in P$, such that I fulfills a given time budget. Additionally, the intersection of all K paths equals to p_0.

3.1   Candidate Set

In [3], the authors observed that groups do not travel long distances between check-ins. Based on this observation, we included a candidate set generation step that takes into account POI near the *starting POI* of the group. That is, for a given location, we consider K number of nearest neighbors. We measure the quality of the candidate set through coverage, such that, checking whether the *next POI* that the group goes to is in the candidate set or not. Figure 2 presents the coverage obtained at different candidate set sizes. As seen in the figure, as the number of neighbors increases, the coverage gets closer to 1.0. However, high number of neighbors incurs computational cost in scoring phase (through recommender system), therefore we need to limit the number while retaining good enough coverage.

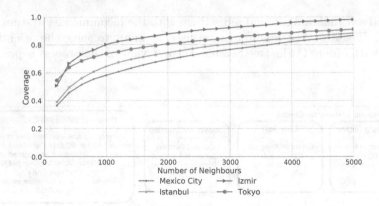

Fig. 2. Coverage of the next venue at different number of neighbors for the cities. As we get a larger set of near venues it is more likely to find the next POI that the group visited next.

3.2   Recommender Systems

We model the group preference for each of the POIs in the candidate set using well-known recommender systems from the literature. We experiment with content-based recommender system [22] based on the content features described in Sect. 4.3, item-based [18], and collaborative filtering model for implicit feedback (iALS) [14].

We build recommendations for groups using two approaches. The first approach is to create a profile for each of the groups using the observed check -ins in the ground-truth dataset. Then, we train the recommender systems to learn the preferences of the group using the history of check-ins. The seconds approach is to build a profile for each user using the known check-ins in the ground-truth dataset. Then, the group recommendation is computed as an aggregation of the predicted score for each of the users in the group. Similar to [2], we use three different aggregation functions: *(i) IndividualSUM* is the sum of the group members' recommendations, *(ii) IndividualMIN* the min value of the group members' recommendations, and *(iii) IndividualFAIR* assigns the mean value of the scores minus one standard deviation.

3.3   Itinerary Construction

Our proposed solution for the orienteering problem uses Monte Carlo Tree Search (MCTS) to efficiently find a promising itinerary. MCTS is built on the idea of combining tree search with random sampling, and finds optimal decision paths [5].

There are different variations of MCTS [5]. We focus on Bandit-Based Methods, as they are a well-known class of sequential decision problems. Specifically, we use Upper Confidence Bound for Trees (UCT), "the most popular algorithm in the MCTS family" [5]. UCT iteratively builds a tree by expanding nodes in each iteration, selecting the most promising node, estimating the expected reward of the selected node, and back-propagating the reward to the parents. This process is repeated until certain computational budget is reached. Finally, starting from the root, we pick the child node with highest expected rewards until we reach a terminal node. In our case, each node represents visiting a POI. The main function of the UCT algorithm is defined in Algorithm 1, and the following is an explanation of each of the functions.

Tree policy expands a node if it is not fully expanded or terminal, otherwise it picks the *best child*, which is the node that requires more attention at the current iteration, and controls the trade between exploration and exploitation. The UCT algorithm picks the *best child* that maximizes the Upper Confidence Bounds defined in Eq. 1.

$$UCT(j) = \bar{X}_j + 2C_p \sqrt{\frac{2 \ln n}{n_j}}, \tag{1}$$

where $\bar{X}_j$ is the average expected reward of a candidate node j calculated during default policy step, C_p controls the exploration-exploitation trade-off. For expected rewards between $[0,1]$, $\frac{1}{\sqrt{2}}$ is recommended [15], n is the total number of visits to all nodes, and n_j the visits to node j.

Default policy simulates an expected reward, if the node is non-terminal. A terminal node is the case when the time budget is exhausted and no more POIs can be visited. In the simplest case, the algorithm picks uniformly random unseen reachable nodes, until a terminal node is found. The expected reward is the average of the nodes visited during simulation. If the node is terminal, the expected reward is score of the node.

Backup is the process of updating the parents of the selected best-child according to the expected reward.

UCTSearch(s_0)
create root node v_0 with starting POI
while *within computational budget* **do**
  $v_j \leftarrow TreePolicy(v_0)$
  $\Delta \leftarrow DefaultPolicy(s(v_j))$
  $Backup(v_j, \Delta)$
end
return *BestChilds*(v_0)

Algorithm 1. The UCT Algorithm.

To generate top-k recommendations, we run the UCT algorithm, select the best itinerary, remove the selected POIs in the itinerary, and run again UCT until K itineraries are generated. We propose three versions of the MCTS method. *MCTS_Simple* selects uniformly random the POIs. The expected reward is defined in Eq. 2.

$$x_j = \frac{\hat{r}_j}{f_{c_j} \cdot \lambda + 1}, \tag{2}$$

where $\hat{r}_j$ is the recommendation score, f_{c_j} the number of occurrences of the venue category, and λ is a category diversity parameter greater than 1.

MCTS_Score replaces the uniformly random sampling with a greedy search, using the score of the recommender system. The expected reward is defined in Eq. 2.

MCTS_Context also replaces the uniformly random sampling with a greedy search, using the score of the recommender system, and contextual information. The expected reward is calculated as Eq. 3.

$$x_j = \frac{\hat{r}_j \cdot popularity(p_j, t_j) \cdot ecp(c_j|c_i)}{f_{c_j} \cdot \lambda + 1}, \tag{3}$$

where $\hat{r}_j$ is the recommendation score, $popularity(v_j, t_j)$ is the proportion of check-ins observed in p_j at time of arrival t_j, $ecp(c_j|c_i)$ is the empirical conditional probability of visiting a venue of type c_j after a venue of type c_i (e.g., going to a coffee shop after a park), f_{c_j} the number of occurrences of the venue category, and λ is a category diversity parameter greater than 1.

4   Data

We start by describing our working dataset, the cleaning steps, and data statistics, followed by presenting data exploration results. Afterwards, we describe the feature engineering.

4.1   Working Dataset

Our approach requires a dataset including both individual users' preferences and group preferences. For this purpose, we collected a dataset of check-ins from *Swarm by Foursquare*, a popular LBSN. *Check-ins* are posts in LBSN with information of the venue that a user visited. Moreover, check-ins may mention other users that are with the user who checked-in. Following the idea of [3], we collected check-ins by searching for public check-ins shared on *Twitter*, a popular social network. Then, we follow a *Snowball* sampling approach [11]. For each of the group check-ins found in Twitter, we search for additional public check-ins in the group members Twitter time-line. This collection has approximately 2.7 M individual check-ins, 1 M group check-ins, 200 K users, 400 K groups, 116 K venues, and 547 categories.

We followed the cleaning steps of [3]. We considered groups with maximum 12 participants, removed users and groups with high geographical dispersion (i.e., according to the standard deviation of their geographical mobility), and removed irrelevant venue categories (i.e., Residence, States & Municipalities, Professional & Other Places and Event, College & University, Travel & Transport, Event, Shop & Service). Table 1 presents statistics of the groups, users, venues, and categories for the Top-4 cities.

Table 1. Statistics for the Top-4 cities in our working dataset.

City	Check-ins	Venues	Categories	Group Check-ins	Group Venues	Group Categories
Istanbul	763 K	32 K	423	205 K	18	362
Izmir	262 K	14 K	373	46 K	5 K	271
Mexico city	179 K	19 K	370	71 K	12 K	340
Tokyo	108 K	19 K	392	7 K	3 K	264

4.2 Data Exploration

LBSN data is rich in contextual features such as time and location. Regarding time, there are seasonality effects for the daytime, day of the week and month. Intuitively, popularity of venues changes depending on the week, the hour, and the season. For example, a restaurant by seaside in Istanbul is more popular during summer where temperatures are around 25 °C, than winter where temperatures are around 5 °C. An example of the effects that the week-day, time, and venue type have on the popularity is shown in Fig. 3.

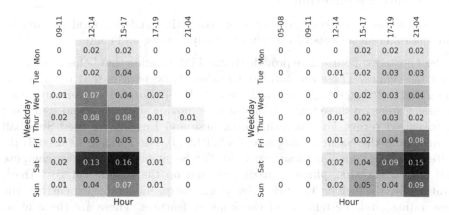

(a) Check-in distribution for museums   (b) Check-in distribution for bars

Fig. 3. Distribution of check-ins per weekday and hour for museums and bars in Istanbul. The rows are the parts of the day, and the columns are days of the week. A darker color represent higher distribution of check-ins.

The dataset contains groups who transit from one venue to another in the same day. These item to item transitions help us to understand popular venue type transitions in a city. Figure 4 shows an extract from the transition matrix for Istanbul.

Current POI Category	Food	Scenic Lookout	Bar	Plaza	Other Great Outdoors	Movie Theater	Historic Site	Nightclub	Park	Athletics & Sports
Food	0.66	0.06	0.08	0.01	0.01	0.04	0.01	0.02	0.01	0.01
Scenic Lookout	0.63	0.1	0.07	0.03	0.04	0.01	0.03	0	0.01	0
Bar	0.43	0.06	0.27	0.02	0.01	0.01	0.01	0.08	0.01	0.01
Plaza	0.57	0.05	0.08	0.04	0.03	0.01	0.07	0.01	0.03	0
Other Great Outdoors	0.54	0.1	0.13	0.02	0.03	0.01	0.04	0.02	0.02	0
Movie Theater	0.87	0.01	0.05	0.01	0.01	0	0	0.01	0	0
Historic Site	0.55	0.12	0.05	0.07	0.04	0.01	0.05	0.01	0.02	0
Nightclub	0.28	0.02	0.1	0	0	0	0	0.5	0	0
Park	0.53	0.11	0.05	0.04	0.04	0.02	0.07	0	0.03	0.01
Athletics & Sports	0.83	0.03	0.04	0	0.02	0.02	0	0	0	0

Next POI Category

Fig. 4. Part of the transition matrix for activity types in Istanbul. The numbers represent the transition probability between two categories. Intuitively, moving from a venue of certain category conditions the next venue category. For example, if the group is in a Park, it is more likely that they will go to a Food place, than to a Nightclub. And, if the group is in a Nightclub, they are more likely to go to a Food place after than a Park.

4.3 Feature Engineering

From the dataset, we construct content, transition and temporal features to represent different aspects of the venues, groups, users, and context.

Content Features: Foursquare provides textual information that helps know more about the venues. These are *tips* left by other users, *phrases* that describe the venue, and *attributes* (e.g., romantic, cozy). We performed conventional pre-processing steps on these textual content by removing stop words and stemming the tokenized terms. For Spanish, Portuguese and English, we used Snowball stemmer from the Natural Language Toolkit by [4]. For Turkish, we used TurkishStemmer[2] and for Japanese, we used TinySegmenter[3]. Then, we compute tf-idf [20] on the tips, phrases, attributes, and on the venues' category tree[4]. Additionally, we include the price tier (e.g., cheap, moderate, expensive), the venue rating, and the number of check-ins as features. These are the content features used by the content-based recommender system.

Transition Features: We use pair-wise transition probabilities between the current POI category c_i and a possible next POI category c_j as transition feature. The empirical conditional probability for the venues categories observed in the training dataset is defined as $ecp(c_j|c_i) = \frac{n_{ij}}{n_i+1}$, where n_{ij} is the number of observed transitions between c_i and c_j, and c_i is the number of check-ins at c_i.

Temporal Features: To include temporal context as a feature, we split the day into 6 segments. 05–08 (Early Morning), 09–11 (Late Morning), 12–14 (Early

[2] https://github.com/otuncelli/turkish-stemmer-python.
[3] http://tinysegmenter.tuxfamily.org/.
[4] https://developer.foursquare.com/docs/resources/categories.

Afternoon), 15–17 (Late Afternoon), 17–19 (Early Evening), 21–04 (Night). The distribution of check-ins for the venue and category in each of the day-segments are added as features.

5 Experiments

5.1 Data Split

For each of the cities presented in Table 1, the dataset is split into training, validation and testing datasets with respect to time order. The training dataset contains the 80% of the data, from the training set we take 5% as validation set. The testing dataset is 20% of the dataset. As the ground truth, we collect itineraries by looking for check-in sequences of the same group on the same day, in the data sets.

5.2 Metrics

Our proposed approach consists of two phases: The recommender systems, which finds relevant POIs to the group; and the itinerary algorithm, which constructs the sequence of POIs according to the constraints. We evaluate the results of these phases independently.

The quality of the group recommender system is measured by evaluating the ranking and the relevance of the recommended POIs. For this, we use normalized discounted cummulative gain. This metric rewards relevant POIs that appear higher in the rank and penalizes relevant POIs that appear lower in the rank. The nDCG@100 is defined as given in Eq. 4.

$$nDCG@100 = \frac{\sum_{i}^{100} \frac{2^{rel_i}-1}{log_2(i+1)}}{IDCG}, \qquad (4)$$

such that rel_i is 1 if the venue was visited. The $IDCG$ is the ideal ordering of the ranking as defined in Eq. 5.

$$IDCG = \sum_{i}^{|REL|} \frac{2^{rel_i}-1}{log_2(i+1)}, \qquad (5)$$

where $|REL|$ is the *next* POIs visited by the group.

The itinerary construction is evaluated on how well the constrains are satisfied while maximizing the predicted preference of the group for the candidate POI. The quality of the itinerary is measured by three metrics. The value of the itinerary, as the sum of the recommendations scores in the itinerary, precision, as defined in Eq. 6, and recall, as defined by Eq. 7.

$$Precision = \frac{|\ \{Visited\ POIs\} \cap \{POIs\ in\ the\ Itinerary\}\ |}{|\ \{POIs\ in\ the\ Itinerary\}\ |} \qquad (6)$$

$$Recall@K = \frac{|\ \{Visited\ POIs\} \cap \{POIs\ in\ the\ Itinerary\}\ |}{|\ \{Visited\ POIs\}\ |} \qquad (7)$$

5.3 Baselines

The recommender systems that are used within this study are compared with each other in terms of nDCG@100 per city. For itinerary construction evaluation, MCTS_Context is compared against MCTS_Simple and MCTS_Score. Additionally, Greedy Algorithm is used another baseline for comparison. In itinerary construction, for each city, the best recommender system to score the candidate sets is used. MCTS_Context, and the three baselines include the penalization over repeating venue types in the itinerary.

Greedy Algorithm. A basic approach to solve the itinerary construction problem is to use a greedy approach. Just like the MCTS, our greedy algorithm also utilizes the recommender system to score the POIs. Starting from a POI, the algorithm asks for top k candidates from the recommender system, filters out the ones that are already included, or break some constraint, and chooses the one that gives the highest score according to the recommender system. The algorithm continues next POI selection until the budget is consumed.

Table 2. Results of evaluating the *Top 100* recommendations of the next POI to visit. The scores are the mean nDCG@100. In bold are the best models for each city. The abbreviation *w/*stands for recommendations filtered to a specific *next* POI category (i.e., Top 100 nearest POI of the given category), and *w/o* when the next category is not specified.

Using next category	Istanbul		Izmir		Mexico city		Tokyo	
	w/o	w/	w/o	w/	w/o	w/	w/o	w/
iALS - Group	**0.546**	**0.700**	**0.550**	**0.705**	**0.531**	**0.716**	0.460	0.666
iALS - IndividualSum	0.395	0.567	0.454	0.643	0.508	0.692	0.455	0.695
Item-to-Item - IndividualSum	0.364	0.532	0.344	0.531	0.418	0.615	0.428	0.655
Item-to-Item - Group	0.502	0.650	0.271	0.441	0.204	0.395	**0.500**	**0.753**
Content - Group	0.269	0.434	0.271	0.441	0.204	0.396	0.282	0.494
Content - IndividualFair	0.246	0.414	0.261	0.430	0.186	0.384	0.145	0.350
Content - IndividualMin	0.246	0.414	0.261	0.430	0.186	0.384	0.145	0.350
Content - IndividualSum	0.246	0.414	0.261	0.430	0.186	0.384	0.145	0.350
Item-to-Item - IndividualMin	0.175	0.321	0.229	0.390	0.212	0.387	0.146	0.363
Item-to-Item - IndividualFair	0.152	0.292	0.132	0.272	0.183	0.338	0.169	0.378
iALS - IndividualMin	0.057	0.206	0.088	0.242	0.134	0.293	0.239	0.409
iALS - IndividualFair	0.034	0.163	0.060	0.163	0.075	0.215	0.112	0.326

5.4 Recommender Systems Evaluation Results

We used Apple's Turi Create[5] recommender module for the implementation of recommender systems. We fine-tuned the parameters using grid search on the validation set. The code for generating the candidate sets, training, and predictions, was executed on a single machine with 16 GB of RAM, 16 cores. Table 2 shows the results of the different recommender systems.

---

[5] https://github.com/apple/turicreate.

Table 3. Mean value of the itineraries. The value of an itinerary is the sum of the POIs recommendations.

	Greedy	MCTS_Context	MCTS_Score	MCTS_Simple
Istanbul	0.318	0.357	0.356	**0.361**
Izmir	0.276	0.317	0.315	**0.322**
Mexico city	0.417	**0.445**	**0.445**	**0.445**
Tokyo	0.320	0.363	0.363	**0.371**

Table 4. Mean precision of the *Top 10* itineraries recommendation. In bold are the highest values for each city.

	Greedy	MCTS_Context	MCTS_Score	MCTS_Simple
Istanbul	0.008	**0.009**	**0.009**	**0.009**
Izmir	0.005	**0.007**	**0.007**	**0.007**
Mexico city	0.024	0.024	0.023	**0.026**
Tokyo	0.044	**0.054**	0.048	0.048

Table 5. Mean recall of the *Top 10* itineraries recommendation. In bold are the highest values for each city.

	Greedy	MCTS_Context	MCTS_Score	MCTS_Simple
Istanbul	**0.011**	**0.011**	**0.011**	0.010
Izmir	**0.013**	**0.013**	0.012	0.012
Mexico city	0.024	0.023	0.023	**0.025**
Tokyo	**0.042**	0.041	0.038	0.037

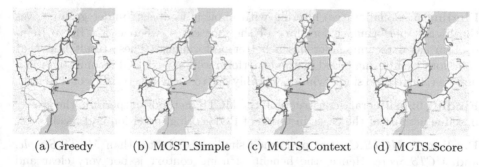

(a) Greedy   (b) MCST_Simple   (c) MCTS_Context   (d) MCTS_Score

Fig. 5. Examples of itineraries generated by different algorithms for a group in Istanbul: Greedy, MCTS_Simple, MCTS_Context, and MCTS_Score. Each line represent one of the Top-10 recommended itineraries.

5.5   Itinerary Construction Evaluation Results

We generate Top-10 itineraries for each of the group tours in the testing set. For all the itineraries we used a budget of 8 hours, and assumed that the group walks to commute between the POIs at a speed of 1.4 m/s. We took the first POI as the first check-in of the group sequence in our testing set, the starting POI is excluded from the evaluations. We experimented with λ values of 1, 2, 3, and 4. The best λ values for each city are 4 for Istanbul, 3 for Izmir, 3 for Mexico City, and 2 for Tokyo.

The results of comparing the greedy baseline with the MCTS models are presented in Table 3 for the itinerary values, Table 4 for average precision, and Table 5 for average recall. Figures 5 presents an example of the Top 10 itineraries generated for a group in Istanbul.

5.6   Findings on Experiments

The following findings summarize the experimental results.

Finding 1: LBSNs provide enough clues to determine group preferences for venues and contextual information.

Finding 2: In Table 2 we observe that for most of the cases, recommendation models constructed with group history perform better than the models including the aggregation of individual recommendations. However, aggregating individual recommendations could be useful for cold start groups (i.e., groups without check-ins history).

Finding 3: Predicting the next POI using a fixed category type (i.e., using next category) improves the quality of the recommender systems considerably.

Finding 4: Except for content based recommender, aggregation of individual recommendations under different functions effect the quality of the recommender systems. Among three aggregation functions, *summation* of individual recommendation scores provide the highest performance.

Finding 5: Collaborative filtering with implicit feedback (under group) gives the highest performance for three of the four cities, whereas for Tokyo item-to-item based recommender has a better performance. These result hint that collaborative filtering based methods perform well in next POI recommendation, however, there is no single winner, possibly due to the characteristics of the cities.

Finding 6: In itinerary construction, the MCTS methods outperform the greedy baseline in most of the cases in terms of POI scoring, precision and recall.

Finding 7: *MCTS_Context* performed similarly or better than *MCTS_Simple*, and *MCTS_Score*. Hence, the benefit of using context is not very clear and possibly depends on the group behavior characteristics per city. For instance, Tokyo dataset contains longer ground truth group itineraries, which enables MCTS_Context to capture and use context information. On the other hand, in

Mexico City data set, the itineraries are much shorter, of length two on average. In such a case, simple reward function performs better in itinerary construction.

Finding 8: Sample constructed itineraries for Istanbul in Fig. 5 include high overlap. Hence, it is hard to conclude about the characteristics of the constructed itineraries. On the basis of the mean itinerary scores given in Table 3, the slight differences in itineraries, especially on the leftmost part of the figures, are due to differences in POI selection with different scores.

6   Conclusions and Future Work

We present an approach for solving the top-k context-aware tour recommendations for groups problem by using MCTS and contextual information. Our solution integrates selection of a candidate set of neighbor POIs, a group recommender system, and Monte Carlo Tree Search. Given a starting POI, building top-k itineraries consists of these steps: selecting and ranking a candidate set of next POIs, scoring the POIs by recommender system, and solving the orienteering problem using MCTS under time budget constraint.

In itinerary construction, MCTS performs better than greedy algorithm in terms of POI scoring and accuracy. On the other hand, benefits of using the contextual information in MCTS_Context in comparison to MCTS_Score and MCTS_Simple is hard to measure due data sparsity.

The main limitation of our work is the difficulty of evaluating the top itineraries for groups using the ground-truth dataset. Even though we collected approximately 300 K group check-ins, the number of publicly available sequences of check-ins remain limited, and most of the sequences are of length two.

We envision further work on several research direction. Understanding what makes different recommender systems to perform better for characteristics of the cities is a potential problem to work on. In order to improve the itinerary construction performance, using category type as constraint during itinerary generation, experimenting with different candidate selection strategies, and using more advanced scoring functions that combine contextual information with recommendations can be further investigated. Additionally, studying itinerary recommendations for groups with no check-ins (cold start) is an interesting follow-up research problem.

Acknowledgments. The authors would like to thank Aristides Gionis, and Levente Kocsis for their input to our work. Frederick Ayala-Gómez was supported by the Mexican Postgraduate Scholarship of the Mexican National Council for Science and Technology (CONACYT). This work is partially supported by TUBITAK under the project grant number 117E566, by the Hungarian Government project 2018-1.2.1-NKP-00008: Exploring the Mathematical Foundations of Artificial Intelligence and by the Momentum Grant of the Hungarian Academy of Sciences.

References

1. Amer-Yahia, S., Roy, S.B., Chawlat, A., Das, G., Yu, C.: Group recommendation: semantics and efficiency. Proc. VLDB Endow. **2**(1), 754–765 (2009)
2. Anagnostopoulos, A., Atassi, R., Becchetti, L., Fazzone, A., Silvestri, F.: Tour recommendation for groups. Data Min. Knowl. Discov. **31**(5), 1157–1188 (2017)
3. Ayala-Gómez, F., Daróczy, B., Mathioudakis, M., Benczúr, A., Gionis, A.: Where could we go? Recommendations for groups in location-based social networks. In: WebSci 2017, pp. 93–102. ACM, New York (2017). https://doi.org/10.1145/3091478.3091485
4. Bird, S., Klein, E., Loper, E.: Natural Language Processing with Python: Analyzing Text with the Natural Language Toolkit. O'Reilly Media, Inc., Sebastopol (2009)
5. Browne, C.B., et al.: A survey of monte carlo tree search methods. IEEE Trans. Comput. Intell. AI Games **4**(1), 1–43 (2012)
6. Castro, J., Barranco, M.J., Rodriguez, R.M., Martinez, L.: Dealing with diversity and novelty in group recommendations using hesitant fuzzy sets. In: Proceedings of the FUZZ-IEEE International Conference on Fuzzy Systems (2017)
7. Chen, G., Wu, S., Zhou, J., Tung, A.K.: Automatic itinerary planning for traveling services. IEEE Trans. Knowl. Data Eng. **26**(3), 514–527 (2014)
8. De Choudhury, M., Feldman, M., Amer-Yahia, S., Golbandi, N., Lempel, R., Yu, C.: Automatic construction of travel itineraries using social breadcrumbs. In: Proceedings of the 21st ACM Conference on Hypertext and Hypermedia, pp. 35–44. ACM (2010)
9. Fan, L., Bonomi, L., Shahabi, C., Xiong, L.: Multi-user itinerary planning for optimal group preference. In: Gertz, M., Renz, M., Zhou, X., Hoel, E., Ku, W.-S., Voisard, A., Zhang, C., Chen, H., Tang, L., Huang, Y., Lu, C.-T., Ravada, S. (eds.) SSTD 2017. LNCS, vol. 10411, pp. 3–23. Springer, Cham (2017). https://doi.org/10.1007/978-3-319-64367-0_1
10. Gionis, A., Lappas, T., Pelechrinis, K., Terzi, E.: Customized tour recommendations in urban areas. In: Proceedings of the 7th ACM International Conference on Web Search and Data Mining, pp. 313–322 (2014)
11. Goodman, L.A.: Snowball Sampling. The Annals of Mathematical Statistics, pp. 148–170 (1961)
12. Guo, Z., Tang, C., Niu, W., Fu, Y., Xia, H., Tang, H.: Beyond the aggregation of its members–a novel group recommender system from the perspective of preference distribution. In: Knowledge Science, Engineering and Management, pp. 359–370 (2017)
13. Hall, W., Tiropanis, T.: Web evolution and web science. Comput. Netw. **56**(18), 3859–3865 (2012)
14. Hu, Y., Koren, Y., Volinsky, C.: Collaborative filtering for implicit feedback datasets. In: Eighth IEEE International Conference on Data Mining, ICDM 2008, pp. 263–272. IEEE (2008)
15. Kocsis, L., Szepesvári, C., Willemson, J.: Improved Monte-Carlo search. University of Tartu, Estonia, Technical report 1 (2006)
16. Lim, K.H., Chan, J., Karunasekera, S., Leckie, C.: Personalized itinerary recommendation with queuing time awareness. In: Proceedings of the 40th International ACM SIGIR Conference on Research and Development in Information Retrieval (SIGIR 2017). Google Scholar (2017)
17. Lim, K.H., Chan, J., Leckie, C., Karunasekera, S.: Personalized trip recommendation for tourists based on user interests, points of interest visit durations and visit recency. Knowledge and Information Systems, pp. 1–32 (2017)

18. Linden, G., Smith, B., York, J.: Amazon. com recommendations: item-to-item collaborative filtering. IEEE Internet Comput. **7**(1), 76–80 (2003)
19. Liu, X., Tian, Y., Ye, M., Lee, W.C.: Exploring personal impact for group recommendation. In: Proceedings of the 21st ACM International Conference on Information and Knowledge Management, pp. 674–683 (2012)
20. Luhn, H.P.: A statistical approach to mechanized encoding and searching of literary information. IBM J. Res. Dev. **1**(4), 309–317 (1957)
21. Qian, Y., Lu, Z., Mamoulis, N., Cheung, D.W.: P-LAG: location-aware group recommendation for passive users. In: Advances in Spatial and Temporal Databases, pp. 242–259 (2017)
22. Ricci, F., Rokach, L., Shapira, B.: Recommender Systems Handbook, 2nd edn. Springer Publishing Company, Incorporated, Heidelberg (2015)
23. Roy, S.B., Das, G., Amer-Yahia, S., Yu, C.: Interactive itinerary planning. In: 2011 IEEE 27th International Conference on Data Engineering (ICDE), pp. 15–26. IEEE (2011)
24. Xiao, L., Min, Z., Yongfeng, Z., Zhaoquan, G., Yiqun, L., Shaoping, M.: Fairness-aware group recommendation with pareto-efficiency. In: Proceedings of the Eleventh ACM Conference on Recommender Systems, pp. 107–115 (2017)
25. Yu, Y., Chen, X.: A survey of point-of-interest recommendation in location-based social networks. In: Workshops at the Twenty-Ninth AAAI Conference on Artificial Intelligence, vol. 130 (2015)
26. Yuan, Q., Cong, G., Lin, C.Y.: COM: a generative model for group recommendation. In: Proceedings of the 20th ACM SIGKDD International Conference on Knowledge Discovery and Data Mining, pp. 163–172 (2014)

A Knowledge-Based Weighted KNN
for Detecting Irony in Twitter

Delia Irazú Hernández Farías[1]([✉]), Manuel Montes-y-Gómez[1],
Hugo Jair Escalante[1], Paolo Rosso[2], and Viviana Patti[3]

[1] Instituto Nacional de Astrofísica, Óptica y Electrónica (INAOE),
Tonantzintla, Mexico
{dirazuherfa,mmontesg,hugojair}@inaoep.mx
[2] PRHLT Research Center, Universitat Politècnica de València, Valencia, Spain
prosso@dsic.upv.es
[3] Dipartimento di Informatica, University of Turin, Turin, Italy
patti@di.unito.it

Abstract. In this work, we propose a variant of a well-known instance-based algorithm: WKNN. Our idea is to exploit task-dependent features in order to calculate the weight of the instances according to a novel paradigm: the Textual Attraction Force, that serves to quantify the degree of relatedness between documents. The proposed method was applied to a challenging text classification task: irony detection. We experimented with corpora in the state of the art. The obtained results show that despite being a simple approach, our method is competitive with respect to more advanced techniques.

Keywords: Instance-based algorithm · WKNN · Irony detection

1 Introduction

Social media are nowadays an important communication channel where people express their opinions, thoughts, and ideas. Analyzing such kind of content is the main aim of Sentiment Analysis (SA). An important challenge to address in SA is the presence of figurative language devices such as irony [12]. The most common definition of irony refers to an utterance by which the words are used with the intention of communicating the opposite of what is literally said [10].

During the last years the interest in detecting the presence of ironic content in social media has grown significantly especially on Twitter. Several computational linguistics approaches have been proposed to deal with irony detection. Mainly, they take advantage of different features attempting to capture the presence of ironic content together with machine learning algorithms. For further details on state-of-the-art irony detection methods see [14]. Moreover, detecting the presence of irony has been the aim of some shared tasks in both languages English [7,22] and Italian [1,2].

© Springer Nature Switzerland AG 2018
I. Batyrshin et al. (Eds.): MICAI 2018, LNAI 11289, pp. 194–206, 2018.
https://doi.org/10.1007/978-3-030-04497-8_16

In this paper, we are proposing an approach for addressing irony detection that exploits a variation of WKNN. Our proposal is about a new schema for calculating the weights. We took advantage of knowledge-based features together with a novel paradigm called Textual Attraction Force (that allows us to calculate the relatedness between two documents) to assign the weights. With the purpose of evaluating the performance of the proposed model, we decided to apply it for detecting irony in Twitter. The performance of the proposed approach was assessed over a set of state-of-the-art corpora. The obtained results outperform those from well-known classifiers validating the usefulness of our method. It is worth mentioning that in spite of the fact that the proposed model is simple, the obtained results are quite encouraging when compared with both traditional classifiers and more sophisticated techniques such as deep learning.

Summarizing, the main contributions of this paper are: (i) We propose a new variant of WKNN that exploits knowledge-based features by means of the Textual Attraction Force in order to classify unlabeled instances according to a weighted voting KNN; and, (ii) We evaluated our approach on irony detection. A set of corpora of the state of the art was used for experimental purposes. The obtained results demonstrate the usefulness of our approach for irony detection.

The paper is organized as follows. Section 2 introduces the basis of the proposed approach. Section 3 describes the knowledge-based features exploited to characterize irony. Section 4 describes the experimental setting and results. Finally, Sect. 5 draws some conclusions.

2   A Knowledge-Based Weighted KNN

kNN (*k-Nearest Neighbors*) is one of the oldest and simplest classification approaches based on the use of the nearest neighbor rule. It assumes that the class of an unlabeled instance is assigned by a majority voting between the classes of its k nearest neighbors. kNN belongs to the family of the instance-based learning algorithms [15]; besides, it has several advantages: it is simple, effective, intuitive, and it has a competitive classification performance in many domains.

With the aim of improving kNN, different variations have emerged, where the main idea is to incorporate a weighting schema, namely WKNN. One approach, often known as *Distance-based WKNN (DWKNN)*, consists in weighting close neighbors more heavily according to their distances to the test instance [6,9].

In this paper, we propose a variation of the WKNN called Knowledge-based Weighted KNN (hereafter KBWKNN). We attempt to exploit a novel paradigm: the Textual Attraction Force (TAF), in order to calculate the weight of the instances. The idea behind TAF is to emulate the Newton's gravitational force where the attraction between two objects depends on their masses and the distance between them. The higher their masses and the lower their distance, the greater the attraction force between them is.

Current text classification approaches are based on finding similar documents, considering that all the instances have the same relevance for building a given classifier. TAF represents a change in the way by which the relationship between documents is calculated. By using TAF for text classification, we could evaluate the relationship between two documents by considering not only their similarity (for example in terms of vocabulary) but also their relevance (or mass). Therefore, defining the mass function is crucial for identifying similar documents. The mass (or relevance) of an instance refers to a subjective and dependent aspect of the problem in hand. In this paradigm, the main hypothesis is that there are differences regarding the relevance of the objects, i.e., some objects are more relevant than others. Therefore, relevant objects have the greatest influence during the classification phase. Given two documents (t_i and t_j), each one having a mass ($mass(t)$) associated with a particular aspect. The TAF between them is calculated as:

$$TAF(t_i, t_j) = \frac{mass(t_i) * mass(t_j)}{d(t_i, t_j)} \tag{1}$$

where: $mass(t)$ Is a mass function related to a particular aspect.
  $d(t_i, t_j)$ Is a distance metric between the documents.

As mentioned before, we are proposing KBWKNN, a variant of WKNN where the weights are determined by the TAF. We are using the term "knowledge-based" due to the fact that in this approach the information regarding the problem in hand is used in order to calculate the TAF between instances, and finally to determine the class of a given instance by considering a weighting schema.

The KBWKNN algorithm is as follows: Given a set of training instances $<t_i, f(t_i)>$, an unlabeled instance t_q, and the set of the k nearest neighbors to t_q (denoted as knn) in the training set, the class of t_q is determined as follows[1]:

$$f(t) \leftarrow \underset{c \in C}{argmax} \sum_{i=1}^{k} TAF(t_q, knn_i) * \partial(c, f(knn_i)) \tag{2}$$

Figure 1 shows a schematic example of the representation of an instance (symbolized as a star) to be classified by kNN in (a) and by using KBWKNN in (b). In both cases k takes as values 3 and 5. By kNN, the assigned class will be the one of the gray circles since in both cases it represents the majority class in the neighborhood. On the other side, in Fig. 1(b) each instance has a different size that represents its relevance in terms of a given particular aspect. The higher the magnitude of a circle, the greater the mass is. Then, the unlabeled instance will be assigned according to the class of the examples having a higher weight, i.e., the one represented by darker circles.

---

[1] Where $\partial(c, f(knn_i))$ returns 1 if $c = f(knn_i)$ and 0 otherwise.

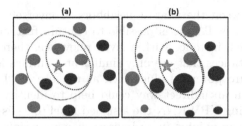

Fig. 1. Schematic representation of an unlabeled instance with its nearest neighbors.

3 KBWKNN for Detecting Irony in Twitter

Irony detection is considered as an special case of text classification, where the aim is to identify ironic texts from non-ironic ones. As mentioned before, different approaches have been proposed to deal with such a complex task. Here, we are proposing to perform irony detection by using the approach described in Sect. 2. It is important to highlight that this is the first time that the Textual Attraction Force paradigm is applied for detecting irony.

Let us to introduce an example[2]:

tw I absolutely LOVE moving house[3].

Nearest neighbors[4]
tw1 #elf #why not moving http://t.co/ZcrJaOwPqZ **
tw2 the_angry_ranga UberFacts For the masses. I love it. **
tw3 #forgive #others #because they #deserve #forgiveness #you #peace love this from @i2imovement #words #wisdom ... http://t.co/MjXdZCZDud **
tw4 love is bliss... **
tw5 Absolutely love waking up to snow *
tw6 I just love the NHS *
tw7 I'm in love with the cocoa **

Provided that *tw* is the instance to be classified. First, its k-nearest neighbors were identified (represented as *tw1 − tw7*). Supposing that a bag-of-words based representation is used, then the instances in the example are within the same neighborhood because they share two noticeable features: having at least a term in common ("love" or "moving") and they are composed by less than six tokens[5]. According to a kNN approach, the class of *tw* is "Nonirony", because of the

---

[2] This example is part of the obtained results when the aforementioned method was applied with a size of k = 7. All the tweets were extracted from dataset developed by [22].

[3] This tweet was labeled with the class "Irony".

[4] The real classes of these tweets are denoted as follows: "*" represents the class "Irony" while "**" is denoted as "Nonirony".

[5] Hashtags, mentions, emoji, and url were not considered in the bag-of-words model.

majority voting between the nearest neighbors. Conversely, taking advantage of a paradigm such the TAF it is possible to capture further information that allows to provide each instance with a given relevance, therefore, it is possible to discard those instances that are only similar at a surface level. For instance, being able to capture potential clues such as the use of "LOVE" to stress a subjective opinion in an indirect way could help to identify ironic content. In this case, by exploiting KBWKNN the assigned label to tw is "Irony", since the instances expressing an opinion in an indirect way are weighted more heavily.

In order to exploit KBWKNN for irony detection, it is needed to harness domain-related knowledge that allows us to capture specific aspects of the use of irony in Twitter. Attempting to take advantage of various factors that could be useful for characterizing irony in tweets, we defined a set of six different mass functions described below.

3.1   Mass Functions for Detecting Irony

Our set of mass functions aim to cover different aspects related to such interesting linguistic phenomenon. Below we introduce each of the mass functions evaluated.

- *Structural.* Unlike the case of spoken communication, where enunciation, stress and tone of voice help us to communicate effectively the sense of our utterances, in written communication it is needed to make use of lexical marks to point out the intended meaning of a text. In order to calculate the *mass_Structural.* of a tweet t, we considered the frequency of five aspects: Str_i: Uppercase characters; Str_{ii}: Words in uppercase; Str_{iii}: Punctuation marks[6]; Str_{iv}: Hashtag, mention, emoticons, and emoji; and, Str_v: Internet Slang[7] terms. Once we have these frequencies, we determine the value of *mass_Structural* as follows:

$$sumStr = Str_i + Str_{ii} + Str_{iii} + Str_{iv} + Str_v$$

$$mass\_Structural(t) = \begin{cases} sumStr + 1 & \text{if } sumStr > 0 \\ 1 & otherwise \end{cases} \quad (3)$$

- *Sentiment.* Irony can be used to reveal an evaluative judgment. Thus, the sentiment score of a tweet may help to characterize ironic instances. We attempt to capture such value focusing especially on the positive sense of each tweet. It has been recognized the important role of positive words for masking the ironic intention [11]. To determine the *mass_Sentiment* we used the Hu&Liu lexicon (HL) [13]. It is a well-known resource developed for opinion mining that includes more than six thousand of words divided into two groups: positive and negative. Formula 4 shows how to calculate the *mass_Sentiment.*

$$mass\_Sentiment(t) = 1 + \frac{Pos - Neg}{Len} \quad (4)$$

---

[6] We consider five different punctuation marks: ".", ",", ":", "!", and "?".

[7] We used a list of terms defined in https://en.wiktionary.org/wiki/Appendix:English_internet_slang.

where: *Pos*   It refers to the number of positive terms in the tweet t.
  Neg   It refers to the number of negative terms in t.
  Len   It refers to the length in words of t.

– *Emotions.* Affective information plays a key role for irony comprehension-communication. We defined two mass functions attempting to capture information related to emotions:

  • *Categorical Model of Emotions.* Several theories propose different sets of basic or fundamental emotions. In our approach, we adopted the eight basic emotions considered in the Plutchik model [18]: *anger, anticipation, disgust, fear, joy, sadness,* and *trust.* In particular, we took advantage of EmoLex [16], a lexical resource containing more than fourteen thousand words labeled according to the Plutchik's model of emotions. First, we compute *eCatScore* that captures how many words in the tweet t are associated to an emotion category in EmoLex, then *mass_EmotCat* is calculated as:

$$mass\_EmotCat(t) = \begin{cases} \log_{10}(eCatScore) + 1 & \text{if } eCatScore > 0 \\ 1 & Otherwise \end{cases} \quad (5)$$

  • *Dimensional Model of Emotions.* We employ information regarding dimensional models of emotions by taking advantage of SenticNet 3 (SN3) [5]. It contains 30,000 concepts associated with the four dimensions of the Hourglass of Emotions: Pleasantness (Pl), Attention (At), Sensitivity (Sn), and Aptitude (Ap). In [4], the authors propose a formula for calculating a polarity measure in terms of the affective dimensions in the Hourglass of Emotions. We decided to take advantage of this formula in order to calculate *mass_EmotDim*:

$$mass\_EmotDim(t) = \sum_{i=1}^{n} \frac{Pl(c_i) + |At(c_i)| - |Sn(c_i)| + Ap(c_i)}{3N} \quad (6)$$

  where: c_i Is an input concept
    N Is the total number of concepts of the tweet t

– *Lexical Cohesion.* Comprehending irony involves getting the literal sense of the words and then understanding the figurative intention behind them [8]. Often, a way to achieve an ironic sense in an utterance is to use words that are semantically unrelated. We attempt to quantify the degree of lexical cohesion (*mass_LexicalCohesion*) in a tweet by exploiting word embeddings-based[8] similarity scores between words in a sentence. We calculated the similarity score for each pair of words in a tweet, then the maximum value is kept. The final value is determined as follows:

$$mass\_LexicalCohesion(t) = 1 + \max_{i,j \in tweet} (sim(w_i, w_j)) \quad (7)$$

---

[8] We used the embeddings pre-trained on the Google News corpus. https://code.google.com/archive/p/word2vec/.

- *Cognitive Aspects.* Understanding irony involves different cognitive processes such as the ones devoted to text comprehension. Several factors influence such process. One of them is related to lexical sophistication. A measurable aspect related to this issue is "Word Concreteness", which can be calculated for each term in a sentence to indicate a ratio of how abstract or concrete a word is. In [21] Skalicky and Crossley, analyzing different features to identify satiri-cal[9] and non-satirical reviews, found that the language in satirical reviews was more concrete than in non-satirical ones, i.e., words with more specific meaning were used in satirical reviews. In other words, they observed that satirical sentences have higher levels of word concreteness. Inspired by those findings, we decided to experiment with such feature for detecting irony in tweets. We exploited a lexical resource developed by Brysbaert et al. [3]. First, we measure the overall word concreteness (denoted as *tweetConcreteness*) of a tweet as the sum of all the ratings of the words contained on it. Then, the mass_CognitiveAspects is calculated as follows:

$$mass\_CognitiveAspects(t) = 1 + tweetConcreteness \qquad (8)$$

4   Experiments and Results

4.1   Evaluation Datasets

For evaluation purposes, we took advantage of four different corpora for irony detection covering different aspects such as collection criteria and balance degree. Below we briefly introduce each of the corpora we used:

- *TwMohammad2015.* Mohammad et al. [17] collected a set of tweets labeled with hashtags pertaining to the 2012 US presidential elections. Those data where manually annotated considering several aspects such as: sentiment, emotions, purpose and style; the latter being the one including the presence of irony. The distribution of classes in *TwMohammad2015* is 532 ironic and 1,397 non-ironic tweets.
- *TwRiloff2013.* A set of more than three thousands of tweets compose the dataset created by Riloff et al. [20]. They followed a mixed approach consider-ing first a set of tweets tagged with the #sarcasm[10] and #sarcastic hashtags. Then, after removing the aforementioned hashtags, those tweets were man-ually annotated on the presence of sarcastic content. *TwRiloff2013* contains 474 ironic tweets and 1,689 nonironic ones.
- *TwReyes2013.* Reyes et al. [19] retrieved a set of tweets by using four hashtags: #irony, #education, #humor, and #politics. The first hashtag was used in order to get ironic instances relying on the idea that the author of a tweet

---

[9] Satire is strongly related to verbal irony, providing a detailed definition of such a concept is beyond of the scope of this work.

[10] In computational linguistics, irony is often considered as an umbrella term that covers also sarcasm.

self-annotate her ironic intention. *TwReyes2013* is composed by 40,000 tweets equally distributed in four classes[11].

– *TwVanHee2018*. In the framework of the *SemEval-2018 Task 3: Irony Detection in English tweets*[12] shared task, a dataset containing more than four thousand tweets was developed. Van Hee et al. [22] collected a set of tweets labeled with a set of hashtags: #irony, #sarcasm and #not. As a second step, the tweets were manually annotated attempting to minimize the noise. A total of 2,222 ironic and 2,396 nonironic tweets composed this dataset.

4.2 Experimental Setting

A preprocessing phase was applied to the data attempting to reduce the high dimensionality of the feature space. We filtered out all the mentions, hashtags, url, emoticons, emoji, and stop words for each tweet. Additionally, all data were converted to lowercase. Finally, each instance was represented as a binary bag-of-words vector. The vocabulary of each dataset was built according to a single criterion: we kept the words with a minimum frequency. In the case of *TwMohammad2015*, *TwRiloff2013*, and *TwVanHee2018* we considered all terms that appear more than twice in the training set. While in the *TwReyes2013* the minimum frequency was fixed as twenty.

In order to calculate the mass functions defined in Sect. 3.1, we used the original content of each tweet, i.e. any kind of preprocessing was applied during this phase. In this case it is crucial to avoid losing important information that could be discarded. For example by converting the text to lowercase, some of the factors considered for calculating *mass_Structural* cannot be captured.

We experimented with the mass functions previously defined but adding a criterion before calculating the TAF. The idea is to compensate the imbalance degree by assigning a greater weight for neighbors belonging to the minority class. For each mass described above we applied the following criterion: if the instance belongs to the minority class, the mass is recalculated as follows[13]:

$$massFuction\text{-}Modified(t) = e^{massFunction(t)} \qquad (9)$$

For the sake of the readability, the same acronyms defined in Sect. 3.1 will be used for introducing the obtained results.

In addition, we decided to combine all the masses together into a single one *mass_Combination* by means of the sum of all the masses *sumMass(t)* considering also the imbalance degree between the classes by means of the amount of instances per class (denoted as *nClass*).

$$mass\_Combination = \begin{cases} sumMass(t) * \frac{nClass_i * 100}{nClass_i + nClass_j} & \text{if } t \in Class_j \\ sumMass(t) * \frac{nClass_j * 100}{nClass_i + nClass_j} & \text{if } t \in Class_i \end{cases} \qquad (10)$$

---

[11] We performed three different binary classifications by combining each of the nonironic classes with the ironic one. From now on, these experiments will be referred as *TwReyes2013-Edu*, *TwReyes2013-Hum*, and *TwReyes2013-Pol*.

[12] https://competitions.codalab.org/competitions/17468.

[13] Where *massFunction* can be any of the functions defined in Sect. 3.1.

We experimented with three variations of the k-nearest neighbor classifier: kNN, DWKNN (we used the inverse of the distance for calculating the weights), and KBWNN. Concerning the size of k, we assessed the performance of our proposed approach with three different values: 3, 5, and 7. As distance function for finding the neighbors we used the Cosine Distance calculated as $D(t_i, t_j) = 1 - simCos(t_i, t_j)$.

Furthermore, for comparison purposes we also experimented with standard classifiers previously used in irony detection. The Scikit-learn implementation of Naïve Bayes (NB), Decision Tree (DT), and Support Vector Machine (SVM)[14] was used. All the experiments were carried out in a 5-fold cross-validation setting.

4.3 Results

In Table 1 we present the results achieved by applying different classification approaches. Underlined values are used to point out that the achieved outcome is higher than using one of the standard classifiers. Bold values highlight the best obtained result for the "Irony" class in each dataset.

We compared the performance of our approach against to standard classifiers. There are many cases where KBWKNN outperforms at least one classifier in terms of Macro F-score. Often, the obtained results with our method overcome those of DT. Following the evaluation schema of the SemEval-2018 Task 3, we also present the results in terms of the class of interest, i.e., the ironic one. All the experiments involving each individual mass in KBWKNN outperforms the outcomes achieved by kNN and DWKNN in most of the corpora used for evaluation purposes. On the other hand, some of these results also improve the performance of the other classifiers.

Concerning to *TwMohammad2015* and *TwVanHee2018*, the best performance for the class of interest achieved by exploiting the *mass_CognitiveAspects*. In the case of *TwRiloff2013*, and *TwReyes2013*, the best performance was obtained when all the masses were combined into a single one. Overall, the best improvement in terms of the ironic class is observed on the experiments involving data where a crowd-sourcing process is part of the corpora construction. Similarly, a difference was observed in [11], where the proposed irony detection model was evaluated with different state-of-the-art corpora featured by different collection and annotation methodologies, and the performance was different depending on the methodology applied for developing the corpora. Such difference represents an interesting aspect that deserves to be further investigated. It is possible that the annotation methodology exploited for developing corpora for irony detection affects the consistency of data, especially when we compare, on the one hand, corpora developed via self-tagging, where irony-related hashtags used by ironists to express their intention to be ironic are taken as class labels, and, on the other hand, manually annotated corpora, which involve external annotators tagging the ironic intention of tweets written by others.

---

[14] The default configuration of parameters in the classifiers was applied.

Table 1. Results obtained in macro F-score and F-score for the "Irony" class.

Method	k	TwMohammad2015 F-score	F_{Irony}	TwRiloff2013 F-score	F_{Irony}	TwVanHee2018 F-score	F_{Irony}	TwReyes2013-Edu F-score	F_{Irony}	TwReyes2013-Hum F-score	F_{Irony}	TwReyes2013-Pol F-score	F_{Irony}
Naive Bayes		0.529	0.257	0.637	0.403	0.612	0.597	0.865	0.814	0.851	0.858	0.895	0.896
Decision Tree		0.522	0.279	0.577	0.33	0.582	0.565	0.83	0.824	0.821	0.818	0.884	0.882
Support Vector Machine		0.539	0.311	0.645	0.435	0.601	0.57	0.862	0.863	0.858	0.861	0.907	0.909
kNN	3	0.519	0.242	0.597	0.339	0.574	0.53	0.82	0.814	0.83	0.823	0.884	0.882
kNN	5	0.511	0.211	0.589	0.314	0.584	0.546	0.825	0.82	0.832	0.825	0.884	0.882
kNN	7	0.505	0.18	0.594	0.319	0.594	0.557	0.83	0.825	0.834	0.828	0.885	0.883
DWKNN	3	0.522	0.247	0.596	0.338	0.578	0.534	0.827	0.822	0.839	0.832	0.887	0.886
DWKNN	5	0.511	0.212	0.585	0.309	0.588	0.55	0.833	0.828	0.839	0.833	0.887	0.886
DWKNN	7	0.505	0.181	0.588	0.311	0.596	0.558	0.835	0.831	0.839	0.833	0.884	0.884
KBWKNN mass_CognitiveAspects	3	0.49	0.387	0.576	0.415	0.534	0.641	0.781	0.793	0.798	0.789	0.859	0.87
KBWKNN mass_CognitiveAspects	5	0.432	0.415	0.534	0.425	0.478	0.654	0.733	0.759	0.758	0.746	0.826	0.847
KBWKNN mass_CognitiveAspects	7	0.373	0.419	0.479	0.415	0.441	0.655	0.69	0.73	0.732	0.718	0.8	0.83
KBWKNN mass_Sentiment	3	0.497	0.388	0.58	0.412	0.543	0.641	0.8	0.807	0.815	0.805	0.871	0.879
KBWKNN mass_Sentiment	5	0.532	0.348	0.617	0.428	0.568	0.638	0.816	0.821	0.819	0.812	0.88	0.886
KBWKNN mass_Sentiment	7	0.527	0.384	0.595	0.42	0.535	0.652	0.801	0.809	0.808	0.798	0.87	0.878
KBWKNN mass_Structural	3	0.497	0.374	0.59	0.419	0.54	0.622	0.794	0.802	0.807	0.798	0.865	0.874
KBWKNN mass_Structural	5	0.455	0.404	0.577	0.436	0.502	0.645	0.772	0.788	0.785	0.776	0.845	0.86
KBWKNN mass_Structural	7	0.421	0.406	0.55	0.436	0.473	0.652	0.745	0.77	0.765	0.756	0.828	0.848
KBWKNN mass_LexicalCohesion	3	0.492	0.385	0.581	0.415	0.543	0.64	0.802	0.809	0.816	0.806	0.871	0.88
KBWKNN mass_LexicalCohesion	5	0.526	0.349	0.618	0.432	0.567	0.638	0.814	0.82	0.818	0.81	0.879	0.885
KBWKNN mass_LexicalCohesion	7	0.517	0.39	0.595	0.427	0.353	0.652	0.8	0.808	0.807	0.798	0.87	0.879
KBWKNN mass_EmotCat	3	0.505	0.377	0.59	0.422	0.555	0.641	0.803	0.811	0.819	0.811	0.875	0.882
KBWKNN mass_EmotCat	5	0.518	0.374	0.612	0.432	0.558	0.637	0.81	0.817	0.817	0.809	0.877	0.884
KBWKNN mass_EmotCat	7	0.519	0.372	0.599	0.426	0.547	0.654	0.803	0.813	0.812	0.804	0.872	0.879
KBWKNN mass_EmotDim	3	0.496	0.387	0.583	0.418	0.541	0.641	0.8	0.808	0.815	0.805	0.871	0.879
KBWKNN mass_EmotDim	5	0.532	0.348	0.618	0.428	0.567	0.637	0.816	0.821	0.82	0.812	0.88	0.886
KBWKNN mass_EmotDim	7	0.525	0.389	0.603	0.43	0.538	0.653	0.8	0.809	0.808	0.798	0.87	0.878
KBWKNN mass_Combination	3	0.513	0.38	0.586	0.416	0.574	0.533	0.827	0.823	0.839	0.833	0.885	0.884
KBWKNN mass_Combination	5	0.522	0.371	0.597	0.439	0.582	0.545	0.835	0.832	0.839	0.834	0.888	0.888
KBWKNN mass_Combination	7	0.517	0.372	0.583	0.428	0.586	0.55	0.836	**0.835**	0.842	**0.837**	0.887	**0.888**

We are also interested in comparing the performance of the proposed approach with the state of the art. Table 2 shows such information. Previous results by exploiting an irony detection model (called "emotIDM") on the *TwMohammad2015*, *TwRiloff2013*, and *TwReyes2013* corpora are found in [11][15]. The proposed approach, KBWKNN outperforms the results of *TwMohammad2015* and *TwRiloff2013*. Conversely, in the case of the experiments related to the *TwReyes2013*, our proposed approach did not achieve higher performance than "emotIDM". Regarding the comparison with *TwVanHee2018*, we reported the three best official results achieved during the shared task.

Deep learning-based approaches were exploited by the three best ranked systems[16]. As it can be noticed, our three best results are higher than the one obtained by the 3rd best ranked system. It serves to validate that even if our method is simple, it is able to obtain competitive results against more sophisticated techniques.

Table 2. Comparison of our results with the state of the art.

	emotIDM	Our approach		
Dataset	SVM	Mass1	Mass2	Mass3
TwMohammad2015	0.011	0.419	0.415	0.406
TwRiloff2013	0.134	0.439	0.436	0.432
TwReyes2013-Edu	0.892	0.835	0.832	0.823
TwReyes2013-Hum	0.89	0.837	0.834	0.833
TwReyes2013-Pol	0.888	0.888	0.886	0.885
TwVanHee2018		Our approach		
Ranking position	F_{Irony}		Mass	F_{Irony}
1st	0.705		Mass1	0.655
2nd	0.671		Mass2	0.654
3rd	0.650		Mass3	0.653

5   Conclusions

In this paper, we introduce a variation of WKNN by exploiting knowledge-based information together with a novel paradigm, called Textual Attraction Force. The proposed approach was evaluated in an special case of text classification: irony detection. This is the first time that such a complex task is addressed by

---

[15] The authors reported the performance of their model in terms of F-measure considering both classes together. Attempting to compare our results, we carried out experiments by exploiting the aforementioned model but instead of considering an overall performance, we are reporting only the performance in terms of the ironic class.

[16] For further details on the shared task see [22].

exploiting this kind of approach. We have performed several experiments over a set of state-of-the-art corpora. Across most of the experiments carried out, it can be concluded that using knowledge-based information for calculating the TAF between two instances (and then using this value as the weight of the instances in KBWKNN), despite being a simple model exploiting a traditional representation (bag-of-words) together with domain-dependent features not only improves the classification performance of well-known machine learning algorithms for irony detection, but also validates the usefulness of using novel paradigms (more intuitive and easier to interpret) to find similar documents in text classification related tasks. As future work, it could be interesting to further explore different ways to calculate the mass functions as well as comparing our results against deep learning techniques.

Acknowledgments. This research was funded by CONACYT project FC 2016-2410. The work of P. Rosso has been funded by the SomEMBED TIN2015-71147-C2-1-P MINECO research project. The work of V. Patti was partially funded by Progetto di Ateneo/CSP 2016 (IhatePrejudice, S1618_L2_BOSC_01).

References

1. Barbieri, F., Basile, V., Croce, D., Nissim, M., Novielli, N., Patti, V.: Overview of the Evalita 2016 sentiment polarity classification task. In: Proceedings of Third Italian Conference on Computational Linguistics, vol. 1749. CEUR-WS.org (2016)
2. Basile, V., Bolioli, A., Nissim, M., Patti, V., Rosso, P.: Overview of the Evalita 2014 sentiment polarity classification task. In: Proceedings of the First Italian Conference on Computational Linguistics, pp. 50–57 (2014)
3. Brysbaert, M., Warriner, A.B., Kuperman, V.: Concreteness ratings for 40 thousand generally known English word lemmas. Behav. Res. Met. **46**(3), 904–911 (2014)
4. Cambria, E., Hussain, A.: Sentic Computing, vol. 1. Springer, Cham (2015). https://doi.org/10.1007/978-3-319-23654-4
5. Cambria, E., Olsher, D., Rajagopal, D.: SenticNet 3: a common and common-sense knowledge base for cognition-driven sentiment analysis. In: Proceedings of AAAI Conference on Artificial Intelligence, pp. 1515–1521 (2014)
6. Dudani, S.A.: The distance-weighted k-nearest-neighbor rule. IEEE Trans. Syst., Man, Cybern. SMC **6**(4), 325–327 (1976)
7. Ghosh, A., et al.: SemEval-2015 task 11: sentiment analysis of figurative language in Twitter. In: Proceedings of the 9th International Workshop on Semantic Evaluation, pp. 470–478 (2015)
8. Giora, R., Fein, O.: Irony: context and salience. Metaphor. Symb. **14**(4), 241–257 (1999)
9. Gou, J., Du, L., Zhang, Y., Xiong, T.: A new distance-weighted k-nearest neighbor classifier. J. Inform. Comp. Sci. **9**(6), 1429–1436 (2012)
10. Grice, H.P.: Logic and conversation. In: Cole, P., Morgan, J.L. (eds.) Syntax and Semantics: Volume 3: Speech Acts, pp. 41–58. Academic Press, San Diego (1975)
11. Hernández Farías, D.I., Patti, V., Rosso, P.: Irony detection in Twitter: the role of affective content. ACM Trans. Internet Technol. **16**(3), 19:1–19:24 (2016)

12. Hernández Farías, D.I., Rosso, P.: Irony, sarcasm, and sentiment analysis. chapter 7. In: Pozzi, F.A., Fersini, E., Messina, E., Liu, B. (eds.) Sentiment Analysis in Social Networks, pp. 113–127. Morgan Kaufmann (2016)
13. Hu, M., Liu, B.: Mining and summarizing customer reviews. In: Proceedings of the 10th SIGKDD International Conference on Knowledge Discovery and Data Mining, pp. 168–177 (2004)
14. Joshi, A., Bhattacharyya, P., Carman, M.J.: Automatic sarcasm detection: a survey. ACM Comput. Surv. **50**(5), 73:1–73:22 (2017)
15. Mitchell, T.M.: Machine learning and data min. Com. ACM **42**(11), 30–36 (1999)
16. Mohammad, S.M., Turney, P.D.: Crowdsourcing a word-emotion association lexicon. Comput. Intell. **29**(3), 436–465 (2013)
17. Mohammad, S.M., Zhu, X., Kiritchenko, S., Martin, J.: Sentiment, emotion, purpose, and style in electoral tweets. Inf. Process. Manag. **51**(4), 480–499 (2015)
18. Plutchik, R.: The nature of emotions. Am. Sci. **89**(4), 344–350 (2001)
19. Reyes, A., Rosso, P., Veale, T.: A multidimensional approach for detecting irony in Twitter. Lang. Resour. Eval. **47**(1), 239–268 (2013)
20. Riloff, E., Qadir, A., Surve, P., Silva, L.D., Gilbert, N., Huang, R.: Sarcasm as contrast between a positive sentiment and negative situation. In: Proceedings of the Conference on Empirical Methods in Natural Language Processing, pp. 704–714. ACL (2013)
21. Skalicky, S., Crossley, S.: A statistical analysis of satirical Amazon.com product reviews. Eur. J. Humour Res. **2**, 66–85 (2015)
22. Van Hee, C., Lefever, E., Hoste, V.: SemEval-2018 task 3: irony detection in English tweets. In: Proceedings of the 12th International Workshop on Semantic Evaluation, SemEval-2018. ACL, June 2018

Model for Personality Detection Based on Text Analysis

Yasmín Hernández[1(✉)], Carlos Acevedo Peña[2], and Alicia Martínez[2]

[1] Instituto Nacional de Electricidad y Energías Limpias, Gerencia de
Tecnologías de la Información, Reforma 113, 62490 Cuernavaca, Mexico
yasmin.hernandez@ineel.mx
[2] Tecnológico Nacional de México, CENIDET, Interior Internado Palmira,
62490 Cuernavaca, Mexico
{carlos.acevedo,amartinez}@cenidet.edu.mx

Abstract. Personality is a unique trait which distinguish people from each
other. It is a set of individual differences in thinking, feeling and behaving of
people, and it affects interaction, relationships and environment of people.
Personality can be useful to several tasks like education, training, marketing and
personnel recruitment. Several methods to detect personality have been proposed and there are several psychological models proposing different personality
dimensions. Previous research states that personality can be detected by means
of text analysis. We have built a model for personality detection based on
statistical analysis of language and DISC model. As fundamental components of
the model, we built a linguistic corpus with personality annotations and a corpus
of words related to personality. To build the model, we conducted a study where
120 individuals participated. The study consisted in filling a personality test and
writing some paragraphs. We trained several machine learning algorithms with
data from the study, and we found Sequential Minimal Optimization algorithm
achieved best results in classification.

Keywords: DISC model · Personality linguistic corpus · Machine learning
Personality detection · Text analysis

1 Introduction

Personality refers to individual differences in characteristic patterns of thinking, feeling
and behaving [1]. To know personality is a way to understand how the diverse parts of
a person come together as a whole, since it is a combination of characteristics and
behavior of an individual when dealing with different situations [2]. Moreover, personality can influence people choices in several things such as websites, books, music
and films [3]. Personality affects the way we interact with other people, our relationships and the environment around us. Personality has been shown to be relevant to
many types of interactions, and also to be useful in predicting job satisfaction, professional relationships success, and even preference for different user interfaces [4].

Personality is important for several processes such as: personnel recruitment,
psychological therapies conduction, tutoring o teaching students, and health advising,

I. Batyrshin et al. (Eds.): MICAI 2018, LNAI 11289, pp. 207–217, 2018.
https://doi.org/10.1007/978-3-030-04497-8_17

among others. Therefore, several applications could benefit from personality insights. That is why many organizations are paying attention to know of personality.

Previous work on personality and interfaces showed that users are more receptive and have more confidence in interfaces and information presented from the perspective of their own personality traits. Namely, introvert people prefer messages presented from the perspective of an introvert individual. If the personality of a user can be predicted from their social media profile, online marketing among other applications can use it to personalize their messages and their presentation [4].

Several models of personality have been proposed, such as the Big Five model [5], the PEN model [6] or the DISC model [7, 8]. Typically, in order to identify personality, it is necessary for the individual to undergo a psychological assessment or a personality test based on a personality model.

Researchers have tried to obtain information about the personality of people through direct means such as the revised Eysenck personality questionnaire [9] which mostly consist in items that individuals try to apply themselves. However, indirect methods, such as linguistic analysis, can be used to detect personality [4]. Several approaches proposing indirect methods use semi-supervised multi-label classification [10], data mining techniques, machine learning, with demographic and text attributes [11].

Since personality is considered to be stable over time and throughout different situations, specialized psychologists are able to infer the personality profile of a subject by observing the subject's behavior. One of the sources of knowledge about the behavior of individuals is written text [9].

Much research in personality prediction has been conducted, however most of them are focused on English language and they are based on the Big Five model. We are interested in predict personality through Spanish text analysis and with base on the DISC model of personality. Therefore, we have developed a model to detect personality through text analysis considering DISC personality model.

We decided to use DISC model of personality because it is considered a simple model since it needs short time to assess, the result can be obtained easily, and can provide adequate information regardless if people conducting the survey are knowledgeable in psychology [12].

Linguistic resources are a fundamental part of our model. We built a set of Spanish personality words and a linguistic corpus with DISC personality annotations for Spanish. These resources will be useful in the identification of the personality through the analysis of texts. In order to build the linguistic resources, we conducted a study of personality with 120 people participating. The study consisted on to have participants answering a general questionnaire and the DISC test, and also writing a text on a general topic they selected. Answers and text were handwritten.

The rest of the paper is organized as follows: Sect. 2 presents background and related work, Sect. 3 describes briefly the DISC personality model, Subsect. 3.3 presents our approach to build the corpus, and finally conclusions and future work are presented in Sect. 4.

2 Background and Related Work

Personality is a very important player in many tasks and fields. Consequently, there is a need to know the personality of a person in easy ways which allow to make decisions promptly. Several psychological models of personality have been proposed, such as the Big Five model [5], the PEN model [6] or the DISC model [7, 8]. The most studied and used model is the Big Five model.

We are interested in detecting personality through linguistic analysis and in the DISC personality model. In this section, we briefly describe DISC model of personality, and also we present some relevant and recent research in personality detection.

2.1 Background

The DISC model of personality [7, 8] proposes four dimensions or unique characteristics of the personality. The four personality traits proposed by DISC Model are: *Dominance*, *Influence*, *Steadiness*, and *Compliance* [13], and they denote the basic behavioral styles. The first dimensions represent assertiveness and receptiveness, and the other two axes represent openness and control. The personality of individual lies between these dimensions [8, 13]. Although these traits represent existing characteristics in every person in certain extent, there is a predominant factor which explain the personality.

A high *Dominance* factor describes someone with an independent attitude and a motivation to succeed on their own terms. High-D's people have the strength to work well under pressure, and are always ready to take on responsibility [8, 14, 15].

When *Influence* stands out as a major factor in a DISC profile, that profile describes someone with a positive attitude to other people, and the confidence to demonstrate that attitude. Influential people are comfortable in social situations, and interact with other people in an open and expressive way [8, 14, 15].

Steadiness is related to natural pace of a person and her reactions to changes. This factor describes a reticent and careful person. Compared to *Dominance* or *Influence*, a person whose major factor is *Steadiness* will tend to be far less open or direct. Usually Steady people respond to events, rather than take pro-active steps themselves. As it implies, steady people are consistent and reliable in their approach. Indeed, they prefer to operate in situations that follow established patterns, and to avoid unplanned developments. Because of this, people with high *Steadiness* tend to be quite resistant to change, and will take time to adapt to new situations [8, 14, 15].

Compliance factor is connected to accuracy, organization and attitudes to authority. A person who shows high compliance in their DISC profile has a concern for practicality and detail. The key to this factor lies in attitudes to authority. High-C's are concerned with working within rules, and they are often described as rule-oriented. They are also concerned with accuracy and structure, and understanding the ways things work [8, 14, 15].

In this model, personality is established by a test, which is answered by individuals. The test consists of 28 groups of four adjectives. In order to assess the personality, individuals have to choose the adjective that identifies them the most and the adjective that identifies them the less.

Besides the dominant factor, the personality of an individual can be assigned to one of 15 classic personality combinations or patterns proposed by DISC Model: *developer, result oriented, inspirational, creative, promoter, persuader, counselor, appraiser, specialist, achiever, agent, investigator, objective thinker, perfectionist,* and *practitioner.*

2.2 Related Work

Social networks have enable people to express themselves in many ways. Mainly, people write what they think and what they do. In this sense, a way to identify the personality has been through the analysis of the behavior of users of social networks. For example, in a study some features are introduced to capture information about the social behavior of a set of users of *Twitter.* This study found a correlation between these characteristics and personality traits from Big Five Model. Sixty people participated in this study [16].

A recent study explores the relationship between personality and content preferences of books [17].

An important work is the dataset called *Essay.* This dataset consists of flow of consciousness essays written by volunteers in a controlled environment related with Big Five personality traits. Authors used the Linguistic Inquiry and Word Count (LIWC) functions to determine the correlation between dataset essays and personality [5, 18, 19].

In order to identify the personality through the analysis of texts, linguistic resources are a primordial component. In this way, several linguistic corporá have been constructed. For example, MRC sociolinguistic database is a database containing 150,000 words and their linguistic and psychological characteristics [10].

Other research focusing text classification to predict personality analized English and Indonesian tweets. The result is the MyPersonality corpus which consists of 10,000 updates of 250 users. Users are labeled in Big Five personality dimensions [2].

In a recent work, the method to extract personality traits from stream of consciousness essays using a convolutional neural network is proposed [20].

Another important research built a 200,000-word corpus and it is used for predicting the personality of the author of a text, as a result the attribution of authorship in 145 authors achieves results with an accuracy of 50% [21].

Unlike of aforementioned research, which work with texts in English and they are based on the Big Five personality model, in this paper, we propose a model based on the DISC model of personality to detect personality in Spanish texts.

3 Model for Personality Detection

In order to detect personality, we propose a model based on DISC model of personality and linguistic analysis. Proposed model comprises: (i) a set of words denoting personality, (ii) a corpus of Spanish texts labeled with a personality dimension of DISC model, and (iii) a machine learning algorithm for classification. Figure 1 shows a block diagram representing the model.

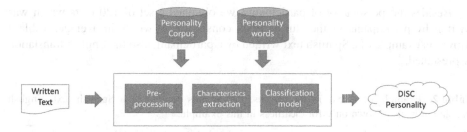

Fig. 1. Model for personality detection. Personality words in a text. Each verb and adjective is assigned with TF-IDF weights accordingly with the personality words set. This process is a part of the construction of the linguistic corpus and it is repeated for each text gathered in the study.

In order to build the model and linguistic resources, we conducted a study for knowing personality of participants and to relate it with writing features. In this section we describe our approach to build the model for personality detection.

3.1 Personality Study to Gather Data

The basis of the proposed model is a personality study. The study consisted of three parts. In the first part, participants answered questions about personal information, such as: gender, age, schooling, marital status, occupation, preferred social networks, and number of friends in such social networks. In the second part, participants filled out the test of the DISC personality model. In the third part, participants hand wrote some paragraphs about any topic. They were free to choose any topic, however some topics were suggested such as hobbies and what they did last day. These written texts are the basis to construct the model and the linguistic resources.

We had 120 college students participating in the study, 49 females and 71 males, they are between 20 and 30 years old. We asked them for age range rather than precise age. The questionnaires and texts were hand written by participants. Texts were transcript to electronic texts. After evaluating the personality tests, we obtained the following results: the most frequent personality factor was *Steadiness*, with 61 people, the second more frequent factor was *Influence* with 25 people, the third more frequent factor was *Compliance* with 19 people, and the less frequent is the factor *Dominance* with 15 people. Table 1 shows the resultant personality for the 120 participants.

Table 1. Personality results in the study. Participants answered a personality test based on the DISC model. The most pronounced factor for each participant is shown.

DISC factors	Female	Male	Total
Dominance	8	7	15
Influence	9	16	25
Steadiness	25	36	61
Compliance	7	12	19
	49	71	120

Besides the personality of participants, we obtained a set of 120 texts which was written by participants in the study. Texts consist of 90 words in average. Table 2 shows an example of a Spanish text written by a participant, also the English translation is presented.

Table 2. Example of a text in the corpus. Participants handwrote a Spanish text. English translation is also shown only for clearness in this example.

Spanish text written by a participant	English translation
"Mi pasatiempo favorito es tocar mi saxofón, me gusta la música, también me gusta ensayar con pistas, cantar. Mis principales metas son: terminar mis estudios, ahora me encuentro cursando el 9o semestre de la carrera de ingeniería en sistemas computacionales en el Instituto Tecnológico de Iguala. El día de ayer hice un poco de aseo en la casa, me gusta tener las cosas en orden"	"My favorite hobby is to play saxophone, I like music, I also like to practice with tracks, to sing. My main goals are: to finish my studies, now I am studying the 9th semester of the computer systems engineering career at the Technological Institute of Iguala. Yesterday, I did some cleaning in the house, I like to have things in order"

As we mention this study is the basis to build linguistic resources, for the time being, we only use the personality and the written text. The demographic data will be use in a future analysis.

3.2 Personality Words Set

Words that people choose when write or speak can describe their personality [2]. In this way, when people write, they express themselves in different ways from other people, and these differences correspond to their very individual personality traits and moods [9].

As follows, we built a set of words which denotes personality by analyzing all the texts gathered in the study. Participant's texts were grouped by personality dimension; in this way, instead to have 120 texts, we have four documents. The analysis includes to count how many verbs and adjectives appear in each document. Stop words were eliminated and all words were lemmatized. Personality words were chosen according their number of occurrences in the texts and their TF-IDF weights.

Term frequency–inverse document frequency, TF-IDF, is a numerical statistic that is intended to reflect how important a word is to a document in a collection of documents. This value increases proportionally to the number of times a word appears in the document and is offset by the frequency of the word in the collection, which helps to adjust for the fact that some words appear more frequently in general [22].

After the analysis, we have a personality words set consisting of 505 words, 346 verbs and 159 adjectives, and these words are labeled with weights TF-IDF for each personality dimension, which is the importance of that word in the personality document. Table 3 shows some examples from of the personality words set.

Table 3. Examples of personality words. The complete set includes 505 words: 346 verbs and 159 adjectives. The importance of the word (TF-IDF) and the occurrences of that word for each personality factor is shown.

Spanish word	English translation	Type	Occurrences				TF-IDF weights			
			D	I	S	C	D	I	S	C
Favorito	Favorite	Adjective	4	7	25	5	0.0076	0.0059	0.0113	0.0069
Feliz	Happy		2	1	4	1	0.0038	0.0008	0.0018	0.0014
Principal	Main		3	14	25	4	0.0057	0.0118	0.0113	0.0055
Estudiar	To study	Verb	1	4	10	2	0.0019	0.0034	0.0045	0.0027
Gustar	To like		4	15	24	9	0.0076	0.0127	0.0108	0.0124
Sentir	To feel		3	3	3	1	0.0057	0.0025	0.0014	0.0014

3.3 Personality Corpus Construction

The personality word set is used in the construction of the corpus of personality. Separately, we analyzed the 120 texts from study participants. We looked for verbs and adjectives which are in the personality words. We assign to each verb and adjective their TF-IDF weights for each personality dimension as shown in Fig. 2.

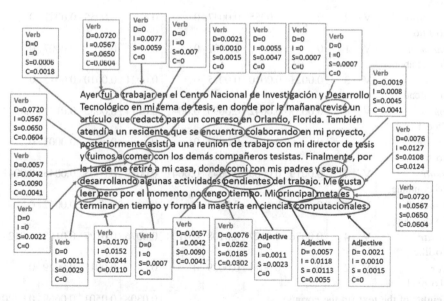

Fig. 2. Personality words in a text. Each verb and adjective is assigned with TF-IDF weights accordingly with the personality words set. This process is a part of the construction of the linguistic corpus and it is repeated for each text gathered in the study.

In this way, every verb and adjective in each text was assigned with their TF-IDF weights within the personality words set. A weight for each personality dimension is calculated. Then, all the TF-IDF weights of these words are added to obtain the TF-IDF weights for that text in the corpus. Table 4 shows the process of calculating of weights for a text.

Table 4. Resultant TF-IDF weights for each word in a text. These weights are used for obtaining TF-IDF weights in the corpus. A weight for each personality dimension is calculated.

Word-lemma	Tp	Occ	Word weight in personality doc				Word weight in current text			
			D	I	S	C	D	I	S	C
Principal -Main	A	4	0.0057	0.0118	0.0113	0.0055	0.0057	0.0118	0.0113	0.0055
Computacional -Computer	A	3	0.0021	0.0010	0.0015	0	0.0021	0.0010	0.0015	0
Trabajar -To work	V	2	0	0.0077	0.0059	0	0	0.0077	0.0059	0
Revisar -To check	V	1	0	0	0.0007	0	0	0	0.0007	0
Redactar -To write	V	1	0	0	0.0007	0	0	0	0.0007	0
Atender -To serve	V	2	0	0	0.0006	0.0018	0	0	0.0006	0.0018
Encontrar -To find	V	2	0	0.0055	0.0047	0	0	0.0055	0.0047	0
Colaborar -To collaborate	V	1	0	0	0.0007	0	0	0	0.0007	0
Asistir -To attend	V	3	0.0021	0.0010	0.0015	0	0.0021	0.0010	0.0015	0
Comer -To eat	V	4	0.0057	0.0042	0.0090	0.0041	0.0114	0.0085	0.0181	0.0082
Retirar -To live	V	1	0	0	0.0007	0	0	0	0.0007	0
Seguir -To continue	V	4	0.0019	0.0008	0.0045	0.0041	0.0019	0.0008	0.0045	0.0041
Desarrollar -To develop	V	1	0	0	0.0022	0	0	0	0.0022	0
Gustar -To like	V	4	0.0076	0.0127	0.0108	0.0124	0.0076	0.0127	0.0108	0.0124
Leer - To read	V	2	0	0.0011	0.0029	0	0	0.0011	0.0029	0
Weights of the text (in the corpus)							0.0308	0.0501	0.0669	0.0320

This process was repeated for each word in the text. After this process, we had a linguistic corpus with personality annotations. Is it worth to mention, we had several versions of the corpus, until we get a corpus which allows a better performance within the classification model described in the next subsection. We found verbs that are repeated many times behaves as a stop-word, therefore we eliminate them from the personality words set, examples are: to go, to finish and to play. Each entry in our final corpus consists of word, number of times the word appeared in each personality document, TF-IDF weights for each personality dimension and the personality. Some examples of entries in the corpus are shown in Table 5.

Table 5. Examples of Corpus entries. The corpus consists of 120 texts labeled with their DISC personality. Written texts were gathered via a study.

No. of words	Dominance TF-IDF	Influence TF-IDF	Steadiness TF-IDF	Compliance TF-IDF	Personality label
86	0.1173	0.0739	0.0812	0.0860	Dominance
59	0.1099	0.0681	0.0841	0.0792	Dominance
81	0.0571	0.1135	0.0811	0.0472	Influence
107	0.0955	0.1857	0.1505	0.1075	Influence
71	0.0436	0.0472	0.0797	0.0464	Steadiness
88	0.0766	0.0776	0.1078	0.0604	Steadiness
67	0.0413	0.061	0.0637	0.1325	Compliance
76	0.0966	0.0970	0.0963	0.1238	Compliance

3.4 Classification Model

As we have mention, we want to build a model for personality detection. We base our proposal on data gathered by means of study and rely on machine learning techniques. We evaluated several classifier algorithms through several metrics.

Machine learning is devoted to construct algorithms and models that can learn from data and make data-driven predictions. These models are built from sample inputs. A classifier algorithm is a machine learning algorithm which looks for relationships between unlabeled objects and a set of correctly labeled objects in order to correctly classify the unlabeled objects [23]. In literature, there are several classifier algorithms to match with available data and applications. In order to build our model, we evaluated the algorithms: ZeroR, k-Nearest-Neighbor, Naive Bayes, DTable, RepTree, Support Vector Machines, and Multilayer Perceptron.

The Support Vector Machines (SVM) algorithm achieved the best performance in the classification process over the other algorithms. A stratified ten times ten-fold cross-validation technique was used because it is the standard evaluation technique in situations where only limited data is available [24], we also used percentage split. We evaluated such algorithms by means of accuracy, recall, and F-measure metrics. We compare these metrics in order to select the best performance model. We used several versions of the corpus until we achieve the best performance in classification. The accuracy, recall, and F-measure metrics of SVM algorithm is shown in Table 6.

Table 6. Performance of classification model. The model is based on support vector machines algorithms.

Evaluation	Precision	Recall	F-measure	Instances correctly classified
Ten-fold cross-validation	41.6%	53.3%	41.2%	53.3%
percentage split	67.9%	75%	67.8%	75%

4 Conclusions and Future Work

Personality traits impacts every aspect of humans. It is an important player in decisions and success. To know personality of people is useful to understand diverse aspects such as collective behaviors and preferences of people, such a way personality could be useful to select candidates for a job or to plan a focused marketing.

We are interested in detecting personality through linguistic analysis and in the DISC personality model. In this paper, we proposed a model for personality detection with three main components: (i) a set of personality Spanish words, (ii) a linguistic corpus for Spanish language with personality annotations as proposed by DISC model, and (iii) a classification model based on SVM algorithm.

To build the model and the linguistic resources, we conducted a study where 120 people participated. In this study, participants were asked to fill a DISC personality test, to write some paragraphs, and to answer a demographic questionnaire.

We evaluate several classification algorithms in order to achieve a classification model with a good performance. Additionally, we had several versions of the linguistics corpus.

This work is still in progress, there is a lot of work to be done in order to produce conclusive results. For example, the analysis of the patterns that are obtained from the personality tests since these are obtained when selecting adjectives in the personality test, one could find an important relation between the adjectives used in the test and the personalities. Additionally, we could improve the linguistic resources, gathering more texts from a diverse group of people regarding age or schooling, in order to achieve a richer set of words that would tell us more about each personality.

Acknowledgments. This research has been partially funded by European Commission and CONACYT, through the SmartSDK project.

References

1. Kazdin, A.E. (ed.): Encyclopedia of Psychology: 8 Volume Set. American Psychological Association and Oxford University Press (2000). http://www.apa.org/pubs/books/4600100.aspx?tab=3
2. Pratama, B.Y., Sarno, R.: Personality classification based on Twitter text using Naïve Bayes, KNN and SVM. In: 2015 International Conference on Data Software Engineering, pp. 170–174 (2015)

3. Cantador, I., Fernández-Tobías, I., Bellogín, A.: Relating personality types with user preferences in multiple entertainment domains. In: CEUR Workshop Proceedings, vol. 997 (2013)
4. Golbeck, J., Robles, C., Edmondson, M., Turner, K.: Predicting personality from Twitter. In: Proceedings of the 2011 IEEE International Conference on Privacy, Security Risk Trust IEEE International Conference on Socity Computing PASSAT/SocialCom 2011, pp. 149–156 (2011)
5. Tupes, E.C., Christal, R.E.: Recurrent personality factors based on trait ratings. J. Pers. **60** (2), 225–251 (1992)
6. Eysenck, H.J.: Dimensions of Personality. Transaction Publishers, Piscataway (1950)
7. Marston, W.M.: Emotions of Normal People. Harcourt, Brace and Company, New York (1928)
8. A&S Ltd.: What is DISC? https://www.discusonline.com/disc/what-is-disc.php. Accessed 01 Jan 2017
9. Achaerandio, Y.S., Navarro, C., Sáez, A.M., Viñuela, P.I.: A system for personality and happiness detection. Ijimai **2**(5), 8–16 (2014)
10. Lima, A.C.E.S., de Castro, L.N.: A multi-label, semi-supervised classification approach applied to personality prediction in social media. Neural Netw. **58**, 122–130 (2014)
11. Wald, R., Khoshgoftaar, T., Sumner, C.: Machine prediction of personality from Facebook profiles. In: Proceedings of the 2012 IEEE 13th International Conference on Information Reuse and Integration, IRI 2012, pp. 109–115 (2012)
12. Yuniar, I., Agung, A.A.G.: Personality Assessment Website using DISC. In: 2016 International Conference on Information Management and Technology, no. November, pp. 72–77 (2016)
13. DISC profile: Perfil en el lugar de trabajo [PDF file] (2013). https://www.discprofile.com/DiscProfile/media/Everything-DiSC/Workplace-Spanish-report.pdf. Accessed 05 Jan 2017
14. D. Insight: The DISC Insights Web Development Team. https://www.discinsights.com/. Accessed 05 Jan 2017
15. D. P. 4U: DISC Behavioral Styles. http://www.discprofiles4u.com/. Accessed 05 Jan 2017
16. Adali, S., Golbeck, J.: Predicting personality with social behavior. In: 2012 IEEE/ACM International Conference on Advances in Social Networks Analysis and Mining, pp. 302–309 (2012)
17. Annalyn, N., Bos, M.W., Sigal, L., Li, B.: Predicting personality from book preferences with user-generated content labels. IEEE Trans. Affect. Comput. **3045**(c), 1–12 (2018)
18. Pennebaker, J., King, L.: Linguistic styles: language use as an individual difference. J. Personal. Soc. **77**(6), 1296–1312 (1999)
19. Pennebaker, J.W., Booth, R.J.: LIWC. http://liwc.wpengine.com/. Accessed 10 Jan 2017
20. Majumder, N., Poria, S., Gelbukh, A., Cambria, E.: Deep learning-based document modeling for personality detection from text. IEEE Intell. Syst. **32**(2), 74–79 (2017)
21. Luyckx, K., Daelemans, W.: Personae: a corpus for author and personality prediction from text. In: Sixth International Conference Language Resources and Evaluatio, LR 2008, no. May 2017, pp. 2981–2987 (2008). A European Language Resources
22. Leskovec, J., Rajaraman, A., Ullman, J.D.: Mining of Massive Datasets, 2nd edn. Cambridge University Press, Cambridge (2014)
23. Loyola, O., Medina, M.A., García, M.: Inducing decision trees based on a cluster quality index. IEEE Latin Am. Trans. **13**(4), 1141–1147 (2015)
24. Witten, I.H., Frank, E., Hall, M.: Data Mining: Practical Machine Learning Tools and Techniques. 2nd edn. Morgan Kaufmann Publishers, Burlington (2005)

Analysis of Emotions Through Speech Using the Combination of Multiple Input Sources with Deep Convolutional and LSTM Networks

Cristyan R. Gil Morales[✉] and Suraj Shinde

everis AI Digital Lab, 06600 Mexico City, Mexico
{cgilmora,suraj.shinde}@everis.com

Abstract. Understanding emotions expressed in speech by a person is fundamental in having a better interaction between humans and machines. Many algorithms have been developed to solve this problem before. They have been tested on different datasets, some of these datasets were recorded by actors under ideal recording conditions and some others were recorded from people's opinion on some video streaming platform. Deep learning has shown very positive results in recent years and the model presented here follows this approach. We propose the use of Fourier transformations as the input of a convolutional neural network and Mel frequency cepstral coefficients as the input of an LSTM neural network. Finally, we concatenate the outputs of both models and obtain a final classification for five emotions. The model is trained using the MOSEI dataset. We also perform data augmentation by using time variations and pitch changes. Our model shows significant improvements over state-of-the-art algorithms.

Keywords: Speech Emotion Recognition · CNN · LSTM
Deep learning · MFCC · stft

1 Introduction

The emotions classification through speech, also called Speech Emotion Recognition (SER), is one of the biggest challenges in terms of audio analysis. However, accomplishing it accurately implies some great advantages, for example, if our desire is to enable machines to detect emotions, it would definitely be very useful, since changing actions could be proposed to change the mood of the person resulting in better user experience [13].

The task of analyzing only the audio to recognize emotions becomes very complex due to the fact that emotions are something subjective. This affects how emotions are perceived as well as how these emotions are labeled. Previously we used datasets that contained data from actors in very controlled situations, which could lead to good results. The disadvantage is that when testing in real

I. Batyrshin et al. (Eds.): MICAI 2018, LNAI 11289, pp. 218–226, 2018.
https://doi.org/10.1007/978-3-030-04497-8_18

situations the models did not respond correctly. However, now a days there is a tendency to use data from people expressing their opinions in a natural way, normal people expressing themselves in real situations [5].

Another aspect being considered is the use of multimodal analysis, since it gives us more information about the user. Nonetheless, in this work we hypothesise that only by using the audio of people's opinions we can achieve good results in the classification of emotions without using the other aspects of multimodal analysis such as facial expressions. With the advantage of combining our speech analysis model in a multimodal system could improve classification accuracy.

The rest of the paper has the following structure. Section 2 is devoted to presenting the related work associated with our research objectives. Section 3 describes the dataset used for testing our approach. Section 4 describes the main results achieved. Finally, Sect. 5 provides some conclusions and directions for future work.

2   Related Work

The selection of characteristics is an important task for recognizing emotions in speech in traditional machine learning methods. The most common methods of extraction include the pitch frequency feature, the energy-related feature, the formant feature, the spectral feature, etc. After the features were extracted, they were used to train models such as Bayesian networks [18], Hidden Markov Model [11], Gauss Mixed Model (GMM) [12] and multi-classifier fusion [20]. The primary advantage of this method is that it could be trained without very large amount of data. Some disadvantages include, difficulty in judging the quality of the features, probable loss of some key features, decreasing the accuracy of recognition and difficulty to ensure that good results can be achieved with different datasets.

Many of the current works that seek to solve the recognition of emotions through speech have used datasets recorded by actors in real recording conditions. Their results have been improved when using Deep Learning techniques which have shown a better performance.

The main difference between the traditional methods of machine learning and Deep Learning models is the ability to extract high-level features [1] that help us achieve better results. The recognition of emotions through speech by using Deep Belief Network (DBN) models can capture non-linear features, and such models have shown improvement in classification over performance baselines that do not employ deep learning.

Several approaches based on deep learning have shown promising results with databases such as EmoDB [3] or IEMOCAP [4]. Mao et al. [10] proposed learning affect-salient features for SER using CNN, which leads to stable and robust recognition performance. A CNN was trained on by Lee et al. [9] taking as an input the sort time fourier transform (stft) of the audio, such combination was capable of reaching the accuracy of 86% percent on the EmoDB.

Prasomphan et al. [16] the emotion was detected by using information inside the spectrogram, then using the Neural Network to classify the emotion again on the EmoDB database, and got the accuracy of 83% trained on five emotions.

Fayek [6] provided a method to augment training data, but the accuracy is less than 61% on ENTERFACE database and SAVEE database. In [22] a ConvNet was employed and combined with DTPM (Discriminant Temporal Pyramid Matching) for automatic feature learning the final result was an accuracy of 86% on EmoDB. Niu et al. [14] proposed a very interesting way on performing data augmentation of the audios based on Retinal Image Principle on the IEMOCAP database. Lee et al. [8] proposed Attention Networks for multimodal representation learning between speech and text data for emotion classification in the MOSEI dataset, reaching out the overall accuracy of 83%.

As shown in this section there are plenty of works trying to solve the SER problem using IEMOCAP or EmoDB which can be consider small if compared with MOSEI, however due to the recent release of the MOSEI dataset just a few works [2,8,15,17,19], can be found in the literature and most of them use a multimodal approach.

In this work we use audios of people's opinion from the MOSEI dataset, instead of the multimodal approach used in [8]. We perform data augmentation in order to compare the results of the original data with the augmented data (expecting to have better results on the last data), preprocess it with short time fourier transform (stft) and Mel Frequencies Ceptral Coeficient (MFCC). Finally the model was trained to classify emotions with a combination of LSTM and CONVNET.

3   Data Description

Our work is based on a standardized emotion dataset, called CMU-MOSEI [21], from the CMU Multimodal Data SDK. This dataset contains video segments that were collected from YouTube wherein the speaker is providing their review of a movie that they have seen. The segments have been labeled by humans for 6 different emotions, including the null case. These labels are: Anger, Disgust, Fear, Happiness, Sadness, and Surprise. Each segment can have any combination of emotion labels, or no labels at all. In addition, for each emotion label there is a corresponding regression value in the range of [0, 3]. This means that every video segment can be characterized with an emotion as well as the intensity of that emotion. The CMU-MOSEI dataset [21] provides pre-processed features and a means to align features.

We decided to use the raw data instead of the preprocessed information that came with the dataset with the purpose of extracting the information directly from the source. Also due to the small quantity of segments labeled with the surprise emotion, we decided to exclude it from our experiments.

3.1 Data Preprocessing

In this work we transform the audio using short time fourier transform (stft) with window of 512, the number of points used to calculate the fourier transform or nfft with a value of 512, and number of overlapping of 384. In addition we also preprocess the audio to obtain the Mel Frequencies Ceptral Coefficients (MFCC) using the librosa python library with a sampling rate of 22050 and 20 MFCCs [5]. We also performed data augmentation were we increased the time stretch by a factor of 1.1 and decreased by a factor of 0.9. Then we made a modification in the pitch shift. [7]

4   Experiments and Results

In this paper we propose a Deep Learning architecture where we combine the ability to extract characteristics of a convolutional neural network and the ability to recognize temporal patterns of a LSTM network, both receiving inputs with different preprocessing. On the one hand, the matrices resulting from applying Short Time Fourier Transform are input into the convolutional network, while the resulting matrices from the Mel Frequencies Ceptral Coefficients are feed into the LSTM network. The architecture can be found in Fig. 1.

4.1 Experiment with the Original Data

The original data is composed of more than 23 thousand of samples of which 70% were used for training, 15% for validation and the other 15% for testing as we can see on Table 1.

Table 1. Dataset statistics.

Dataset	
Training set size	16,198
Validation set size	3471
Test set size	3471

The model was trained approximately 200 epochs, with a batch size of 64 using a SGD optimizer with a learning rate of 0.001 and the error was calculated using by mse function. The results of testing the model on the validation dataset are shown in Table 2, the values are expressed in percentages for better visualization.

The results are quite good, however, with the increase in the data, a better result is expected.

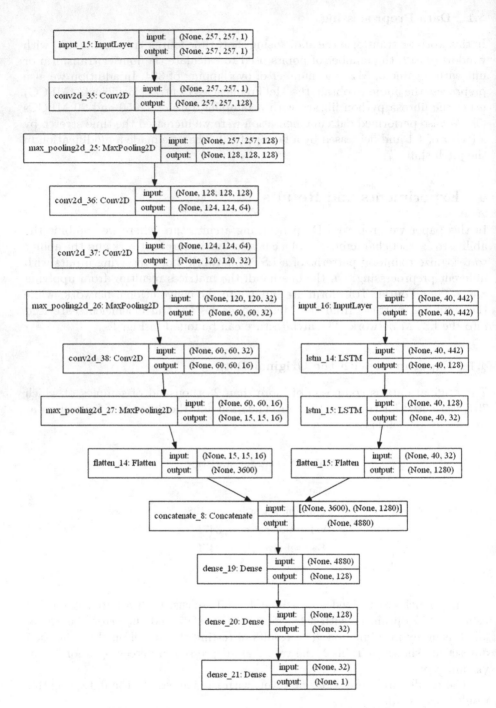

Fig. 1. Hybrid deep neural networks for integrating MFCC and STFT.

Table 2. Confusion matrix original data

%	Anger	Disgust	Fear	Happy	Sad
Anger	82.0	2.0	3.0	10.0	3.0
Disgust	4.0	85.0	1.0	8.0	2.0
Fear	5.0	2.0	75.0	15.0	3.0
Happy	9.0	0.0	0.0	89.0	2.0
Sad	2.0	3.0	1.0	15.0	79.0

4.2 Experiment with the Augmented Data

After data augmentation, we obtained a final dataset with 115700 samples, of which 70% were used for training, 15% for validation and the other 15% for testing as we can see on Table 3.

Table 3. Dataset statistics.

Dataset	
Training set size	80,990
Validation set size	17,355
Test set size	17,355

The model was trained approximately 200 epochs, with a batch size of 64 using a SGD optimizer with a learning rate of 0.001 and the error was calculated using by mse function. The results of testing the model on the validation dataset are shown in Table 4, the values are expressed in percentages for better visualization.

Table 4. Confusion matrix augmented data

	Anger	Disgust	Fear	Happy	Sad
Anger	91.0	0.0	0.0	6.0	1.0
Disgust	2.0	91.0	0.0	4.0	1.0
Fear	3.0	1.0	83.0	9.0	2.0
Happy	2.0	0.0	0.0	94.0	1.0
Sad	2.0	1.0	0.0	8.0	87.0

One of the best results obtained in "First Workshop and Grand Challenge on Computational Modeling of Human Multimodal Language" was the paper [7], which uses a multimodal analysis approach on the same dataset used in the experiments carried out in this work. The comparison of these results with those obtained by us are shown in Table 5.

Table 5. Comparison table.

Emotion	Convolutional attention networks	Ours
Anger	85.0	91.0
Disgust	88.0	91.0
Fear	94.0	83.0
Happy	92.0	94.0
Sad	88.0	87.0

As we can see our results compete with one of the best results obtained in the ACL contest, although they used multimodal analysis and we only used speech.

5 Conclusions

We have shown that we can analyse speech in an isolated way, i.e perform unimodal analysis of emotions. We have also shown how we used classical techniques for extracting characteristics such as Short Time Fourier Transform (STFT) and Mel Frequencies Ceptral Coefficients (MFCC). The results were fed into a model that combines the computational power of CNNs and LSTMs to achieve impressive results in terms of accuracy.

We have also shown that increasing the amount of the data is beneficial to the performance of our model, increasing the data in a rate of 5 times the original size boosts the accuracy and let us to outperform the results of similar works. Our results may seem low compared to bimodal and trimodal results, but the fact that our unimodal behaves quite well with only one variable analysed is an indication that integrating this type of model in multimodal analysis may bring greater benefits to the final result.

References

1. Bengio, Y., Courville, A.C., Vincent, P.: Unsupervised feature learning and deep learning: a review and new perspectives. CoRR abs/1206.5538 (2012). http://arxiv.org/abs/1206.5538
2. Blanchard, N., Moreira, D.M., Bharati, A., Scheirer, W.J.: Getting the subtext without the text: scalable multimodal sentiment classification from visual and acoustic modalities. CoRR abs/1807.01122 (2018)
3. Burkhardt, F., Paeschke, A., Rolfes, M., Sendlmeier, W., Weiss, B.: A database of German emotional speech, vol. 5, pp. 1517–1520 (2005)
4. Busso, C., et al.: IEMOCAP: interactive emotional dyadic motion capture database. Lang. Resour. Eval. **42**, 335–359 (2008)

5. Davletcharova, A., Sugathan, S., Abraham, B., James, A.P.: Detection and analysis of emotion from speech signals. CoRR abs/1506.06832 (2015). http://arxiv.org/abs/1506.06832
6. Fayek, H.M., Lech, M., Cavedon, L.: Towards real-time speech emotion recognition using deep neural networks. In: 2015 9th International Conference on Signal Processing and Communication Systems (ICSPCS), pp. 1–5, December 2015
7. Ko, T., Peddinti, V., Povey, D., Khudanpur, S.: Audio augmentation for speech recognition. In: INTERSPEECH (2015)
8. Lee, C.W., Song, K.Y., Jeong, J., Choi, W.Y.: Convolutional attention networks for multimodal emotion recognition from speech and text data. CoRR abs/1805.06606 (2018)
9. Lim, W., young Jang, D., Lee, T.: Speech emotion recognition using convolutional and recurrent neural networks. 2016 Asia-Pacific Signal and Information Processing Association Annual Summit and Conference (APSIPA), pp. 1–4 (2016)
10. Mao, Q., Dong, M., Huang, Z., Zhan, Y.: Learning salient features for speech emotion recognition using convolutional neural networks. IEEE Trans. Multimedia **16**, 2203–2213 (2014)
11. Mao, X., Chen, L., Fu, L.: Multi-level speech emotion recognition based on HMM and ANN. In: 2009 WRI World Congress on Computer Science and Information Engineering. vol. 7, pp. 225–229, March 2009
12. Neiberg, D., Elenius, K., Laskowski, K.: Emotion recognition in spontaneous speech using GMMS. In: INTERSPEECH (2006)
13. Niu, Y., Zou, D., Niu, Y., He, Z., Tan, H.: A breakthrough in speech emotion recognition using deep retinal convolution neural networks. CoRR abs/1707.09917 (2017)
14. Niu, Y., Zou, D., Niu, Y., He, Z., Tan, H.: Improvement on speech emotion recognition based on deep convolutional neural networks. In: Proceedings of the 2018 International Conference on Computing and Artificial Intelligence, pp. 13–18. ICCAI 2018. ACM, New York, NY, USA (2018). https://doi.org/10.1145/3194452.3194460
15. Pham, H., Manzini, T., Liang, P.P., Póczos, B.: Seq2seq2sentiment: Multimodal sequence to sequence models for sentiment analysis. CoRR abs/1807.03915 (2018)
16. Prasomphan, S.: Improvement of speech emotion recognition with neural network classifier by using speech spectrogram. In: 2015 International Conference on Systems, Signals and Image Processing (IWSSIP), pp. 73–76, September 2015
17. Sahay, S., Kumar, S.H., Xia, R., Huang, J., Nachman, L.: Multimodal relational tensor network for sentiment and emotion classification. CoRR abs/1806.02923 (2018)
18. Ververidis, D., Kotropoulos, C.: Fast and accurate sequential floating forward feature selection with the bayes classifier applied to speech emotion recognition. Signal Processing **88**(12), 2956–2970 (2008). http://www.sciencedirect.com/science/article/pii/S0165168408002120
19. Williams, J., Kleinegesse, S., Comanescu, R., Radu, O.: Recognizing emotions in video using multimodal DNN feature fusion. In: Proceedings of Grand Challenge and Workshop on Human Multimodal Language (Challenge-HML), pp. 11–19. Association for Computational Linguistics, July 2018
20. Wu, C.H., Liang, W.B.: Emotion recognition of affective speech based on multiple classifiers using acoustic-prosodic information and semantic labels. IEEE Trans. Affect. Comput. **2**(1), 10–21 (2011)

21. Zadeh, A., Liang, P.P., Poria, S., Vij, P., Cambria, E., Morency, L.P.: Multi-attention recurrent network for human communication comprehension. CoRR abs/1802.00923 (2018)
22. Zhang, S., Zhang, S., Huang, T., Gao, W.: Speech emotion recognition using deep convolutional neural network and discriminant temporal pyramid matching. IEEE Trans. Multimedia **20**(6), 1576–1590 (2018)

Robustness of LSTM Neural Networks for the Enhancement of Spectral Parameters in Noisy Speech Signals

Marvin Coto-Jiménez[1,2]([✉]) [iD]

[1] PRIS-Lab, Escuela de Ingeniería Eléctrica, San Pedro, Costa Rica
marvin.coto@ucr.ac.cr
[2] Universidad de Costa Rica, San José, Costa Rica

Abstract. In this paper, we carry out a comparative performance analysis of Long Short-term Memory (LSTM) Neural Networks for the task of noise reduction. Recent work in this area has shown the advantages of this kind of network for the enhancement of noisy speech, particularly when the training process is performed for specific Signal-to-Noise (SNR) levels.

For application in real-life environments, it is important to test the robustness of the approach without the a priori knowledge of the SNR noise levels, as classical signal processing-based algorithms do. In our experiments, we conduct the training stage with single and multiple noise conditions and perform the comparison of the results with the specific SNR training presented previously in the literature.

For the first time, results give a measure on the independence of the training conditions for the task of noise suppression in speech signals, and shows remarkable robustness of the LSTM for different SNR levels.

Keywords: Deep learning · LSTM · MFCC
Neural networks · Speech enhancement

1 Introduction

Speech signals are often affected by additive noise, reverberation and other distortions in real-world environments. Communication devices and applications of speech technologies may be affected in their performance [2,23,28,29] with such noise added to the speech information.

During the past decades, speech enhancement algorithms have been presented to suppress or reduce such distortions and preserve or enhance perceived signal quality [14]. Several recent algorithms for the task of enhancing speech signals are based on deep neural networks (DNN) [4,5,13,22]. The most common approach is that of learning mapping features from noisy speech into the features

---

Supported by the University of Costa Rica.

I. Batyrshin et al. (Eds.): MICAI 2018, LNAI 11289, pp. 227–238, 2018.
https://doi.org/10.1007/978-3-030-04497-8_19

of the corresponding clean speech, using autoencoders based on perceptrons or recurrent neural networks (RNNs).

Among the new types of RNNs, the Long Short-Term Memory Network (LSTM) has succeeded in mapping features derived from the spectrum, usually Mel-Frequency Cepstrum Coefficients (MFCC). These features have been used widely in speech-related tasks because automatic speech recognition systems are frequently based on them.

In this work, we extend previous experiences of speech enhancement with LSTM by measuring its robustness, considering more than one levels of noise with a single network. Benefits from this type of speech enhancement can be applied to more realistic tasks in mobile phones, VoIP, speech recognition, and devices for hearing-impaired listeners [15].

1.1 Related Work

Several techniques for enhancement of speech signals based on DNN have been presented in the past few years. Typically, these techniques rely on the enhancement of spectral features, such as MFCC and its discrete derivatives. For example, MFCCs plus its first and second derivatives are used in [1,18,24].

The deep learning approaches have been successful in outperformed classical methods based on signal processing when the speech signals contain noise of different types with various signal-to-noise radio (SNR) [16,21,26], or reverberant speech [8,19]. Also, the advantage in reducing the musical artifact commonly present in speech enhancement classical algorithms has been observed [30].

The principal method for enhancing the signals using deep learning is to apply the networks as regression models, mapping the noisy parameters of the speech into the corresponding clean parameters [28,29].

LSTM networks for speech enhancement have been presented previously in [3], using MFCC as features, for the case of applying one LSTM network for enhancement of each noise type and SNR level. Even though the LSTM outperforms other deep networks in this task, the training process for its successful implementation requires single specific noise conditions, and a priori knowledge of the SNR during test procedure. The robustness of LSTM networks with several SNR levels has also been tested for Voice Activity Detection in [25], with relevant success.

In the present paper, we consider a more realistic scenario for noise reduction, where the networks are trained with more than one level of noise, to measure the capacity of the LSTM networks to enhance speech signals without the a priori information of the SNR level in test sets. The process is trained, validated and tested with examples from the Carnegie Mellon University speech database. Several objective measures are used to test the results, which show the capacity of the LSTM in robust noise reduction.

The rest of this paper is organized: Sect. 2 gives the background and context of the problem of denoising and the LSTM, Sect. 3 describes the experimental setup, Sect. 4 presents the results with a discussion, and finally, in Sect. 5, we present the conclusions.

2   Background

2.1   Problem Statement of Speech Enhancement and Robustness

In the field of speech enhancement, we can assume that a noisy signal, y, is the sum of a speech signal, x, and noise d, given by:

$$y(t) = x(t) + d(t) \qquad (1)$$

In the spectral domain, the formulation of the problem becomes:

$$Y_k(n) = X_k(n) + D_k(n), \qquad (2)$$

where k is the frequency index and n the time-segment index. In most methods, $x(t)$ is considered uncorrelated to $d(t)$, and signal processing-based enhancement algorithms estimate $X_k(n)$ from the power spectral domain of $x(t)$ and $d(t)$. In deep learning-based approaches, $x(t)$ can be estimated using algorithms that learn an approximated function $f(\cdot)$ between the noisy and clean data of the form:

$$\hat{x}(t) = f(y(t)). \qquad (3)$$

The precision of the approximation $f(\cdot)$ usually depends on the amount of training data and the algorithm selected. Previous attempts has estimated $f(\cdot)$ for each type of noise and SNR. It means that for N SNR levels, there is a set of N deep neural networks trained and then applied separately to estimate the set $f_{SNR_{-1}}(\cdot), f_{SNR_{-2}}(\cdot), \cdots, f_{SNR_{-N}}(\cdot)$.

A robust application of noise reduction can provide a single network capable of enhance several SNR levels. It means there is no need to have a priori knowledge of the SNR level presented at the input of the network, because an estimation of $f_R(\cdot)$ can be done for more than one scenario. With this robust network, it is expected to have

$$f_{SNR_{-1}}(\cdot) \approx f_R(\cdot) \qquad (4)$$

$$f_{SNR_{-2}}(\cdot) \approx f_R(\cdot)$$

$$\vdots \quad \vdots$$

$$f_{SNR_{-N}}(\cdot) \approx f_R(\cdot),$$

and for any signal at the input with a given SNR, the regression performed should be similar to those of the network trained with the specific level.

2.2   Long Short-Term Memory Neural Networks

Several kinds of neural networks have been tested for classification and regression purposes over the past several decades. Recently, new kinds of networks organized in many layers, known as Deep Neural Networks (DNN) achieved good results in many problems of a wide spectrum of applications. From the emergence of RNNs, which can store information by feedback connections between neurons in the hidden layers to themselves or others neurons in the same layer [7,31], new possibilities in modeling the dependent nature of sequential information have been opened.

With the aim of expanding the capabilities of RNN by storing information in the short and the long term, LSTM networks presented in [17] have introduced a set of gates within memory cells that control the access, storing and propagation of values over the network. LSTM networks presented encouraging results in speech recognition, music composition and handwriting synthesis, which heavily depends on previous states of the information [10,11,17].

To accomplish the task of preserving values in the long-term and the short-term, the LSTM has four gates that controls the operations of input, output, and erasing the memory. More details on the training procedure and the mathematical modeling of the LSTM can be found in [9].

2.3   Denoising with Deep Neural Networks

The idea of training neural networks in speech enhancement and noise reduction was first introduced several decades ago for binary input patterns, corrupted by randomly flipping a fraction of the input bits. Other than binary inputs, acoustic coefficients were modeled with a single layer a few years later. Neither the computer capabilities nor the algorithms were adequate for including more hidden layers or considering much larger sets of data [16].

For noise reduction and other regression-based tasks, parameters of the networks are found using training data in order to minimize the average reconstruction of the input, that is, to have output $f(y)$ as close as possible to the uncorrupted signal x [27].

One of the recent architectures of neural networks that have achieved considerable success is called a denoising autoencoder, consisting of two steps: the first one is the encoder, which performs a mapping f that transforms an input vector y into a representation h in the hidden layers. The second step is the decoder, which mapped back the hidden representation into a vector $\hat{x}$ in input space.

During the training stage, noise corrupted features are presented at the inputs of the denoising autoencoders, while the corresponding clean features became the outputs. The training algorithm adjusts the parameters of the network in order to learn the complex relationships between them. Modern computers allow the training of many hidden layers and larger datasets.

3   Experimental Setup

In order to test the robustness of LSTM network, the experimental setup, from data generation to evaluation, can be summarized in the following steps:

1. Noisy database generation: Files containing White noise were generated and added to each audio file in the database for a given signal-to-noise ratio (SNR). Five noise levels were added, in order to cover a range from light to heavy noise levels for each noise type.
2. Feature extraction and input-output correspondence: A set of parameters was extracted from the noisy, and the clean audio files. Those from the noisy files were used as inputs to the networks, while the corresponding clean features were the outputs.
3. Training: During training, using forward pass and back-propagation through time algorithm, the weights of the networks were adjusted as the noisy and clean utterances were presented at the inputs and at the outputs. A total of 900 utterances (about 80% of the total database) were used for training. Details and equations of the algorithm followed can be found in [12].
4. Validation: After each training step, the sum of squared errors were computed within the validation set of 182 utterances (about 15% of the total database), and the weights of the network updated in each improvement.
5. Test: A subset of 50 randomly selected utterances (about 5% of the total amount of utterances of the database) was chosen for the test set, for each noise level. These utterances were not part of the training process, to provide independence between the training and testing.

   In the following subsections, further details of the main experimental setup are given.

3.1   Database

In our work, we chose the SLT voice from the CMU ARCTIC databases [20], designed for speech research. The whole set of 1132 sentences were used to randomly define the training, validation and test sets. In our work, we chose the female SLT voice, and the whole set of 1132 sentences were used to randomly define the training (849 sentences), validation (233 sentences) and test sets (50 sentences) for each noise level.

3.2   Feature Extraction

The audio files of the noisy and the clean database were downsampled to 16 kHz, 16 bits, to extract parameters using the Ahocoder system [6]. A frame size of 160 samples and frame shift of 80 samples were used to extract 39 mfcc, f_0 and energy of each sentence.

3.3 Evaluation

To evaluate the results given by the different enhancement methods, we use the following well-known measures:

- Euclidean Distance between MFCC: This measure is computed between each of the 39 vectors of clean and enhanced speech in the test set. For a vector $\mathbf{x}$ of MFCC, and the corresponding enhanced $\hat{\mathbf{x}}$, the distance is computed as:

$$Eu(\mathbf{x_j}, \hat{\mathbf{x}_j}) = \left(\sum_{i=1}^{n} (x_{j_i} - \hat{x_{j_i}})^2 \right)^{\frac{1}{2}}, \tag{5}$$

where n is the number of frames in the test sentences, and $j \in [1, 39]$ the index of the MFCC.
- Mean Absolute Distance between MFCC: Computed as

$$MAD(\mathbf{x_j}, \hat{\mathbf{x}_j}) = \frac{1}{39} \sum_{j=1}^{39} \frac{1}{n} \sum_{i=1}^{n} |x_{j_i} - \hat{x_{j_i}}| \tag{6}$$

We use MAD as a unique measure for each experiment, opposed to Euclidean Distance which is applied on each MFCC.
- Frequency Domain Segmental SNR (SegSNRf): Is a frame-based measure, calculated by averaging the frame level SNR estimates, following the equation:

$$\text{SegSNRf} = \frac{10}{N} \sum_{i=1}^{N} \log \left[\frac{\sum_{j=0}^{L-1} S^2(i,j)}{\sum_{j=0}^{L-1} (S(i,j) - X(i,j))^2} \right] \tag{7}$$

where $X(i,j)$ is Fourier transform coefficient of frame i at frequency bin j, and $S(i,j)$ is the corresponding coefficient for the processed speech. N is the number of frames and L the number of frequency bins. The values are limited to the interval $[-20, 35]$ dB.

Additionally, we show contours of MFCC coefficients to illustrate the result for the different experiments.

3.4 Experiments

For the purpose of testing the robustness of LSTM networks in noise reduction, we train several sets of networks to directly map the noisy features to clean features. The experiments, which contemplates the training of different SNR levels and its comparison with the base system of a single SNR training, are described following the nomenclature:

- Five Levels: The LSTM network were trained using sentences containing all the SNR levels at the input and the corresponding clean sentences at the output. For evaluation purposes, test sets of every SNR level were presented at the trained network, and the evaluation measures applied to the output.

– Three levels: The LSTM network were trained using sentences containing three SNR levels at the input. This case required different networks for the experimentation with each level, to consider a target SNR and the two neighbor SNR. In this paper, we did not consider the training with distant or randomly selected SNR.

   • SNR-10: Given that SNR-10 is the lower level in our experiments, the closer SNR are SNR-5 and SNR0. So the network was trained with SNR-10, SNR-5 and SNR0.
   • SNR-5: The same network of the previous case was applied.
   • SNR0: The training procedure was performed with sentences of SNR-5, SNR0 and SNR5 at the input.
   • SNR5: The training procedure was performed with sentences of SNR0, SNR5 and SNR10 at the input.
   • SNR10: Given that SNR10 is the upper limit of noise level we considered in the experiments, c.

– Two levels: The LSTM network were trained using sentences containing two SNR levels at the input, without particular consideration of a higher or lower level in comparison to the target SNR.

   • SNR-10: The training procedure was performed with sentences of SNR-10 and SNR-5 at the input.
   • SNR-5: The same network of the previous case was applied.
   • SNR0: The training procedure was performed with sentences of SNR0 and SNR5 at the input.
   • SNR5: The same network of the previous case was applied.
   • SNR10: The training procedure was performed with sentences of SNR10 and SNR5 at the input.

– One level (base system): One network for each SNR was trained. This is the case analyzed in previous references, and we consider the base results for comparison.
– None: The evaluation measures were applied to the noisy sentences.

   The LSTM architecture for the networks was defined by trial and error. Initially, we considered a single hidden layer with 50 units and then increased

Table 1. Mean absolute distance (MAD) between MFCC enhanced coefficients in the test set. Lower values represent better results (higher similarity)

Training levels of SNR	Test levels				
	SNR-10	SNR-5	SNR0	SNR5	SNR10
None (noisy)	0.24	0.23	0.21	0.20	0.19
One	0.13	0.13	0.12	0.11	0.11
Two	0.13	0.12	0.12	0.11	0.11
Three	0.13	0.12	0.12	0.11	0.11
Five	0.13	0.12	0.12	0.11	0.11

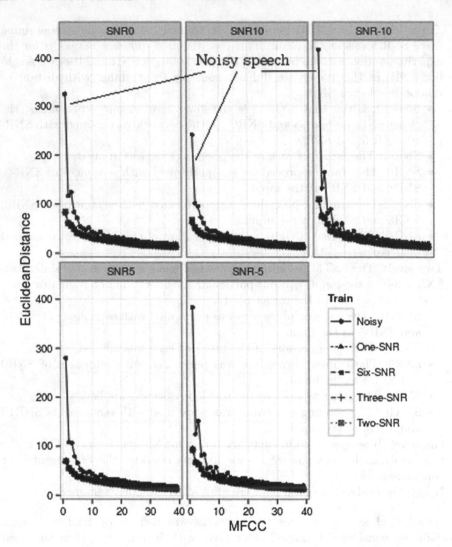

Fig. 1. Comparison of Euclidean Distance between the MFCC of the Noisy Speech and LSTM-enhanced

the size with steps of 50 units, up to three hidden layers with 300 units in each layer. The final selection consisted of a network with three layers containing 100, 100 and 100 units in each one.

This network gave the best results in the trial experiments, and also had a manageable training time, considering that we use 20 LSTM networks in this work. The training procedure was accelerated by a NVIDIA GPU system, taking about 7 h to train each LSTM.

4   Results and Discussion

The results present the measures according to the number of SNR levels described in the precious section. Table 1 summarizes the MAD between the MFCC of clean sentences in the test set and the corresponding enhanced version obtained with the LSTM networks.

These MAD results shows independence of this measure to the amount of SNR levels considered during training. It means that a single LSTM can enhance every noise level with similar success as the training of single levels. The Euclidean distance presented in Fig. 1 also shows the robustness of the LSTM in terms of the distance from each MFCC to those of the clean speech. The capacity of the networks to improve the MFCCs of the noisy speech is considerable for every noise level, and every training condition enhance the speech similarly.

Figure 2 illustrates the evolution of the first MFCC coefficient in three SNR levels and compare the case of training with specific SNR with the training with six SNR. It also shows several important results: The Noisy speech with SNR-10 almost degrade the first MFCC completely, but the LSTM networks restore most

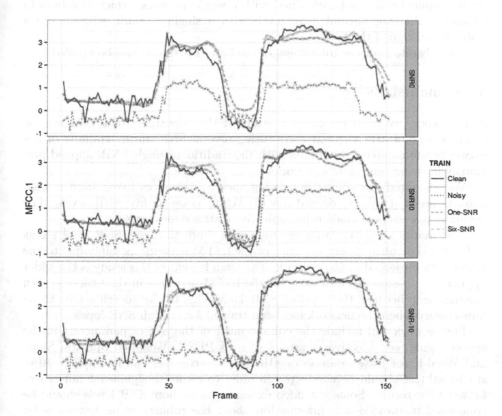

Fig. 2. Comparison of the Trajectory of first MFCC

Table 2. SegSNRf between MFCC enhanced coefficients in the test set. Lower values represent better results

Training levels of SNR	Test levels				
	SNR-10	SNR-5	SNR0	SNR5	SNR10
One	−9.98	−9.95	−9.84	−9.53	−8.27
Two	−9.99	−9.95	−9.88	−9.54	−8.50
Three	−9.99	−9.99	−9.91	−9.59	−8.52
Five	−9.99	−9.96	−9.85	−9.51	−8.27

of the information. But, as the other SNR levels, some of the characteristics of the parameters are smoothed in the denoising process. The differences between the training conditions are not clearly distinguishable, a result that is consistent with those of the previous measures.

Finally, the SegSNRf presented in Table 2 provides additional information about the robustness of the LSTM networks for denoising speech signals. In every case the results are very close. This means that the denoising process in future applications can be obtained with a single network, which is robust in terms of enhancing several noise levels with a single network, without the a priori knowledge of the SNR.

The advantage of the robustness of the LSTM is that it can be applied.

5   Conclusions

In this work, we have presented a study of the robustness of LSTM for the enhancement of MFCC parameters, using several SNR during training of the networks and compare the results with the traditional single SNR, applied previously in speech enhancement works.

We evaluated the comparison using speech utterances taken from a well-known speech database, degraded with White noise at five SNR levels. The evaluations were performed using objective distance measures.

The results show the training stage with more than one SNR level seems to benefit the enhancement capacity of the LSTM network, possibly due to the larger data presented at the network. The main benefit of this study is the wider application possibilities that arise for the LSTM networks in the task of speech enhancement because the a priori SNR knowledge of the speech seems to be unnecessary when the network has been trained for enough SNR levels.

Future work will include the enhancement of the rest of parameters in the speech signal, so objective measures such as PESQ, Weighted Spectral Slope and Word Error Rate in an automatic speech recognizer can be applied. Also, statistical test should be conducted in order to establish significant differences between the results. Some extended experiments in more SNR levels should be conducted to provide new information about the robustness on heavier noise conditions.

Acknowledgments. This work was supported by the Universidad de Costa Rica.

References

1. Abdel-Hamid, O., Mohamed, A.R., Jiang, H., Penn, G.: Applying convolutional neural networks concepts to hybrid NN-HMM model for speech recognition. In: Acoustics, Speech and Signal Processing, pp. 4277–4280. IEEE (2012)
2. Bagchi, D., Mandel, M.I., Wang, Z., He, Y., Plummer, A., Fosler-Lussier, E.: Combining spectral feature mapping and multi-channel model-based source separation for noise-robust automatic speech recognition. In: 2015 IEEE Workshop on Automatic Speech Recognition and Understanding (ASRU), pp. 496–503. IEEE (2015)
3. Coto-Jiménez, M., Goddard-Close, J., Martínez-Licona, F.: Improving automatic speech recognition containing additive noise using deep denoising autoencoders of LSTM networks. In: Ronzhin, A., Potapova, R., Németh, G. (eds.) SPECOM 2016. LNCS (LNAI), vol. 9811, pp. 354–361. Springer, Cham (2016). https://doi.org/10.1007/978-3-319-43958-7_42
4. Deng, L., et al.: Recent advances in deep learning for speech research at Microsoft. In: ICASSP, vol. 26, p. 64 (2013)
5. Du, J., Wang, Q., Gao, T., Xu, Y., Dai, L.R., Lee, C.H.: Robust speech recognition with speech enhanced deep neural networks. In: Association (2014)
6. Erro, D., Sainz, I., Navas, E., Hernáez, I.: Improved HNM-based vocoder for statistical synthesizers. In: Association (2011)
7. Fan, Y., Qian, Y., Xie, F.L., Soong, F.K.: TTS synthesis with bidirectional LSTM based recurrent neural networks. In: Association (2014)
8. Feng, X., Zhang, Y., Glass, J.: Speech feature denoising and dereverberation via deep autoencoders for noisy reverberant speech recognition. In: 2014 IEEE International Conference on Acoustics, Speech and Signal Processing (ICASSP), pp. 1759–1763. IEEE (2014)
9. Gers, F.A., Schraudolph, N.N., Schmidhuber, J.: Learning precise timing with LSTM recurrent networks. J. Mach. Learn. Res. **3**(Aug), 115–143 (2002)
10. Graves, A., Fernández, S., Schmidhuber, J.: Bidirectional LSTM networks for improved phoneme classification and recognition. In: Duch, W., Kacprzyk, J., Oja, E., Zadrożny, S. (eds.) ICANN 2005. LNCS, vol. 3697, pp. 799–804. Springer, Heidelberg (2005). https://doi.org/10.1007/11550907_126
11. Graves, A., Jaitly, N., Mohamed, A.R.: Hybrid speech recognition with deep bidirectional LSTM. In: 2013 IEEE Workshop on Automatic Speech Recognition and Understanding (ASRU), pp. 273–278. IEEE (2013)
12. Greff, K., Srivastava, R.K., Koutník, J., Steunebrink, B.R., Schmidhuber, J.: LSTM: a search space odyssey. IEEE Trans. Neural Netw. Learn. Syst. **28**(10), 2222–2232 (2017)
13. Han, K., He, Y., Bagchi, D., Fosler-Lussier, E., Wang, D.: Deep neural network based spectral feature mapping for robust speech recognition. In: Association (2015)
14. Hansen, J.H., Pellom, B.L.: An effective quality evaluation protocol for speech enhancement algorithms. In: Fifth International Conference on Spoken Language Processing (1998)
15. Healy, E.W., Yoho, S.E., Wang, Y., Wang, D.: An algorithm to improve speech recognition in noise for hearing-impaired listeners. J. Acoust. Soc. Am. **134**(4), 3029–3038 (2013)

16. Hinton, G., et al.: Deep neural networks for acoustic modeling in speech recognition: the shared views of four research groups. IEEE Sign. Process. Mag. **29**(6), 82–97 (2012)
17. Hochreiter, S., Schmidhuber, J.: Long short-term memory. Neural Comput. **9**(8), 1735–1780 (1997)
18. Huang, J., Kingsbury, B.: Audio-visual deep learning for noise robust speech recognition, pp. 7596–7599. IEEE (2013)
19. Ishii, T., Komiyama, H., Shinozaki, T., Horiuchi, Y., Kuroiwa, S. (eds.): In: Interspeech, pp. 3512–3516 (2013)
20. Kominek, J., Black, A.W.: The CMU Arctic speech databases. In: Fifth ISCA Workshop on Speech Synthesis (2004)
21. Kumar, A., Florencio, D.: Speech enhancement in multiple-noise conditions using deep neural networks. arXiv preprint arXiv:1605.02427 (2016)
22. Maas, A.L., Le, Q.V., O'Neil, T.M., Vinyals, O., Nguyen, P., Ng, A.Y.: Recurrent neural networks for noise reduction in robust ASR. In: Association (2012)
23. Narayanan, A., Wang, D.: Ideal ratio mask estimation using deep neural networks for robust speech recognition, pp. 7092–7096. IEEE (2013)
24. Seltzer, M.L., Yu, D., Wang, Y.: An investigation of deep neural networks for noise robust speech recognition, pp. 7398–7402. IEEE (2013)
25. Sertsi, P., Boonkla, S., Chunwijitra, V., Kurpukdee, N., Wutiwiwatchai, C.: Robust voice activity detection based on LSTM recurrent neural networks and modulation spectrum. In: 2017 Asia-Pacific Signal and Information Processing Association Annual Summit and Conference (APSIPA ASC), pp. 342–346. IEEE (2017)
26. Vincent, E., Watanabe, S., Nugraha, A.A., Barker, J., Marxer, R.: An analysis of environment, microphone and data simulation mismatches in robust speech recognition. Comput. Speech Lang. **46**, 535–557 (2017)
27. Vincent, P., Larochelle, H., Lajoie, I., Bengio, Y., Manzagol, P.A.: Stacked denoising autoencoders: learning useful representations in a deep network with a local denoising criterion. J. Mach. Learn. Res. **11**(Dec), 3371–3408 (2010)
28. Weninger, F., Geiger, J., Wöllmer, M., Schuller, B., Rigoll, G.: Feature enhancement by deep lstm networks for asr in reverberant multisource environments. Comput. Speech Lang. **28**(4), 888–902 (2014)
29. Weninger, F., Watanabe, S., Tachioka, Y., Schuller, B.: Deep recurrent de-noising auto-encoder and blind de-reverberation for reverberated speech recognition. In: 2014 IEEE International Conference on Acoustics, Speech and Signal Processing (ICASSP), pp. 4623–4627. IEEE (2014)
30. Xu, Y., Du, J., Dai, L.R., Lee, C.H.: An experimental study on speech enhancement based on deep neural networks. IEEE Sign. Process. Lett. **21**(1), 65–68 (2014)
31. Zen, H., Sak, H.: Unidirectional long short-term memory recurrent neural network with recurrent output layer for low-latency speech synthesis. In: 2015 IEEE International Conference on Acoustics, Speech and Signal Processing (ICASSP), pp. 4470–4474. IEEE (2015)

Tensor Decomposition for Imagined Speech Discrimination in EEG

Jesús S. García-Salinas[✉], Luis Villaseñor-Pineda,
Carlos Alberto Reyes-García, and Alejandro Torres-García

Biosignals Processing and Medical Computation Laboratory,
Language Technologies Laboratory, Instituto Nacional de Astrofísica Óptica y
Electrónica, Cholula, Puebla, México
{jss.garcia,villasen,kargaxxi,alejandro.torres}@inaoep.mx

Abstract. Most of the researches in Electroencephalogram(EEG)-based
Brain-Computer Interfaces (BCI) are focused on the use of motor
imagery. As an attempt to improve the control of these interfaces, the
use of language instead of movement has been recently explored, in the
form of imagined speech. This work aims for the discrimination of imagined words in electroencephalogram signals. For this purpose, the analysis of multiple variables of the signal and their relation is considered
by means of a multivariate data analysis, i.e., Parallel Factor Analysis
(PARAFAC). In previous works, this method has demonstrated to be
useful for EEG analysis. Nevertheless, to the best of our knowledge, this
is the first attempt to analyze imagined speech signals using this approach. In addition, a novel use of the extracted PARAFAC components is
proposed in order to improve the discrimination of the imagined words.
The obtained results, besides of higher accuracy rates in comparison
with related works, showed lower standard deviation among subjects sug-
gesting the effectiveness and robustness of the proposed method. These
results encourage the use of multivariate analysis for BCI applications in
combination with imagined speech signals.

Keywords: Tensor decomposition · Brain Computer Interface
Imagined speech · Electroencephalogram

1 Introduction

A Brain-Computer Interface (BCI) is a system which allows the interaction of a
person with the environment through the analysis of the brain signals in order to
generate an action [7,25]. Different instruments are used to acquire brain signals.
In particular, the use of electroencephalograms (EEGs) is of great interest due
to their simple operation and price.

An imagined speech based BCI requires processing the brain signal in order to
detect the brain activity related to a specific task. Previous works have proposed
methods to address the feature extraction and analysis. Some of them attempted

© Springer Nature Switzerland AG 2018
I. Batyrshin et al. (Eds.): MICAI 2018, LNAI 11289, pp. 239–249, 2018.
https://doi.org/10.1007/978-3-030-04497-8_20

to represent the EEG as multidimensional data and analyzed this representation in many different ways [12–16, 26]. Nevertheless, the relation among different dimensions of the brain signal, i.e., frequency, time and space, had not been considered in most of the approaches. This work aims to find a new characterization which integrates these three dimensions of the EEG signal to find appropriate patterns for imagined speech discrimination.

With a multivariate data analysis, a better representation of the imagined speech is expected due to the extraction of interdependent relationships among the frequency, time and space of the brain signal [4–6, 22]. In the particular case of imagined speech, neurophysiological information of the signals is not well known. Hence, the analysis considering all the information from the EEG signals is essential. This is not the case, for example, with the motor imaging task. In this task, the associated neurophysiological areas are identified, allowing the proposal of successful methods based on spatial patterns discrimination.

This work explores the Parallel Factorial Analysis (PARAFAC) to decompose the EEG signals, and with this representation demonstrate the improvement in the discrimination of imagined words. The PARAFAC is a generalization of Principal Component Analysis (PCA) [2], and it is based on the decomposition of multi-way data into a sum of rank-one tensors in polyadic form [18].

The rest of the paper is organized as follows: Sect. 2 discusses related works that use tensor decomposition analysis in EEG signals. Section 3 will explain the proposed method. In Sect. 4 experimental results are reported. And finally, in Sect. 5 the conclusions and future work are discussed.

2   Related Works

Previous works used tensor decomposition analysis to reconstruct the brain activity, in search of alpha and theta band patterns [1, 19]. The results of these works evidenced that multiple dimensions of EEGs can be properly analyzed by means of tensor decomposition. The first work proposed that a typical structure of the resting state EEG consists of two PARAFAC components. The latter work, found alpha and theta activation patterns in rest and a mental task EEGs respectively, using the components extracted by PARAFAC.

In [13] an EEG classification is proposed, in which a dataset with three tasks, i.e., left/right imagined movement and random words generation, is analyzed. A least squares projection and a hidden states algorithm are proposed. This method achieved similar results as the related works. Nevertheless, a novel tensor configuration was analyzed, considering either the time and the task as a dimension of the tensor.

Other approaches have adapted tensor decomposition analysis to improve classification results. For example, [15] used the class labels and regularization constraints on the tensor factorization in order to find a robust discriminative subspace. Different EEG datasets were tested, the first one involved a motor imagery task, in which the subjects imagined the movement of their right or left hand. In the second dataset, two tasks were performed, geometric figures perception and arithmetic problems solving.

In [16] a general tensor discriminant analysis was performed. This approach was also proposed for image classification in [22]. For this purpose, the dataset of [15] was used, and another one was added. In this new dataset, the subjects were required to recall the meaning and pronunciation of a previous known word, the aiming was to analyze a memory task.

Following the approaches in [14–16] proposed a phase interval value to measure the phase difference between the EEG channels. This value was used to discriminate between two imaginary movements, i.e., the right or left hand, and the features were obtained by means of tensor decomposition analysis.

In [26], both left and right hand imagined movements were analyzed by proposing a tensor decomposition. For this purpose, a slice oriented decomposition from a tensor was developed to discriminate imagined movement. Moreover, a tensor averaging is proposed as a preprocessing step in order to find an invariable feature structure between recordings.

In comparison with previous works, [12] proposed an extension to a multiclass classification of EEG signals. Moreover, it was applied to imaginary movement and steady-state visual evoked potentials. A major issue in this approach is the redundancy of extracted components. To solve this, a non-redundant tensor decomposition was developed.

The method proposed in this work performs a tensor decomposition analysis, in order to find components related to different imagined words and to find discriminative features between them. These components represent the features with which a multi-class classifier is trained and evaluated. To the best of our knowledge, this is the first attempt to use tensor decomposition analysis to address the problem of imagined speech discrimination.

3   Proposed Method

3.1   Introduction

The proposed method is summarized in Fig. 1. This flowchart represents the process applied to each epoch of the imagined words per subject. First, from each epoch of a subject, Continuous Wavelet Transform (CWT) is applied to generate a three-dimensional tensor. From this tensor, a PARAFAC analysis is applied and the obtained components are ordered and labeled to the corresponding imagined word. The components are disposed of in such a way that the components of each mode are concatenated making an element-wise correspondence of each component. This process is repeated for every epoch of every imagined word. Finally, the components are disposed into a clustering method to obtain representative prototypes of the data. These prototypes are used to represent the original set of components as a histogram. The method was implemented in MATLAB® in the R2017b version. A detailed explanation of each step is presented below.

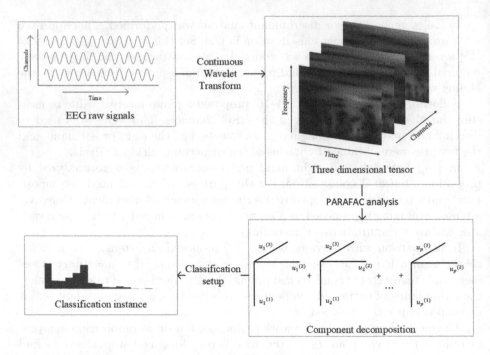

Fig. 1. Feature extraction for one epoch.

3.2 Feature Extraction

The Short time Fourier Transform (STFT) is a common method for frequency analysis. It is an extension of Fourier transform which attempts to analyze the frequency components of a signal in steps of time. A major drawback of this method is the lack of temporal resolution for high-frequency components and lack of spectral resolution for low-frequency components [27]. To solve this issue, the Wavelet Transform (WT) was proposed, which offers a simultaneous localization in time and frequency domain [21]. It is based on the scaling and translation of a basis function named as mother wavelet.

The frequency feature extraction a Morlet Continuous Wavelet Transform (CWT) was applied following the implementation of [17]. This feature extraction is performed for each epoch of the imagined word. Thus, each epoch was represented as a three-dimensional tensor.

A tensor is a multi-dimensional or N-way array, where N is the tensor order. The order N of a tensor is the number of dimensions and it is also referred to as way or mode [8]. A tensor $\mathcal{X}$ can be seen in Fig. 2, where the modes are time, channels and frequency, and e represents the epoch number. The mathematical notation used in this work follows the descriptions of [18].

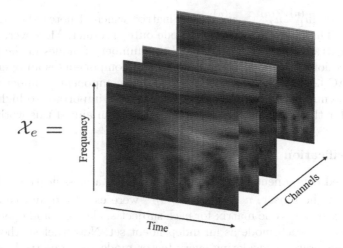

Fig. 2. Tensor example.

3.3 Parallel Factor Analysis

Multilinear algebra introduces methods for tensor decomposition analysis. One of them is the Parallel Factor Analysis (PARAFAC) or Canonical Decomposition (CP), it was simultaneously proposed by [3,10], based on [11]. This method performs an Alternated Least Squares (ALS) approximation of each tensor mode projection. Unlike other component extraction methods, PARAFAC is able to disregard some constraints, as orthogonality or independence.

In Fig. 3, a three dimensional tensor is decomposed into P components in each dimension.

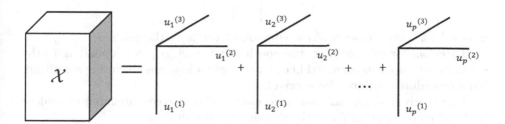

Fig. 3. Tensor decomposition into P factors [18].

Let $\mathcal{X}$ be a N-dimensional tensor, the PARAFAC decomposition is represented in Eq. 1.

$$\mathcal{X} = \sum_{p=1}^{P} u_p^{(1)} \circ u_p^{(2)} \circ \cdots \circ u_p^{(N)} \tag{1}$$

where $U^{(n)} = [u_1^{(n)}, u_2^{(n)}, ..., u_P^{(n)}]$ is a matrix which denotes the components for mode n. The operator $\circ$ is the n-mode outer product. Moreover, P are the number of extracted components and N the number of modes in the tensor.

The selection of an appropriated number of components is an open problem in PARAFAC [2]. For the proposed method, the components number were fixed to match the number of channels of the signal. It is important to highlight that the search for the best components number is not an aim of this work.

3.4   Classification Setup

The extracted components in U were disposed into a classification scheme. For each mode of the tensor, $n \times p$ vectors $u_k^{(n)}$ were extracted and concatenated into a matrix $A_{s,C_i,e}$ (one matrix for each subject s, class C_i, and epoch e). This scheme considers each mode as an independent set. Nevertheless, these components were generated considering every tensor mode and preserved information of the relation among the different modes.

$$A_{s,C_i,e} = \begin{bmatrix} u_1^{(1)} & u_2^{(1)} & \cdots & u_p^{(1)} \\ u_1^{(2)} & u_2^{(2)} & \cdots & u_p^{(2)} \\ \vdots & \vdots & \ddots & \vdots \\ u_1^{(n)} & u_2^{(n)} & \cdots & u_p^{(n)} \end{bmatrix}$$

Each factor $u_k^{(n)}$ is a vector with the form

$$u_k^{(n)} = \begin{bmatrix} u_k^{(n)}(I_1) \\ u_k^{(n)}(I_2) \\ \vdots \\ u_k^{(n)}(I_m) \end{bmatrix}$$

where I_m are the elements of the component vector in the mode n.

Later on, for each subject, the epoch matrices $A_{s,C_i,e}$ corresponding to the same class C_i are concatenated in order to apply a k-means clustering algorithm. This procedure continues for every class.

Once each class has an associated cluster, they are concatenated to achieve a global representation D_s of the signals for the s subject.

The next step is to analyze the elements of every matrix $A_{s,C_i,e}$ and, by means of Euclidean distance, to find and replace each row with the closest prototype in the global representation D_s. This process will transform the matrix A in a sequence of elements of D_s.

This sequence of elements is later transformed into a histogram, which is associated with a class C_i and become a classification instance. The histogram generation, gives as result the conversion of each epoch e in the matrices $A_{s,C_i,e}$ into one single histogram.

Finally, a linear SVM classifier is applied to the set of histograms of each subject. Each classifier was evaluated using a 10-fold cross-validation strategy. This process is summarized in Fig. 4.

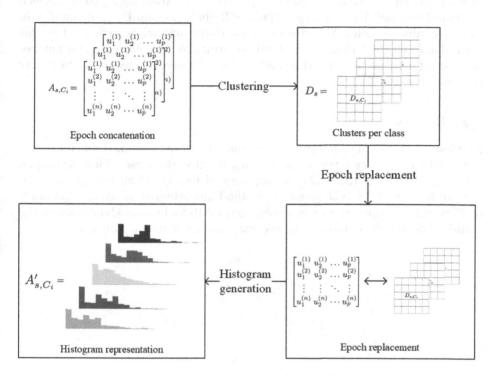

Fig. 4. Classification setup for one subject.

4   Experiments and Results

4.1   Datasets

The first dataset was taken from [24]. The EEG of twenty-seven native Spanish speaking subjects was recorded from fourteen channels (AF3, AF4, F3, F4, F7, F8, FC5, FC6, P7, P8, T7, T8, O1, O2) at 128 Hz sample rate. The data consists of five imagined speech Spanish words ("Arriba", "Abajo", "Izquierda", "Derecha", "Seleccionar"), which are translated in English as ("Up", "Down", "Left", "Right", "Select"), repeated thirty three times each one, with a rest period between repetitions. These recordings were taken in a controlled environment without sound nor visual noise. However, the acquisition protocol main drawback is that the words were presented to the users in a sequential sequence, to reduce bias a random selection of the words would be preferable.

The second dataset was introduced by [20], the EEG of fifteen native Spanish speaking subjects were obtained through six electrodes located in F3, F4, C3, C4, P3, P4 at a 1024 Hz sample rate. The vocabulary consists of six Spanish words ("Arriba", "Abajo", "Izquierda", "Derecha", "Adelante", "Atrás"), which are translated as ("Up", "Down", "Left", "Right", "Forward", "Backward"), in forty recording epochs. Nevertheless, only thirty-nine trials were used in this work due to the fact that public database available, contains only this number of epochs for subjects five and six. Also, the data were downsampled to 128 Hz to match the first database.

4.2 Results

Previous works in imagined speech discrimination have obtained different average results for twenty seven subjects using the first database, [23] reporting an accuracy of 60.11 ± 12.71, [24] an accuracy of 68.45 ± 15.89 and [9] an accuracy of 63.97 ± 13.24. The proposed method has achieved an average accuracy of 75.90 ± 7.13. These results were obtained by 10-fold cross-validation, and the standard deviation represents the accuracy variation among subjects.

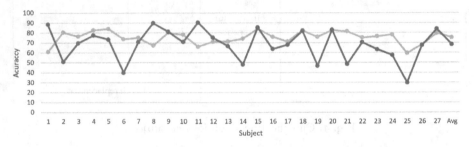

Fig. 5. Results comparison per subject of the proposed work (blue) and [24] (orange) (Color figure online).

The results obtained by the proposed method, besides the higher accuracy, also showed a lower standard deviation among subjects. This is expected due to the expensive feature extraction, which extends the vectorization and matricization methods. Thus, the proposed method is computationally slower but more robust in comparison with previous methods over the first database. The proposed method results per subject, using a cluster number k of 250, are shown in Table 1.

In Table 2 the results of the proposed method are compared with the database proposed by [20]. Nevertheless, such work had only numerically reported the first three subjects, the remaining subjects are presented graphically and followed a similar distribution. For this database, a number of 6000 clusters k is proposed.

These results achieved a great accuracy improvement. In [20], the low accuracy was related to the randomization of the stimuli in the acquisition protocol

Table 1. Proposed method accuracies in [24] database

Subject	Accuracy	Subject	Accuracy	Subject	Accuracy
1	61.03	11	66.10	21	81.32
2	80.59	12	71.03	22	75.04
3	76.40	13	71.32	23	76.43
4	82.61	14	74.08	24	78.13
5	84.12	15	84.23	25	59.41
6	73.82	16	76.03	26	67.90
7	75.04	17	71.10	27	79.38
8	67.24	18	82.43	Avg	75.90 ± 7.13
9	80.00	19	75.88		
10	78.20	20	82.46		

Table 2. Accuracies comparison in [20] database

Subject	Proposed method	[20]
1	66.28	19.31
2	56.20	19.58
3	56.66	19.92
Avg	59.70 ± 5.7	19.60 ± 0.25

and to the use of basic spectral features extraction. Nevertheless, the presence of imagined speech discriminative data in the EEG signals were not discarded.

5 Conclusions

The proposed method allowed the extraction of components which consider inner relations among three different variables in the EEG signal, i.e., time, frequency and space. These components were analyzed in order to successfully discriminate multiple imagined words in two different databases.

In comparison with previous works, a higher classification accuracy rate in both imagined speech databases was achieved. Obtained results suggest that the PARAFAC extracted features allow the discrimination of imagined speech in EEGs. Moreover, these results support the discrimination performance of PARAFAC on motor imagery discrimination.

A statistical One-way ANOVA test of the proposed method and [24] showed that there was a significant effect $[F(1, 52) = 4.48, p = 0.0391]$. In the same way, the analysis of the proposed method and [20] showed a significant effect $[F(1, 4) = 147.9, p = 0.0003]$.

A considerable drawback of the method is the computational cost, which may difficult the direct implementation over a BCI system. Therefore, additional analyses are required to implement a faster analysis based on tensor decomposition.

Acknowledgments. The present work was partially supported by CONACyT (scholarship 487560). Also, the authors thank the support of the Italian Foreign Affairs and Cooperation Ministry, and the International Cooperation for Development Mexican Agency for the project MX14MO06.

References

1. Barzegaran, E., Vildavski, V.Y., Knyazeva, M.G.: Fine structure of posterior alpha rhythm in human EEG: frequency components, their cortical sources, and temporal behavior. Sci. Rep. **7**(1), 8249 (2017)
2. Bro, R.: PARAFAC: tutorial and applications. Chemometr. Intell. Lab. Syst. **38**(2), 149–171 (1997)
3. Carroll, J.D., Chang, J.J.: Analysis of individual differences in multidimensional scaling via an N-way generalization of "eckart-young" decomposition. Psychometrika **35**(3), 283–319 (1970)
4. Cichocki, A.: Tensor decompositions: a new concept in brain data analysis? J. SICE Control Measur. Syst. Integr. Special Issue; Measur. Brain Funct. Bio-Signals **7**, 507–517 (2011). 7 (05 2013)
5. Cichocki, A., et al.: Tensor decompositions for signal processing applications: from two-way to multiway component analysis (2015). https://doi.org/10.1109/MSP. 2013.2297439
6. Cong, F., Phan, A.H., Lyytinen, H., Ristaniemi, T., Cichocki, A.: Classifying healthy children and children with attention deficit through features derived from sparse and nonnegative tensor factorization using event-related potential. In: Vigneron, V., Zarzoso, V., Moreau, E., Gribonval, R., Vincent, E. (eds.) LVA/ICA 2010. LNCS, vol. 6365, pp. 620–628. Springer, Heidelberg (2010). https://doi.org/ 10.1007/978-3-642-15995-4_77
7. Dalhoumi, S., Dray, G., Montmain, J.: Knowledge transfer for reducing calibration time in brain-computer interfacing. In: Proceedings - International Conference on Tools with Articial Intelligence, ICTAI (2014)
8. Devarajan, K.: Matrix and Tensor Decompositions, pp. 291–318. Springer, Boston (2011)
9. García-Salinas, J.S., Villaseñor-Pineda, L., Reyes-García, C., Torres-García, A.A.: Selección de parámetros en el enfoque de bolsa de características para clasificación de habla imaginada en electroencefalogramas. Res. Comput. Sci. **140**(140), 123–133 (2017)
10. Harshman, R.A.: Foundations of the PARAFAC procedure: models and conditions for an "explanatory" multimodal factor analysis. In: UCLA Working Papers Phonetics, vol. 16, no. 10, pp. 1–84 (1970)
11. Hitchcock, F.L.: The expression of a tensor or a polyadic as a sum of products. J. Math. Phys. **6**(1–4), 164–189 (1927). https://doi.org/10.1002/sapm192761164
12. Ji, H., Li, J., Lu, R., Gu, R., Cao, L., Gong, X.: EEG classification for hybrid brain-computer interface using a tensor based multiclass multimodal analysis scheme. Comput. Intell. neurosci. **2016**, 51 (2016)

13. Lee, H., Kim, Y.D., Cichocki, A., ChoI, S.: Nonnegative tensor factorization for continuous EEG classification. Int. J. Neural Syst. **17**(04), 305–317 (2007). https://doi.org/10.1142/S0129065707001159
14. Li, J., Zhang, L.: Phase interval value analysis for the motor imagery task in BCI. J. Circ. Syst. Comput. **18**(08), 1441–1452 (2009). https://doi.org/10.1142/S0218126609005861
15. Li, J., Zhang, L.: Regularized tensor discriminant analysis for single trial EEG classification in BCI. Pattern Recogn. Lett. **31**(7), 619–628 (2010). https://doi.org/10.1016/j.patrec.2009.11.012
16. Li, J., Zhang, L., Tao, D., Sun, H., Zhao, Q.: A prior neurophysiologic knowledge free tensor-based scheme for single trial EEG classification. IEEE Trans. Neural Syst. Rehabil. Eng. **17**(2), 107–115 (2009). https://doi.org/10.1109/TNSRE.2008.2008394
17. Lilly, J.M., Olhede, S.C.: Higher-order properties of analytic wavelets. Trans. Sig. Proc. **57**(1), 146–160 (2009). https://doi.org/10.1109/TSP.2008.2007607
18. Lu, H., Plataniotis, K., Venetsanopoulos, A.: Multilinear Subspace Learning: Dimensionality Reduction of Multidimensional Data. Machine Learning & Pattern Recognition Series. Chapman & Hall/CRC Press, London/Boca Raton (2013)
19. Miwakeichi, F., Martínez-Montes, E., Valdés-Sosa, P.A., Nishiyama, N., Mizuhara, H., Yamaguchi, Y.: Decomposing EEG data into space-time-frequency components using parallel factor analysis. NeuroImage **22**(3), 1035–1045 (2004). https://doi.org/10.1016/j.neuroimage.2004.03.039
20. Pressel Coretto, G.A., Gareis, I.E., Rufiner, H.L.: Open access database of EEG signals recorded during imagined speech. In: Proceedings SPIE, vol. 10160 (2017). https://doi.org/10.1117/12.2255697
21. Sifuzzaman, M., Islam, M.R., Ali, M.Z.: Application of wavelet transform and its advantages compared to Fourier transform. J. Phys. Sci. **13**, 121–134 (2009)
22. Tao, D., Li, X., Wu, X., Maybank, S.J.: General tensor discriminant analysis and Gabor features for gait recognition. IEEE Trans. Pattern Anal. Mach. Intell. **29**(10), 1700–1715 (2007). https://doi.org/10.1109/TPAMI.2007.1096
23. Torres-García, A.A., Reyes-García, C.A., L., L.V.P., Ramirez, J.: Analisis de señales electroencefalograficas para la clasificacion de habla imaginada. Revista mexicana de ingeniería biomedica **34**, 23–39 (2013)
24. Torres-García, A.A., Reyes-García, C.A., Villaseñor-Pineda, L., García-Aguilar, G.: Implementing a fuzzy inference system in a multi-objective EEG channel selection model for imagined speech classification. Expert Syst. Appl. **59**, 1–12 (2016). https://doi.org/10.1016/j.eswa.2016.04.011
25. Wolpaw, J.R., Birbaumer, N., McFarland, D.J., Pfurtscheller, G., Vaughan, T.M.: Brain-computer interfaces for communication and control. Clin. Neurophysiol.: Official J. Int. Fed. Clin. Neurophysiol. **113**(6), 767–91 (2002)
26. Zhao, Q., Caiafa, C.F., Cichocki, A., Zhang, L., Phan, A.H.: Slice oriented tensor decomposition of EEG data for feature extraction in space, frequency and time domains. In: Leung, C.S., Lee, M., Chan, J.H. (eds.) ICONIP 2009. LNCS, vol. 5863, pp. 221–228. Springer, Heidelberg (2009). https://doi.org/10.1007/978-3-642-10677-4_25
27. Zhu, X., Kim, J.: Application of analytic wavelet transform to analysis of highly impulsive noises. J. Sound Vibr. **294**, 841–855 (2006)

Robotics and Computer Vision

A New Software Library for Mobile Sensing Using FIWARE Technologies

Alicia Martinez[1](✉), Hugo Estrada[2], Fernando Ramírez[1], and Miguel Gonzalez[3]

[1] Computer Science Department, TECNM/CENIDET, Cuernavaca, Morelos, Mexico
{amartinez, fernandoramirez16c}@cenidet.edu.mx
[2] Innovation Department, INFOTEC, Mexico City, Mexico
hugo.estrada@infotec.mx
[3] Computer Science Departament, ITESM Mexico's State, Ciudad López Mateos, Mexico
mgonza@itesm.mx

Abstract. Current concept of Next Generation of Internet brings new technologies to the common life of people, such as Internet of Things, Cloud Computing or Big Data. In this new context, smartphones are considered one of the most relevant devices to produce information about the users and their environment. The smartphones are a basic technology for applications for mobility, health, of security. At present, several platforms have been proposed, which consider the development of Future Internet applications, such as the case of FIWARE, RAMI or IDS. However, no generic modules can be found to perform the sensing of data context information from smartphones. In this paper, we present a software library to collect context data from different sensors of smartphones and also, it permits to send that data to the FIWARE Cloud using standard protocols. This new library enables programmers to produce novel applications using the concept of human-as-a-sensor using smartphones.

Keywords: NGSI library · Generic module · FIWARE · Internet of Things

1 Introduction

At present time, most of next generation Internet technologies are related to Internet of Things (IoT) [1] concept. Furthermore, it incorporating new technologies as cloud computing or Big Data. In this context, several platforms have been developed in order to support IoT such as: Smart+Connected Communities [2], Amazon Web Services IoT [3], IBM Watson IoT Platform [4], RAMI [5], IDS [6], FIWARE [7], etc. From all these, FIWARE is one of the most well founded platform for future Internet. FIWARE is compose by a set components, called Generic Enablers, which are components developed for horizontal purposes that could be used to implement robust applications in very different applications domains. In most of the cases, developing new applications with these platforms imply specialized knowledge by developers, for example

© Springer Nature Switzerland AG 2018
I. Batyrshin et al. (Eds.): MICAI 2018, LNAI 11289, pp. 253–263, 2018.
https://doi.org/10.1007/978-3-030-04497-8_21

Open Stack [8] technologies, and also different development languages, for example, Angular [9] to define the front-end of the applications.

In FIWARE, and in most of the current future Internet platforms, one of the main technologies for Internet of Things applications is the smartphones. These applications use the resources of smartphones to produce data about the user context; this is the case of applications as Waze, Google Maps, Uber, etc. However, even the relevance of smartphones as data sources, the review of FIWARE Generic Enabler catalogue demonstrate that there is not a generic component for mobile devices that enables the context information management based on the standard NGSI [10]. This is a relevant lack because the increasing of the concept of human as a sensor based on mobile devices.

In this paper, a new generic component has been developed to capture context data from smartphones sensors and to send this data to the FIWARE cloud based on the NGSI standard [10]. The proposed library can be a useful tool for those applications that take data from smartphones for security, health or mobility domains, isolating the effort of developers to automate the sending of information of sensors to the cloud. This component was developed in the context of the European Project SmartSDK that has the objective of developing data models and reference architectures for smart services based on FIWARE.

The rest of the paper is structured as follow: Sect. 2 shows the basic concepts for this research work. Section 3 describes the NGSI library. Section 4 presents the tests and results. Finally Sect. 5 presents conclusions and future work.

2   Background

2.1   Internet of Things Platforms

The IoT platforms can be defined as, services designed to facilitate the information society through the interconnection of (physical and digital) things based on information and communication technologies running in the cloud [11]. This definition makes explicit the strong link between IoT and the cloud computing for offering platforms that permit the deployment of software applications that manages large scale data context information coming for things connected to internet.

Currently, the market of IoT Platforms is increasing and current trends indicated that more that 80% of companies believe in the IoT as a relevant area to develop business. The connectivity M2M (Machine-to-Machine) is one of the basis for the IoT platforms, which is focused in establish the conditions that enables devices to communicate among them in a bidirectional manner using a telecommunications network [12]. In this context, some IoT platforms are oriented to manage the wireless communication among sensors. PAVENET [13] is a wireless sensor network testbed, which supports the development of wireless sensor network technology. PAVENET has four characteristics: hardware level modularization, dual-CPU architecture, hard real-time transaction support, and network layering APIs. This features help developers to develop not only application functions, but also wireless communication functions. LiteOS [14] is a multithreaded software platform for wireless sensor networks application development. This platform offer functionalities to define hierarchical file system and a wireless shell for user interaction. It also offer a kernel support for dynamic loading and native execution of multithreaded applications, and an online debugging, dynamic memory, and file system assisted communication stacks. This platform has been developed for non-experts developers to create wireless sensors networks.

RAMI [5] is a new Reference Architecture for Industry 4.0. RAMI is an abstract model for the participants of Industry 4.0 compatible with communication and networking. RAMI has a structure compatible with other IoT approaches and a special focus on the manufacturing industry and their specific additional needs. Furthermore, RAMI allows step-by-step migration from current systems.

IDS Platform [6] consists of a community of more than 30 large German and European Companies that cooperate in a pre-competitive, publicly funded innovation project, which involves 11 Fraunhofer institutes for developing IDS reference architecture. For Industry 4.0 IDS forms a Semantic Model bridge between Shop & Office Floor. This Data-driven Service Architecture defines governance, data services, the functional domain and security through the following basic principles: on demand interlinking, linked light semantics, security with industrial containers and certified roles. FIWARE [7] is a future Internet platform that provides a rather simple yet powerful set of APIs that ease the development of Smart Applications in multiple vertical sectors. The specifications of these APIs are public and royalty-free. Besides, an open source reference implementation of each of the FIWARE components is publicly available to community.

At the present, FIWARE is an open source API platform, and it has released 42 standardized software components aimed at helping startups and enterprise build the next generation of smart applications and services for cities, industries, e-health or agribusiness. FIWARE provides a set of tools for different functionalities. It is an innovation ecosystem for the creation of new applications and Internet services. It is especially useful in terms of smart cities, as it ensures the interoperability and the creation of standard data models.

2.2 FIWARE Ecosystem

At the present, FIWARE is an open source API platform, and it has released 42 standardized software components aimed at helping startups and enterprise build the next generation of smart applications and services for cities, industries, e-health or agribusiness. FIWARE provides a set of tools for different functionalities. It is an innovation ecosystem for the creation of new applications and Internet services. It is especially useful in terms of smart cities, as it ensures the interoperability and the creation of standard data models.

From the beginning, FIWARE was designed as an open and standard platform, based on open source to foster the creation of the needed standards to develop smart services and applications in different domains [15], such as smart cities, logistics, energy, agriculture, smart industry, etc. [7].

The standard proposed by FIWARE for context data has been designed to manage the entire life cycle of the context information, including updates, queries, registrations and subscriptions. This standard communication protocol, called FIWARE-NGSI [10], permits developers to use different kind of sensors with different protocols, and unify all the communication of data to be stored in the FIWARE Cloud. In order to create standard information, FIWARE promotes data models for representing context information. The main elements of the NGSI data models are contextual entities, attributes and metadata. In this sense, the data models of FIWARE can be share in the community because they are in a standard format. The FIWARE approach promotes the use of standardized data models for weather, pollution, traffic, alerts, etc.

2.3 Orion Context Broker

The Orion Context Broker is a key component of FIWARE to enable the data context ingestion and also to enable the subscription of applications to the data context. The main concept of the Orion Context Broker is that data context producer can generate information and placed this in the cloud, without a previous knowledge of the users or application that will use the data. For example, the smart mobility software application can generate a traffic alert, which will be sent to the cloud, and later on, the rest of application users can receive the anonymized notification of the alert in their devices. In this context, the Orion Context Broker is a bridge that enable external applications to manage for IoT devices in a transparent and easy manner, hidden the complexity of managing the capture of the data [16]. The Orion Context Broker allows developers to manage all the whole lifecycle of context information including updates, queries, registrations and subscriptions. The Orion Context Broker allows register context elements and manages them through updates and queries. In addition, the context information could be subscribed in order to receive some notification when the context element change [16].

The Context information is represented through values assigned to attributes that characterize those entities relevant to your application. The Context Broker is able to handle context information at large scale by implementing standard REST APIs [16]. The context information may come from many different sources for example: already existing systems, users, through mobile apps, sensor networks, etc. Figure 1 shows the context broker scheme, where the Context Broker can receive context data for very different application domains.

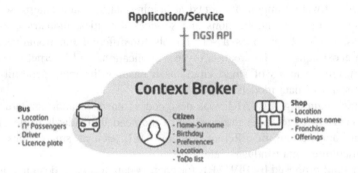

Fig. 1. The Orion Context Broker Schema

2.4 FIWARE NGSI API

The standard NGSI v2 of FIWARE is intended to manage the entire lifecycle of context information, including updates, queries, registrations, and subscriptions [10]. The FIWARE NGSI (Next Generation Service Interface) API defines:

- A data model for context information, based on a simple information model using the notion of context entities.
- A context data interface for exchanging information by means of query, subscription, and update operations.
- A context availability interface for exchanging information on how to obtain context information.

The main elements in the NGSI data model are context entities, attributes and metadata, as shown in Fig. 2. The difference among these elements is that attributes describe the entity and metadata describe attributes.

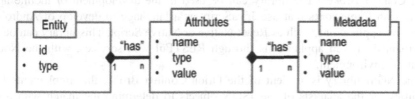

Fig. 2. Elements of the NGSI data model.

Context entities, or simply entities, are the center of gravity in the FIWARE NGSI information model. An entity represents a thing, i.e., any physical or logical object (e.g., a sensor, a person, a room, an issue in a ticketing system, etc.). Each entity has an entity id. Furthermore, the type system of FIWARE NGSI enables entities to have an entity type. Entity types are semantic types; they are intended to describe the type of thing represented by the entity. Each entity is uniquely identified by the combination of its id and type.

Context attributes are properties of context entities. In the NGSI data model, attributes have an attribute name, an attribute type, an attribute value and metadata

- The attribute name describes properties that the attribute value represents in the entity. For example, current speed.
- The attribute type represents the NGSI value type of the attribute value. Note that, FIWARE NGSI has its own type system for attribute values, so NGSI value types are not the same as JSON types.

The attribute value contains: the actual data, optional metadata describing properties of the attribute value like e.g. accuracy, provider, or a timestamp.

Context metadata are used in FIWARE NGSI in several places, one of them being an optional part of the attribute value. Similar to attributes, each piece of metadata has:

- A metadata name, describing the role of the metadata in the place where it occurs. For example, the precision of the name of the metadata indicates the accuracy of a given attribute value.
- A metadata type, describing the NGSI value type of the metadata value. A metadata value containing the actual metadata.

In this research work, the NGSI protocol has been used to communicate the data obtained of the smartphone sensors and for sending this data to the Cloud of the FIWARE ecosystem. This enables to authorize user to use this data from cloud to develop smart services for health, security, mobility scenarios, etc.

3   SDK Library for NGSI in JavaScript Proposal

The NGSI library for JavaScript is a software tool with the aim of transforming JSON entities to NGSI data models, which can be manipulated or operated by the FIWARE Orion Context Broker. The library can be used in the development of mobile applications with frameworks that use JavaScript as a language to develop of Android or IOS native applications, such as React Native or Native Script. This library can be also implemented in web applications through RESTFul web services, with the NodeJS execution environment.

The NGSI library is a client of the Orion Context Broker that implements functionalities for the analysis of the JSON objects to determine the match with a data model, and also, functionalities to transform JSON objects to a NGSI v2 entities.

Currently, several libraries have been developed with a similar functionality. Our proposal is different to other because this library is divided in two modules npm. In this approach, the user can use the library as a unique entity, but also both modules can be used in an independent manner. The library is fully functional and it is currently used in the Smart Security scenario.

3.1   Architecture

The architecture of the NGSI library is composed by two modules npm: ngsi-parser and ocb-sender. These modules can be imported in only one JavaScript project. Figure 3 shows the modules ngsi-parser and ocb-sender of the architecture of the library.

3.2   NGSI-Parser Module

The ngsi-parser module has the objective of analyzing and converting the syntax of a non-structured JSON object or attribute to transform it in a NGSI entity context. Additionally, this module provides the functionality to verify if the entity fulfills with the standard specification of a FIWARE data model. The library verifies if the original JSON structure match with the corresponding FIWARE data model. These data model can be located in the repository dataModels of the account Github of the SmartSDK project.

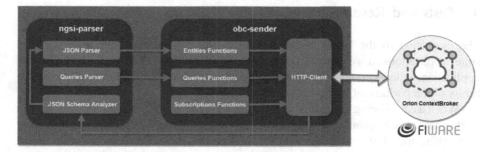

Fig. 3. Architecture of the NGSI library.

The ngsi-parser contains three basic elements to perform the analysis of the JSON objects: (a) The JSON Parser includes the function needed for the analysis and transformation of a non-structured JSON object to one that fulfill with the NGSI standard. (b) The Queries Parser is the responsible element to interpret JSON objects to produce context queries to obtain specific data from the Orion Context Broker, and finally; (c) the Data JSON Schema Analyzer is the responsible to determine if a JSON object fulfill or not fulfill with a data models and also it generates the list of errors in the match between the JSON schema and the data models.

In this research work, three main components are used in the research project presented in this paper

- Orion Context Broker. This component is used to capture context data form the sensors of the Smartphone and it represents the data in the protocol NGSIV2.
- FIWARE data Models. The data models used in this research work are: Alert, Device, Device Model, Road, Road Segment OffStreetParking and Building. These data models are used to standardize the context information capture from the Smartphone.
- API QuantumLep. This component is used to store the data from sensors as time series; in this way it is possible to visualize data in a long period of time.

3.3   OCB-Sender Module

The module ocb-sender has the main objective of manipulating the context information of NGSI context entities and/or FIWARE data models in order to send this information to one instance of the Orion Context Broker.

The ocb-sender module is composed by four elements: first three elements are used to encapsulate the functionalities of the client of the Orion Context Broker: (a) the Entities Functions implements the functions to manipulate the entities of the Orion Context Broker, (b) the Queries Functions considers the functions for personalized queries to the Orion Context Broker, (c) Subscriptions functions implement the functions to manipulate the subscriptions of the Orion Context Broker, (d) the HTTP-Client is the responsible for the connection of the Orion Context Broker. This component is also used for the ngsi-parser to obtain JSON schemas for a repository.

4   Tests and Results

The evaluation of the NGSI library was made through its implementation into two smart applications: a web application, called FIWARE Driving Monitor, and a mobile application, called DrivingApp. These applications enable the monitoring of cars in the limits of the parking of a company.

When a car is being conducted in a wrong way, the application DrivingApp sent an alert to the security guard of the enterprise. Also, it sends alerts of all users near of same area where the alert was generated. Figure 4 shows the detailed architecture of the application where the library was implemented.

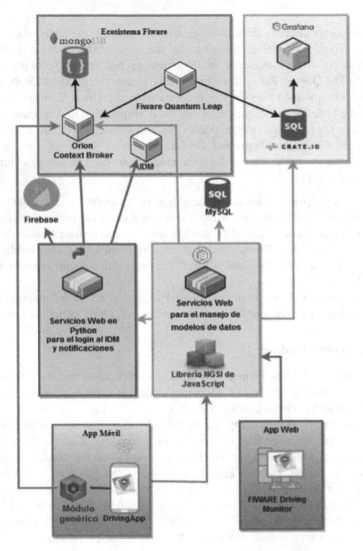

Fig. 4. Detailed architecture of current NGSI library

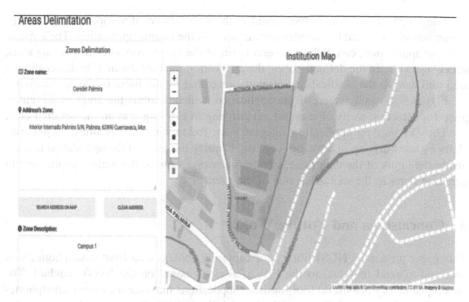

Fig. 5. Screen of the application to delimite areas.

Fig. 6. Module for generating alerts from mobile devices

The architecture of the application uses several components of the FIWARE platform, such as IDM, the Orion Context Broker, Grafana, and Quantum Leap. These components allow us to manage the data context and the store of the information in historic database.

The NGSI library was implemented in the web services developed using node.js. These services are used by the web application and the mobile application. The Driving Monitor application, can delimit several zones of the enter-prise, such as parking slots, street segments. In the defined zones, the application monitor the mobile devices in the area and receive the notifications of alerts generated by the users of the application.

Figure 5 shows a screen of the application, which delimits the areas of the enter-prise for detecting alerts of users of cars driving in a wrong way in the selected area.

The DrivingApp application integrates the following functionalities: to generate alerts by users driving in a wrong way, and to query the users of the application that are inside the limits of the enterprise. Figure 6 shows a view of the mobile application to generate alerts in the mobile device of the user.

5   Conclusions and Future Work

This paper presents a NGSI library for capturing context data from smartphones sensors and to send this data to the FIWARE cloud based on the NGSI standard. The developed library is a useful tool for those applications that take data from smartphones for security, health, mobility domains, etc., isolating the effort of developers to auto-mate the sending of information of sensors to the cloud, which is a common tasks in the mobile applications.

The NGSI library proposal follows the same open approach of FIWARE. Thus, this is available in the FIWARE ecosystem as a specific generic enabler. In this context, any developer can use the library, in order to create applications that take data form smartphones.

The NGSI library is currently being successfully used in security and smart cities for obtaining data and alerts form users. The future work of this work is the devel-opment of more specific domain applications using the proposed NGSI library.

Acknowledgments. This research work has been partially funded by European Commission and CONACYT, through the SmartSDK project.

References

1. PPP: F. I. Led by industry, driven by users Addressing the challenge of Internet development in Europe. https://www.fi-ppp.eu/. Accessed 07 Apr 2017
2. Cisco: Smart + Connected Communities: Changing a City, a Country, the World. http://www.cisco.com/c/dam/en_us/solutions/industries/docs/scc/09CS2326_SCC_BrochureFor West_r3_112409.pdf. Accessed 06 July 2018
3. Amazon: AWS IoT - Amazon Web Services. https://aws.amazon.com/es/iot/. Accessed 06 July 2018
4. Macgillivray, C.: The Platform of Platforms in the Internet of Things. https://www.avnet.com/wps/wcm/myconnect/onesite/97cedd40-2ea0-414e-a091-6d844b70a487/IBM-The-Platform-of-Platforms-IoT.pdf?MOD=AJPERES&attachment=true&id=1488926185691.   Accessed 06 July 2018

5. Reference Architectural Model Industrie 4.0 (RAMI 4.0). https://www.plattform-i40.de/I40/Redaktion/EN/Downloads/Publikation/rami40-an-introduction.pdf?__blob=publicationFile&v=7. Accessed 06 July 2018
6. Industrial Data Space White paper. https://www.fraunhofer.de/content/dam/zv/en/fields-of-research/industrial-data-space/whitepaper-industrial-data-space-eng.pdf. Accessed 06 July 2018
7. FIWARE: https://www.fiware.org/about-us/. Accessed 06 July 2018
8. OpenStack Software: What is OpenStack? https://www.openstack.org/software/. Accessed 06 July 2018
9. AngularJS: https://angularjs.org/. Accessed 06 July 2018
10. FIWARE-NGSI: http://fiware.github.io/specifications/ngsiv2/stable/. Accessed 06 July 2018
11. Pratim Ray, P.: A survey of IoT cloud platforms. Future Comput. Inf. J. 1(1–2), 35–46 (2016)
12. Boswarthick, D., Elloumi, O., Hersent, O.: M2M Communications: A System Approach, p. 727. Wiley, Hoboken (2012)
13. Saruwatari, S., Kashima, T., Minami, T., Morikawa, H., Aoyama, T.: PAVENET: a hardware and software framework for wireless sensor networks. Trans. Soc. Instrum. Control Eng. 1(1), 74–84 (2005)
14. Cao, Q., Abdelzaher, T., Stankovic, J., He, T.: The LiteOS operating system: towards unix like abstraction for wireless sensor networks. In: Proceedings of the 7th International Conference on Information Processing in Sensor Networks, St. Louis, MO, USA, p. 233 (2008)
15. Telefónica. FIWARE. El estándar que necesita el IoT Telefónica. https://iot.telefonica.com/blog/2016/09/es-fiware-estandar-iot. Accessed 06 July 2018
16. FIWARE-Orion Context Broker. https://fiware-orion.readthedocs.io/en/master/. Accessed 06 July 2018

Free Model Task Space Controller Based on Adaptive Gain for Robot Manipulator Using Jacobian Estimation

Josué Gómez[(✉)], Chidentree Treesatayapun, and América Morales

Robotics and Advanced Manufacturing Programm CINVESTAV-Saltillo,
25903 Ramos Arizpe, Mexico
josuegomezcasas@gmail.com

Abstract. A Free Model Task Space Controller (FMTSC) is presented in this paper for an omnidirectional mobile manipulator. However, it is well known the difficulty to know the precsise details of the robotic system and commonly limited for the accuracy of the kinematic and dynamic model, the model based methods are not sufficient, so far. Therefore the use of available information like joints velocities and robot tip velocity allow to estimate the robot Jacobian matrix information, without any requirement of mathematical model. An adaptive Kalman filter is computed to estimate Jacobian to deal with the adaptive robot control in the task space. The control law is developed with the Jacobian estimate for Strong Tracking Kalman Filter (STKF) algortihm. The control algorithm is intended for nonlinear discrete-time system (robot) which provides adpative control gain for taks space controller, designed by Fuzzy Rules Emulated Network Adaptive Gain (FRENAG). The performance of the controller is validated with Kuka youBot mobile manipulator plataform experiments.

Keywords: Free model · Robot manipulator · Jacobian estimate
adaptive Kalman filter · Task space control · Adaptive gain

1 Introduction

Conventionally, there are two frameworks proposed to perform robot manipulators, including model based methods and model free ones. On the other hand, limited for the accuracy of the kinematic and dynamic model, the model based methods are not sufficient, so far. Furthermore, robotic manipulators are nonlinear systems with unknown and changeable parameters, therefore mathematical models of such systems are complex, and quite often exclude some physical phenomena, such as manipulators tip vibration or deformation, as was exposed by Piotr [6]. The aim of this work is to implement a Free Model Task Space Controller (FMTSC) for n degree of freedom (dof) robot manipulator, with a Jacobian estimate based on Strong Tracking Kalman Filter (STKF) to accomplish an object desired position, see [7,8]. Adaptive gain controller is provided by

© Springer Nature Switzerland AG 2018
I. Batyrshin et al. (Eds.): MICAI 2018, LNAI 11289, pp. 264–275, 2018.
https://doi.org/10.1007/978-3-030-04497-8_22

a neuro-fuzzy network based on Fuzzy Rules Emulated Network (FREN) scheme to minimize the error. The Strong Tracking Kalman Filter (STKF) only requires to be fed with joints velocities and robot tip velocity to make the computation of the Jacobian information. Thus, to obtain the velocity of the robot tip is used a Motion Capture System (MOCAPS) in order to send a signal aproximation of the velocity to the adaptive Kalaman filter. The advantages of the proposed control scheme compared with model frameworks is that the controller can be proved without any prior knowledge of the robot model and the parameters are tuned online, so that the Free Model Task Space controller (FMTSC) with Fuzzy Rules Emulated Network Adaptive Gain (FRENAG) ensure the robustness of the control scheme. The Jacobian estimate and controller were validated with an experimental setup. In Sect. 2 is presented the Jacobian estimation methodology. Section 3 provides the control design. In Sect. 4 experimental results are depicted. Finally, Sect. 5 closes with some conclusions.

2   Robotic Jacobian Estimator

During this section is presented the development discrete-time Jacobian matrix for a robot and the estimation of the Jacobian matrix with an adaptive Kalman filter.

2.1   Discrete-Time Jacobian Matrix

The true state of a system is evolved from the state, the system is given by

$$x = f_x(q_1, \ldots, q_n)$$
$$y = f_y(q_1, \ldots, q_n) \tag{1}$$
$$z = f_z(q_1, \ldots, q_n)$$

where $[x, y, z]$ denotes the robot's end effector in 3-D coordinate, f_x, f_y and f_z are geometric functions of joints position $q_1, \ldots, q_n$. The Eq. (1) represents the end effector position in the task space of the robot, in terms of the joint position q_n. The movement of robot manipulator can be considered as a mapping of geometric Jacobian $Jg(q)$, commonly from the joint velocity $\dot{q}(k)$ to the end effector pose $\chi(q)$

$$\chi(q) = \begin{bmatrix} p(q) \\ o(q) \end{bmatrix} \tag{2}$$

where $p(q)$ is the position and $o(q)$ is the orientation of the end effector. In this work is just considered the position of the end effector for the Eq. (2) and can be rewritten as follows

$$\chi(q) = \begin{bmatrix} p(q) \end{bmatrix} \tag{3}$$

the time derivative of end effector position reveals the velocity relationship and $J_g(q) = J_A(q)$

$$\dot{\chi} = J_A(q)\dot{q} \tag{4}$$

in a practical control system, is usually simplified by the first order approximation within a single control interval as

$$\Delta\chi = J_A(q)\Delta q \tag{5}$$

The Eq. (5) can be expressed within discrete time domain

$$\frac{\chi(k+1) - \chi(k)}{Ts} = J_A(q)\left[\frac{q(k) - q(k-1)}{Ts}\right] \tag{6}$$

where $\Delta q = \omega(k)$ is the velocity of the joints, k is the time index and Ts is the sampling time, the Eq. (6) can be written as:

$$\Delta\chi = [Ts \cdot J_A(q(k))]\,\omega(k) \tag{7}$$

the state vector of $x(k+1), y(k+1)$ and $z(k+1)$ are formed by the entries of the robot Jacobian matrix $J_A(q(k))$

$$[Ts \cdot J_A(q(k))] = \begin{bmatrix} \dfrac{\partial fx}{\partial q_1} \cdots \cdots \dfrac{\partial fx}{\partial q_n} \\[2ex] \dfrac{\partial fy}{\partial q_1} \cdots \cdots \dfrac{\partial fy}{\partial q_n} \\[2ex] \dfrac{\partial fz}{\partial q_1} \cdots \cdots \dfrac{\partial fz}{\partial q_n} \end{bmatrix} \cdot Ts \tag{8}$$

and reorganizing the Jacobian matrix in Eq. (8) as a vector $J_v(q(k))$

$$J_v(q(k)) = \begin{bmatrix} \dfrac{\partial fx}{\partial q_1} \cdots \dfrac{\partial fx}{\partial q_n}, \dfrac{\partial fy}{\partial q_1} \cdots \dfrac{\partial y}{\partial q_n}, \dfrac{\partial fz}{\partial q_1} \cdots \dfrac{\partial fz}{\partial q_n} \end{bmatrix}^T \cdot Ts \tag{9}$$

The a measurement matrix $H(k)$ in the Eq. (10) is a block diagonal matrix defined as

$$H(k) = \begin{bmatrix} [\omega_1(k) \ldots \omega_n(k)] & & 0 \\ & \ddots & \\ 0 & & [\omega_1(k) \ldots \omega_n(k)] \end{bmatrix} \tag{10}$$

$\omega_n(k)$ are defined as a velocity of the joints, therfore $H(k)$ is a matrix which contains the robot control signals. For n dof simplifying operations for $\Delta x(k+1)$, $\Delta y(k+1)$ and $z(k+1)$ as,

$$\Delta x(k+1) = \sum_{i=1}^{n} \omega_i(k) \cdot \frac{\partial fx(k)}{q_i(k)}$$

$$\Delta y(k+1) = \sum_{i=1}^{n} \omega_i(k) \cdot \frac{\partial fy(k)}{q_i(k)} \tag{11}$$

$$\Delta z(k+1) = \sum_{i=1}^{n} \omega_i(k) \cdot \frac{\partial fz(k)}{q_i(k)}$$

2.2 Kalman Filter for Jacobian Estimation

In this section is presented an adaptive Kalman filter, also known as STKF, based on the implementation of Li [2,5]. Then a Kalman filter is used to observe the states of the system continuously. A reasonable estimation of the Jacobian at every control interval can be obtained by the following recursive formulation

$$K(k) = P(k)H^T(k)\left[H(k)P(k)H^T(k)\right]^{-1}$$
$$P(k+1) = [I - K(k)H(k)]P(k) \tag{12}$$
$$\hat{J}(k+1) = \hat{J}(k) + K(k)\left[\frac{\Delta\chi}{Ts} - H(k)\hat{J}(k)\right]$$

where $\hat{J}(k)$ is the predicted Jacobian estimate, $P(k)$ is the predicted error covariance, $K(k)$ is the optimal Kalman gain, $P(k+1)$ is the updated error covariance and $\hat{J}(k+1)$ is the updated state estimate. Note that, conventional Kalman filter heavily depends on the exact knowledge of the process and measurement models, or by the case of the robot $\frac{\Delta\chi}{Ts}$ represents the approximation of end effector time derivative. Besides it is important to point out, That the STKF only requires the known values $\omega(k)$ and $\frac{\Delta\chi}{Ts}$ from de robot. The fading factor λ_k can be derived from the innovation sequence $v_k = \frac{\Delta\chi}{Ts} - H(k)\hat{J}(k)$. For STKF is needed to compute the fading factor λ_k and the updated covariance matrix Q_k is including in the updated error covariance matrix $P(k+1)$. This is depicted in (13)

$$P(k+1) = [I - K(k)H(k)P(k)]\lambda_k + Q_k \tag{13}$$

the value of λ_k is restricted for the value of C_k as is expressed in the Eq. (14)

$$\lambda_k = \begin{cases} C_k, & \text{when } C_k \geq 1 \\ 1, & \text{when } C_k < 1 \end{cases} \tag{14}$$

the value of C_k is in terms of the trace value of matrices M_k and N_k

$$C_k = \frac{tr\,[N_k]}{tr\,[M_k]} \tag{15}$$

M_k and N_k can be computed as (16)

$$N_k = V_k - H(k)Q_kH(k)^T$$
$$M_k = H(k)P(k)H(k)^T \tag{16}$$

the matrix V_k is

$$V_k = \begin{cases} v_0 v_0^T & k = 0 \\ \frac{0.95V_{k-1} + v_k v_k^T}{1.95} & k \geq 1 \end{cases} \tag{17}$$

the updated covariance matrix Q_k is

$$Q_k = K(k)\hat{C}(k)K(k) \tag{18}$$

and the estimated covariance of innovation matrix is presented as follow

$$\hat{C}(k) = \frac{1}{N} \sum_{n=k-N+1}^{k} v_n v_n^T \qquad (19)$$

3   Controller Design

3.1   FREN Structure

The architecture of a conventional FREN is illustrated in Fig. 1. Here is given a brief description of FREN's architecture, for a detailed version see [9].

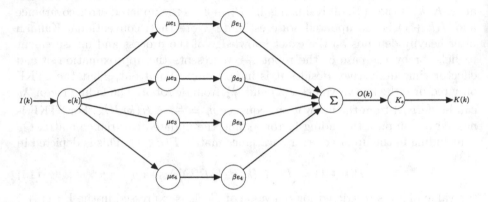

Fig. 1. Artificial Neural Network architecture.

FREN is composed by four layers: *Layer 1*: The error measurement $e(k)$ is the input of this layer which is sent to each node in the next layer directly. Thus, there is no computation in this layer. *Layer 2*: This is called input membership function layer. Each node in this layer contains a membership function corresponding to one linguistic variable (e.g. negative, positive, zero, etc.). The output at the ith node of this layer is calculated by

$$f(k) = \mu_i(e(k)), \qquad (20)$$

where μ_i denotes the membership function at the ith node $(i = 1, 2, ..., N)$. See Fig. 4(a) and (b). *Layer 3*: This layer may be considered as a defuzzification step. It is called the linear consequence (LC) layer, where the initial parameters β_i are intuitively selected and for the case of this work, the values of β_i parameters will remain constant. *Layer 4*: This is the output of the artificial neural network and is calculated as:

$$o(k) = \sum_{i=1}^{N} \beta_i \cdot \mu_i(e(k)) \qquad (21)$$

where N represents the number of linguistic variables. The artificial neural network for the FREN adaptive gain controller is shown in the Fig. 1, its complete structure is fixed as ANN with four hidden layers, and $e(k)$ its input. For the fourth hidden layer which is the adaptive neural netowrk, the inputs is the output of FREN $O_e(k)$. Finally, the control signal is estimated as

$$K(k) = K_e O_e(k) \tag{22}$$

where the parameters K_e remainds constant and the $K(k)$ is the time-varying control gain of FREN controller for $e(k)$, respectively [4, 10].

3.2 Proposed Controller

Figure 2 depicts the block diagram for the controller, where $\chi_d(k)$ is the desired position of the tip robot and $\chi(k)$ is the current position in the task space. The proposed controller is a adaptive gain controller and the output signal $w(k)$ is introduced in the Jacobian estimate algorithm based on Kalman filter. Where the error is defined by the Eq. (23)

$$e(k) = \chi(k) - \chi_d(k) \tag{23}$$

and $K(k)$ is a diagonal matrix that contains the adaptive gians $K_x(k), K_y(k)$ and $K_z(k)$ coming from the output of architecture ANN in Fig. (1), which update the control gain of x, y and z axes for task space control.

$$\nu(k) = -K(k)e(k) \tag{24}$$

the analytic Jacobian pseudoinverse $\hat{J}_A^+(q)$ is

$$\hat{J}_A^+(q) = \hat{J}_A^T(q)(\hat{J}_A(q)\hat{J}_A^T(q))^{-1} \tag{25}$$

where $\hat{J}_A^+(q)$ is build with estimation information of the STKF and now it is possible to calculate the signal of the controller when it is introduced the Eqs. (24) and (25)

$$w(k) = \hat{J}_A^+(q)\nu(k) \tag{26}$$

and the updated control signal is

$$w(k+1) = q(k) + w(k) \cdot Ts \tag{27}$$

The signal in the Eq. (27) allows end effector update to reach the desired position.

4   Experimental Results

4.1   Control Parameters Design

Our research is based on the Kuka youBot mobile manipulator, which has 3 dof for omnidirectional mobile plataform and 5 dof for manipulator arm, [1]. As

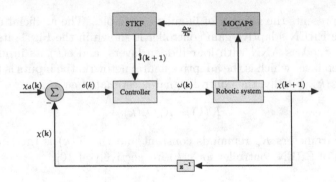

Fig. 2. Control block diagram for adaptive gain using FREN to tune it.

shown throughout Sect. 3, the FRENAG controller needs only updated the gain diagonal matrix $K(k)$ in order to minimize the error, see [3]. The advantage of this control algorithm is that it allows us to reduce the number of adaptation parameters, in comparison with a simple FREN whose number of adaptive control parameters are proportional to the number of membership functions. The objective of this experiment is to verify the performance of the proposed free model robot manipulator. To tune the FRENAG controller for $K_x(k)$, $K_y(k)$ and $K_z(k)$ were used the parameter values in Table 1. The values for βi are tuned intuitively by the experience and remain constant.

Table 1. Value of adaptive FREN gains parameters for $K_x(k)$, $K_y(k)$ and $K_z(k)$.

Name	Parameters	Value for $K_x(k)$ and $K_y(k)$	Value for $K_z(k)$
Positive large	β_{PL}	1.5	1.1
Positive small	β_{NL}	0.7	0.8
Zero	β_{ZE}	0.5	0.7
Negative small	β_{NS}	0.7	0.6
Negative large	β_{NL}	1.5	0.5

The five membership functions used in FRENAG for $K_x(k)$ and $K_y(k)$ are shown in Fig. 3(a), and the five membership functions used for $K_z(k)$ are shown in Fig. 3(b).

4.2   Experimental Set up

By this work the main objective is to test the performance of FMTSC of the robot reaching object position using the estimation of Jacobian by STKF. The proposed FRENAG controller structure guarantees the prescribed performance

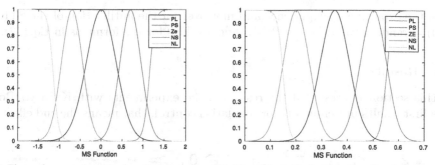

(a) Five rules membership function in terms of e(k) for $K_x(k)$ and $K_y(k)$.

(b) Five rules membership function in terms of e(k) for $K_z(k)$.

Fig. 3. Membership function design.

of end effector position error using a estimation of Jacobian. To know the current $\chi(k)$ that refers the robot's end effector in 3-D coordinate was used Motion Capture System (MOCAPS) which allows to locate in the space rigid bodies as: the robot and objects.

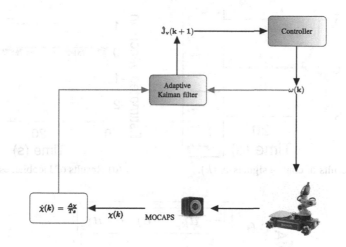

Fig. 4. Free model robot manipulator control using Jacobian estimation.

As is mentioned in Eq. (12) the STKF to estimate the Jacobian needs the velocity of the end effector $\frac{\Delta\chi}{Ts}$. Based on the current position signal $\chi(k)$ coming from the MOCAPS is possible to estimate the velocity of the end effector: $\frac{\Delta\chi}{Ts} = \frac{\chi(k+1)-\chi(k)}{Ts}$. The whole experimental set up can be appreciated in Fig. 4. The home position of the end effector is $[0.1430, 0, 0.6480]$ and the desired position $\chi_d(k)$ is the object, the constant gain $K_e = 0.5$ was intuitively selected according with the experiments experience and moreover for the initilization of the STKF $\hat{P}(0) =$ random values and $\hat{J}(0) =$ random values for the formulas in Eq. (12).

4.3   Results

In this section is presented the results of the experiments with Kuka youBot. The first results presented are for regulation control, that means the end effector

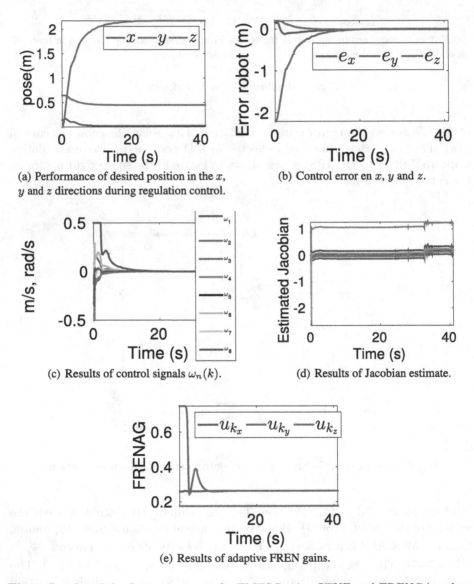

(a) Performance of desired position in the x, y and z directions during regulation control.

(b) Control error en x, y and z.

(c) Results of control signals $\omega_n(k)$.

(d) Results of Jacobian estimate.

(e) Results of adaptive FREN gains.

Fig. 5. Results of the first experiment for FMTSC using STKF and FRENGA, robot reaching object.

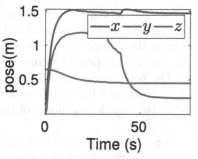

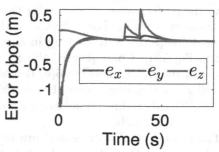

(a) Performance of desired position in the x, y and z directions during tracking of the object.

(b) Control error en x, y and z during change object position.

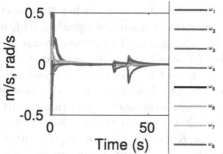

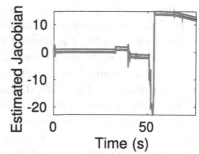

(c) Results of control signals $\omega_n(k)$ during change object position.

(d) Results of Jacobian estimate during change object position.

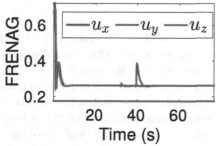

(e) Results of adaptive FREN gains during change object position.

Fig. 6. Results of the second experiment for FMTSC using STKF and FRENGA, during change object position.

robot reaches an object. The Fig. 5(a) depicts the result of the end effector reaching the object position, while the object remains static. During this experiments one can see the satisfactory response for the task space position control and the convergence of the control error $e(k)$ as is shown in Fig. 5(b). The Fig. 5(c) shows the performance of joint velocities $\omega_n(k)$ (or control signals) which lead to zero after a while. The behaivor of the Jacobian estimate as vector $\hat{J}$ it is assumed acceptable and working in agreement with the control and it is portrait in Fig. 5(d). Consequently, the STKF complies an acceptable estimation to control the robot. The Fig. 5(e) demonstrates the performance of FRENAG controller and it is easy to see that $K_x(k)$ and $K_y(k)$ demand high gains values than $K_z(k)$, this is directly related with the need of the mobile plataform of the robot to reach the object.

The second results presented are during change object position. By this experiment after the robot reaches the desired position, the object was moved twice to test the performance of the FMTSC. It is found that the controller is able to drive the Kuka youBot to follow the object and obtain satisfied results. The Figs. 6(a) and (b) show the position of the end effector and control error during change object position, it is easy to appreciate the change when the object is moved. The Fig. 6(c) depicts the joints velocites as a control signals $\omega_n(k)$, it is possible to see during the desired position changes $\omega_n(k)$ presents overpeaks until after a while converge to zero. In Fig. 6(d) can be seen the performance of the estmated Jacobian $\hat{J}$ for object tracking where it is appreciated the variation due to the changes in the desired position (object), however, the adaptation of the STKF work in smooth way to avoid that the Jacobian changes dramatically. The Fig. 6(e) depicts the performance of FRENAG controller for change object position and it is easy to see the variation in $K_x(k)$, $K_y(k)$ and $K_z(k)$ during the changes in the desired position where the gains adapt according with the controller demand.

5   Conclusion

In this work a free model for task space controller of 8 dof omnidirectional mobile manipulator has been proposed. The controller can work without any prior knowledge of the robot model, by means of STKF and FRENAG the parameters of the control scheme are tuned online. Jacobian estimate is computed with avilable information of robot manipulator, the STKF only requires the control signals and measurement output of the robot to calculate a satisfactory performance of the proposed analytical Jacobian vector. The results demonstrate that STKF performs the estimation of Jacobian just tuning the covariance matrix and fading factor. The STKF has a good accuracy and convergence due to an adaptive covariance matrix can capture the instantaneous change of the Jacobian and further improve the tracking of STKF. The FRENAG controller requires the updating of only control gain parameters in the diagonal matrix $K(k)$ and this can be demonstrated the FRENAG controller design and the experimental system validates the performance of proposed algorithm with the convergence of the parameters. As the robot is using a Jacobian estimamte by the second experiment during change object position, the control of the robot does not respond as fast as is expected. For future work is planning to analyze the stability of the FMTSC.

Acknowledgments. The authors would like to thank CONACyT (Project number 257253) for the financial support through this work and Science Basic project Number: 285599) which is called "Toma de decisiones multiobjetivo para sistemas altamente complejos". The first author thanks CONACyT for his PhD scholarship.

References

1. Bischoff, R., Huggenberger, U., Prassler, E.: KUKA youBot-a mobile manipulator for research and education. In: 2011 IEEE International Conference on Robotics and Automation (ICRA), pp. 1–4. IEEE (2011)
2. Braganza, D., Dawson, D.M., Walker, I.D., Nath, N.: A neural network controller for continuum robots. IEEE Trans. Robot. **23**(6), 1270–1277 (2007)
3. Chiang, C.J., Chen, Y.C.: Neural network fuzzy sliding mode control of pneumatic muscle actuators. Eng. Appl. Artif. Intell. **65**, 68–86 (2017)
4. Lee, C.H., Teng, C.C.: Identification and control of dynamic systems using recurrent fuzzy neural networks. IEEE Trans. Fuzzy Syst. **8**(4), 349–366 (2000)
5. Li, M., Kang, R., Branson, D.T., Dai, J.S.: Model-free control for continuum robots based on an adaptive Kalman filter. IEEE/ASME Trans. Mechatron **23**(1), 286–297 (2018)
6. Magdalena, M., Piotr, G.: Advanced neuro-fuzzy system in the task of position-force control of a manipulator. Appl. Mech. Mater. **817** (2016)
7. Mohamed, A., Schwarz, K.: Adaptive Kalman filtering for INS/GPS. J. Geod. **73**(4), 193–203 (1999)
8. Qian, J., Su, J.: Online estimation of image Jacobian matrix by Kalman-bucyfilter for uncalibrated stereo vision feedback. In: Proceedings IEEE International Conference on Robotics and Automation, ICRA 2002, vol. 1, pp. 562–567. IEEE (2002)
9. Treesatayapun, C., Uatrongjit, S.: Adaptive controller with fuzzy rules emulated structure and its applications. Eng. Appl. Artif. Intell. **18**(5), 603–615 (2005)
10. Treesatayapun, C.: Data input-output adaptive controller based on if-then rules for a class of non-affine discrete-time systems: the robotic plant. J. Intell. Fuzzy Syst. **28**(2), 661–668 (2015)

Design and Equilibrium Control
of a Force-Balanced One-Leg Mechanism

Hiram Ponce[(✉)] and Mario Acevedo

Facultad de Ingeniería, Universidad Panamericana, Augusto Rodin 498,
03920 Mexico City, Mexico
{hponce,macevedo}@up.edu.mx

Abstract. The problem of equilibrium is critical for planning, control, and analysis of legged robot. Control algorithms for legged robots use the equilibrium criteria to avoid falls. The computational efficiency of the equilibrium tests is critical. To comply with this it is necessary to calculate the horizontal momentum rotation for every moment. For arbitrary contact geometries, more complex and computationally-expensive techniques are required. On the other hand designing equilibrium controllers for legged robots is a challenging problem. Nonlinear or more complex control systems have to be designed, complicating the computational cost and demanding robust actuators. In this paper, we propose a force-balanced mechanism as a building element for the synthesis of legged robots that can be easily balance controlled. The mechanism has two degrees of freedom, in opposition to the more traditional one degree of freedom linkages generally used as legs in robotics. This facilitates the efficient use of the "projection of the center of mass" criterion with the aid of a counter rotating inertia, reducing the number of calculations required by the control algorithm. Different experiments to balance the mechanism and to track unstable set-point positions have been done. Proportional error controllers with different strategies as well as learning approaches, based on an artificial intelligence method namely artificial hydrocarbon networks, have been used. Dynamic simulations results are reported. Videos of experiments will be available at: https://sites.google.com/up.edu.mx/smart-robotic-legs/.

Keywords: Mechanical design · Legged robots · Equilibrium
Control systems · Artificial hydrocarbon networks
Reinforcement learning

1 Introduction

The problem of equilibrium is critical for planning, control, and analysis of legged robots [1]. Control algorithms for legged robots use the equilibrium criteria to

This research has been funded by Universidad Panamericana through the grant "Fomento a la Investigación UP 2017", under project code UP-CI-2017-ING-MX-03, and partially supported by Google Research Awards for Latin America 2017.

© Springer Nature Switzerland AG 2018
I. Batyrshin et al. (Eds.): MICAI 2018, LNAI 11289, pp. 276–290, 2018.
https://doi.org/10.1007/978-3-030-04497-8_23

avoid falls [2,3]. Regardless of the application, the computational efficiency of the equilibrium tests is critical, as computation time is often the bottleneck of control algorithms. To achieve static equilibrium, i.e. a dynamic wrench equal to zero, the total wrench of gravity and contact forces must therefore be equal to zero. This means that the center of mass (CoM) of the mechanism projects vertically inside the convex hull of the contact points when these points are located in the same horizontal plane. To comply with the previous condition, it is necessary to calculate the horizontal momentum rotation with respect to the CoM whose location, in general, has also to be calculated for every moment. Although for arbitrary contact geometries, more complex and computationally-expensive techniques are required to check equilibrium, some requiring to do it 100–1000 times per second [4].

Designing equilibrium controllers for legged robots is a challenging problem [5]. Inverted pendulums, ballbots, hoppers and cart-pole systems have been revised in literature [5–7]. It has been shown that nonlinear or more complex control systems have to be designed to balance those, which it also complicates the computational cost and demands robust actuators [7,8].

In this paper, we propose a force-balanced mechanism as a building element for the synthesis of legged robots that can be easily balance controlled. The mechanism has two degrees of freedom (DOFs), in opposition to the more traditional one DOF linkages generally used as legs in robotics [9]. As the mechanical system is balanced, the CoM of the whole system is located in an specific location within the mechanism. This facilitates the efficient use of the "projection of the center of mass" criterion with the aid of a counter rotating inertia, and reduces the number of calculations required by the control algorithm to check equilibrium and correct the CoM position. We conducted different experiments to balance the mechanism and to track unstable set-point positions. To do so, we implemented proportional error controllers with different strategies as well as learning approaches, based on an artificial intelligence method namely artificial hydrocarbon networks.

The rest of the paper is organized as follows. Section 2 introduces the design of the one-leg mechanism. Section 3 describes the control and learning strategies, based on artificial hydrocarbon networks, to balance the mechanism. Section 4 reports experimental results. Lastly, conclusions and future work are discussed.

2 Description of the One-Leg Mechanism

The proposed mechanism is presented in Fig. 1. It is a close-loop mechanism formed by two inverted double-pendula balanced by force [10]. The right side pendulum is formed by bars OA and CF, while the left one is formed by bars OB and DG. Both double pendula have a counterweight (green disc) at points F and G, respectively.

This configuration of the mechanism makes its CoM located exactly at point E. Thus, it is very easy to check the equilibrium condition: in this case to maintain the CoM of the system in a vertical position.

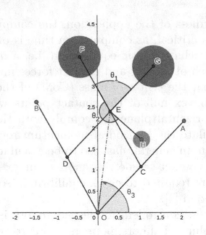

Fig. 1. Proposed one-leg mechanism. (Color figure online)

The mechanism has two DOFs. One DOF is driven by a motor associated to the angle θ_1, which allows to stretch and shrink the leg. The other DOF is driven by a wrench generated by the contact force present at point O and the weight at E, compensated by a torque introduced by a driving motor associated to the angle θ_2, moving a counter inertia defined by the bar EH and a mass.

In this way, the CoM of the whole leg system can be positioned in such a way that the static equilibrium is reached in a tilted pose, where the angle θ_3 is different from 90°.

2.1   Dynamic Model

The dynamic equations of motion of the mechanism have been obtained following a multibody approach. In this case the multibody system is constructed by a group of rigid bodies, which depend on the kinematic constraints and on the applied forces.

If the proposed planar mechanism is made up of b moving rigid bodies, the number of Cartesian generalized coordinates is $n = 3 \times b$. Thus the vector of generalized coordinates for the system can be written as (1) where $\mathbf{q}_b = [x_b, y_b, \phi_b]^T$ for all $b = 1, \dots, 5$.

$$\mathbf{q} = [\mathbf{q}_1, \mathbf{q}_2, \mathbf{q}_3, \mathbf{q}_4, \mathbf{q}_5]^T \tag{1}$$

On the other hand, a revolute kinematic pair introduces a pair of constraints between bodies i and j that in general can be described by (2); where $\mathbf{r}_i$ is the position vector of the CoM of body i, $\mathbf{A}_i$ is its rotation matrix and $\mathbf{s}'_i$ is local coordinates vector that positions the kinematic pair with respect to its local reference frame. The same applies for body j.

$$\Phi = (\mathbf{r}_i + \mathbf{A}_i \mathbf{s}'_i) - (\mathbf{r}_j + \mathbf{A}_j \mathbf{s}'_j) \tag{2}$$

So in this case, the complete set of m kinematic constraints, $m = 12$, dependent on the $n = 15$ generalized Cartesian coordinates at time t, can be expressed as (3).

$$\boldsymbol{\Phi}(\mathbf{q}, t) = 0 \tag{3}$$

The first derivative of Eq. (3) with respect to time is used to obtain the velocities Eq. (4), while the second derivative of Eq. (3) with respect to time yields the accelerations equations as (5); where $\boldsymbol{\Phi}_q$ is the Jacobian matrix, ν and γ are the velocities and accelerations respectively, associated with the terms dependent on time solely.

$$\boldsymbol{\Phi}_q \dot{\mathbf{q}} = \nu \tag{4}$$

$$\boldsymbol{\Phi}_q \ddot{\mathbf{q}} = \gamma \tag{5}$$

The equations of motion for the constrained multibody system are obtained applying the virtual power principle [11], and can be expressed by (6).

$$\mathbf{M}\ddot{\mathbf{q}} + \boldsymbol{\Phi}_q^T \boldsymbol{\lambda} = \mathbf{g} \tag{6}$$

For dynamics analysis, the kinematic constraint equations determine the algebraic configuration, while the dynamical behavior can be defined by the second order differential equations. Therefore, Eqs. (5) and (6) are arranged to form of differential-algebraic equations (DAEs) as (7):

$$\begin{bmatrix} \mathbf{M} & \boldsymbol{\Phi}_q^T \\ \boldsymbol{\Phi}_q & 0 \end{bmatrix} \begin{bmatrix} \ddot{\mathbf{q}} \\ \boldsymbol{\lambda} \end{bmatrix} = \begin{bmatrix} \mathbf{g} \\ \gamma \end{bmatrix} \tag{7}$$

having to solve the positions problem (3) and the velocities problem (4) every specified number of time steps to reduce the accumulation error in numerical simulations to obtain an accurate solution.

3 Control of and Learning the Balancing Task

At first glance, the designing process of the balancing control system for the one-leg mechanism might be seen as a challenging task [5]. Nevertheless, the force-balanced mechanical design provides a simple way to take it over. Thus, in this section, we propose two approaches to solve the balancing problem in the mechanism: (i) designing a controller, and (ii) learning control actions from scratch. We introduce not only these strategies, but also we describe artificial hydrocarbon networks that suit the balancing task under disturbances.

3.1 Artificial Hydrocarbon Networks

Artificial hydrocarbon networks (AHN) algorithm is a supervised learning method proposed by Ponce and Ponce [16]. It is inspired in the inner mechanisms and interactions of chemical hydrocarbon compounds. It aims to model data in packages of information, called molecules, that interact among them to capture the nonlinearities of data correlation. From this interaction, an artificial hydrocarbon compound is built and it can be seen as a net of molecules. If required, more than one artificial hydrocarbon compounds can be added up to finally obtain a mixture.

A molecule is the basic unit of information. It performs an output response $\varphi(x)$ due to an input $x \in \mathbb{R}^k$, as expressed in (8) where $v_C \in \mathbb{R}$ represents a carbon value, $h_{i,r} \in \mathbb{C}$ are the hydrogen values attached to this carbon atom, and d represents the number of hydrogen atoms in the molecule.

$$\varphi(x) = v_C \sum_{r=1}^{k} \prod_{i=1}^{d \leq 4} (x_r - h_{i,r}) \qquad (8)$$

Unsaturated molecules, $d < 4$, can be joined together with other molecules forming chains called compounds. In this work, compounds are made of n molecules: $(n-2)$ CH_2 molecules and two CH_3 molecules, such that the hydrocarbon compound shapes like $CH_3 - CH_2 - \cdots - CH_2 - CH_3$ [17]. CH_d-symbol represents a molecule with d hydrogens. A piecewise function ψ denoted as (9) is associated to the compound; where, $L_t = \{L_{t,1}, \ldots, L_{t,k}\}$ for all $t = 0, \ldots, n$ are bounds where molecules can act over the input space. Different compounds can be selected and added up to form complex structures called mixtures.

$$\psi(x) = \begin{cases} \varphi_1(x) & L_{0,r} \leq x_r < L_{1,r} \\ \cdots & \cdots \\ \varphi_n(x) & L_{n-1,r} \leq x_r \leq L_{n,r} \end{cases} \qquad (9)$$

For training purposes, the least squares error (LSE) is used for obtaining carbon and hydrogen values, while bounds are computed using a gradient descent method with learning rate $0 < \eta < 1$ based on the energy of adjacent molecules [17]. To this end, AHN is trained using the so-called AHN-algorithm. Details can be found in [16–19].

3.2   Artificial Organic Controller

We propose to use an artificial organic controller (AOC) to tackle the balancing problem in the one-leg mechanism, since AOC has been proved to be very effective in uncertain domains [13,20,21]. AOC is an intelligent control system that performs the control law using an ensemble method namely fuzzy-molecular inference (FMI) system [20], as shown in Fig. 2. FMI consists on three steps: fuzzification, fuzzy inference engine, and defuzzification via AHN.

Fuzzification and fuzzy inference engine steps are similar to fuzzy logic. An input x is mapped to a set of fuzzy sets, using membership functions. Then, an inference operation, represented as a fuzzy rule, is applied to obtain a consequence value y_p. Considering, the pth fuzzy rule R_p denoted as (10), inference calculates y_p in terms of an artificial hydrocarbon compound with n molecules, M_j, each one with behavior $\varphi_j(x)$ for all $j = 1, \ldots, n$. The membership value

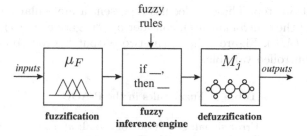

Fig. 2. Block diagram of the fuzzy-molecular inference system.

of y_p is calculated using the min function, expressed as $\mu_\Delta(x_1, \ldots, x_k)$, over the fuzzy inputs.

$$R_p : \text{if } x_1 \in A_1 \wedge \cdots \wedge x_k \in A_k,$$
$$\text{then } y_p = \varphi_j(\mu_\Delta(x_1, \ldots, x_k)) \tag{10}$$

Lastly, the defuzzification step computes the crisp output value y, using the center of gravity approach [20], as expressed in (11).

$$y = \frac{\sum_p \mu_\Delta(x_1, \ldots, x_k) \cdot y_p}{\sum_p \mu_\Delta(x_1, \ldots, x_k)} \tag{11}$$

Thus, two inputs (i.e. DOFs) are expected for the one-leg mechanism, let's say: τ_1 and τ_2. In order to design a controller for the balancing task in this system, we decided to decoupled the two inputs such that τ_1 will regulate the distance R between point E and the origin O, and τ_2 will regulate the angle θ_3 (see Fig. 1).

As shown later, the balancing problem of this decoupled system can be solved using a proportional (P) controller for each degree of freedom. Thus, we design a P-based artificial organic controller (P-AOC) for each variable R and θ_3. First, the error signal $e(t)$ is considered as input to the AOC, with three fuzzy partitions: "negative" (N), "zero" (Z) and "positive" (P). Figure 3 shows the input membership functions used for the both controlled variables.

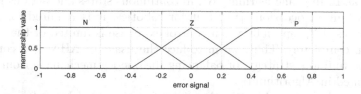

Fig. 3. Membership functions of the input error signal $e(t)$ used for both distance and angle.

In addition, the set of fuzzy rules for P-AOC are summarized in Table 1. Lastly, an artificial hydrocarbon compound of three molecules is proposed for the defuzzification step. These molecules represent a molecular partition of the output signal of the control law $u(t)$, considering: "negative" (M_N), "zero" (M_Z) and "positive" (M_P). Figure 4 shows the two output hydrocarbon compounds, one for each controlled variable.

Table 1. Fuzzy rules in the P-AOC.

Error signal: $e(t)$	Control signal: $u(t)$
N	M_N
Z	M_Z
P	M_P

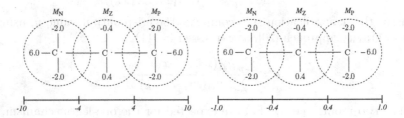

Fig. 4. Artificial hydrocarbon compound for calculating the output signal $u(t)$: (left) for the distance value R, and (right) for the angle θ_3.

3.3 AHN-Based Reinforcement Learning

As an alternative, we also propose a learning approach to tackle the balancing task using reinforcement learning (RL). The aim of RL is to produce/learn a policy π (i.e. a sequence of actions) that is performed by an agent or robot to reach a goal. In machine learning, RL in continuous states and actions has been studied in limited occasions [14]. But for robotics, it is required for learning complex tasks [12], especially when control expertise is limited or the system is not known completely. Thus, we introduce the usage of AHN as a continuous RL method to find a strategy for balancing the one-leg mechanism from scratch, and presented in Algorithm 1.

In this work, we consider the dynamics function model $\hat{f}_\omega$ of a continuous system as (12) where $\mathbf{s}_t \in \mathbb{R}^D$ represents the state and $\mathbf{a}_t \in \mathbb{R}^F$ the action applied in time t.

$$\hat{\mathbf{s}}_{t+1} = \mathbf{s}_t + \hat{f}_\omega(\mathbf{s}_t, \mathbf{a}_t) \tag{12}$$

The dynamics function $\hat{f}_\omega$ is proposed to be parameterized with AHN, where the parameter vector ω represents the hydrogen, carbon and bounds values over the hydrocarbon compound (i.e. $\{h, v_C, L\}$ for all molecules). Then, $\hat{s}_{t+1}$ represents the predicted next state that occurs after the predicted change in state s_t over the time step Δt.

For training the AHN, we use tuples $(s_{t-1}, a_{t-1}) \in \mathbb{R}^{D+F}$ as training inputs and differences $\Delta s_t = s_t - s_{t-1} \in \mathbb{R}^D$ as training outputs, collected in a dataset $\mathcal{D}$. Thus, we train the dynamics function model $\hat{f}_\omega$ by minimizing the error (13) using the gradient descent method.

$$E(\omega) = \frac{1}{2} \sum_{(s_{t-1}, a_{t-1}, s_t) \in \mathcal{D}} \| \Delta s_t - \hat{f}_\omega(s_{t-1}, a_{t-1}) \|^2 \qquad (13)$$

Once the dynamics model is learned, we apply a policy search method using the learned model $\hat{f}_\omega$. For simplicity, we use the random-sampling shooting method described in [22] and employed in [14]. This policy evaluation considers that at each time step t, it generates K candidate action sequences of H actions each. Then, the learned dynamics model is applied to predict the resulting future states from running each action sequence and evaluating it in a prior cost function $c(s_t, a_t)$ that encodes the task. The optimal action sequence is then selected, but only the first action a_t is executed. Then, a replan at the next step is done, as suggested in the model predictive control (MPC) [14]. Lastly, the tuples (s_{t-1}, a_{t-1}) and Δs_t resulting from this policy evaluation are collected in $\mathcal{D}_{\hat{f}}$ and added to the current training dataset $\mathcal{D}$ in order to update the learned model with this new information. Lastly, the proposed method iterates over the dynamics model training, policy search and collection of new training data, until the task is learned.

Algorithm 1. Proposed AHN-based RL method.

Apply random control signals and collect initial training dataset $\mathcal{D}$.
repeat
    Learn the dynamics model $\hat{f}_\omega$ using $\mathcal{D}$.
    $\pi \leftarrow$ Policy search using $\hat{f}_\omega$ and MPC.
    $\mathcal{D}_{\hat{f}} \leftarrow$ Collect new data (s_{t-1}, a_{t-1}) and Δs_t from π.
    Add new data to the training dataset: $\mathcal{D} \leftarrow \mathcal{D} \cup \mathcal{D}_{\hat{f}}$.
until task learned

---

4   Experimental Results

We simulated the one-leg mechanism (Fig. 1). The distances $OA = OB = CF = DG = 30\,cm$, $OC = CA = 15\,cm$, $OD = DB = 15\,cm$, $CE = EF = 15\,cm$, $DE = EG = 15\,cm$, and $EH = 12\,cm$. The physical parameters of the bodies are presented in Table 2.

Then, we applied the two AHN-based approaches for tackling the balancing problem. To compare the performance of the AHN-based strategies, we developed a conventional P-controller like (14) with $K_P = diag\{-10, 2\}$, and a fuzzy

Table 2. Physical parameters of the mechanism bodies

Body	Mass (kg)	Moment of Inertia (kg $\cdot$ m^2)
OA	0.0915	0.0007621
OB	0.0915	0.0007621
CF	0.0915	0.0007621
DG	0.0915	0.0007621
Disc at F	0.0915	0.000097
Disc at G	0.0915	0.000097
Counter inertia EH	0.271	0.0000736

P-controller using input membership functions of Fig. 3, fuzzy rules of Table 1 and output membership functions of Fig. 5.

$$u(t) = K_P e(t) + u(t-1) \tag{14}$$

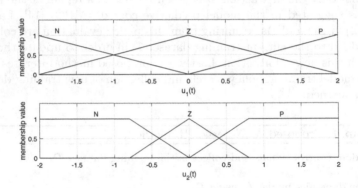

Fig. 5. Membership functions of the output control signal $u(t)$: (top) for the distance value R, and (bottom) for the angle θ_3.

We employ four common performance indices for the analysis [15]: the integral of squared error (15), the integral of absolute error (16), the integral of time multiply squared error (17), and the integral of time multiply absolute error (18); where, e is the error signal at time t.

$$ISE = \int_0^\infty e^2(t)dt \tag{15}$$

$$IAE = \int_0^\infty |e(t)|dt \tag{16}$$

$$ITSE = \int_0^\infty te^2(t)dt \tag{17}$$

$$ITAE = \int_0^\infty t|e(t)|dt \qquad (18)$$

4.1 Set-Point Tracking Control

The first experiment consisted on tracking a set-point using the simulated mechanism. This experiment aims to determine the performance of the P-AOC controller assuming that the design is not influenced by noise. The experiment was run over 10 s with 0.05 s of time step, and the reference was moved several times during the running. Figure 6 shows a comparison between P-AOC, conventional-P and fuzzy-P controllers. As notice, the simplest conventional-P controller performed well when tracking the set-point, and the other controllers also accomplished the task as expected. In addition, P-AOC reaches the settling time 80% more quickly than conventional-P. Quantitatively, Table 3 summarizes the performance of the output response of the controllers. In all cases, performances indices show that P-AOC has the best performance for this particular set-point tracking.

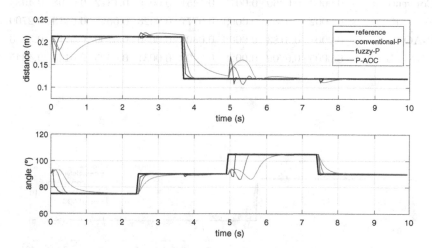

Fig. 6. Comparative output response for the set-point tracking controllers.

4.2 Set-Point Tracking Control with Disturbances

The same experiment was conducted again, but this time the mechanism was subjected to random disturbances of horizontal forces in the range $[-10, 10]$ N over all simulation time. This experiment aims to determine the performance of the P-AOC under disturbances. Figure 7 depicts a comparison between the output responses of the controllers and Table 3 shows the resultant performance indices. Instead all the controllers deal with uncertain scenarios, fuzzy-P and P-AOC are the best. To this end, it is shown that P-AOC obtained the best performance in the experiment.

Table 3. Summary of performance of the controllers.

Controller type	Distance R				Angle θ			
	ISE	IAE	ITSE	ITAE	ISE	IAE	ITSE	ITAE
Set-point tracking control								
Conventional-P	0.0041	0.1119	0.0130	0.3728	0.1193	0.5846	0.4227	2.1505
Fuzzy-P	0.0009	0.0278	0.0032	0.0900	0.0678	0.2828	0.2248	0.9598
P-AOC	0.0008	0.0182	0.0025	0.0536	0.0306	0.1322	0.1060	0.4395
set-point tracking control with disturbances								
Conventional-P	0.0044	0.1341	0.0146	0.5256	0.1278	0.7042	0.4660	2.9003
Fuzzy-P	0.0014	0.0779	0.0058	0.3705	0.0819	0.5204	0.2960	2.1890
P-AOC	0.0013	0.0690	0.0052	0.3409	0.0395	0.3510	0.1528	1.6100
balancing task								
Conventional-P	0.0003	0.0360	0.0013	0.1451	0.0056	0.1322	0.0198	0.4983
Fuzzy-P	0.0002	0.0183	0.0009	0.0777	0.0028	0.0659	0.0115	0.2700
P-AOC	0.0001	0.0183	0.0008	0.0803	0.0030	0.0790	0.0115	0.3156
AHN-RL	0.0005	0.0700	0.0028	0.3490	0.0031	0.0905	0.0144	0.4345

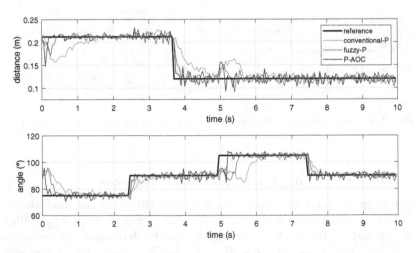

Fig. 7. Comparative output response for the set-point tracking controllers under disturbances.

4.3   Control of Balancing Task

Once the set-point tracking was implemented, the balancing problem was tackled with the same controllers employed so far. Particularly, this experiment was run with instant disturbances of 20 N horizontal force, one per second, during 10 s. The one-leg mechanism has initial conditions at distance 0.21 m and angle 90°. The goal of this experiment is to measure the performance of the P-AOC to handle disturbances and compensate it to return to the initial condition. Figure 8 shows a comparison between the output response of the controllers, and Table 3 summarizes the performance indices for this experiment labeled as *balancing task*. As shown in Fig. 8, conventional-P takes 5s to get in a safe position for instant disturbances, while fuzzy-P and P-AOC only take one second (see time range 1–2 s). However, any of the controllers can deal with these disturbances without diverging.

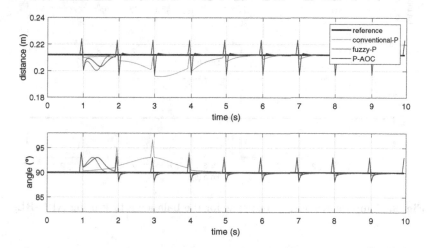

Fig. 8. Comparative output response for the balancing task using control.

4.4   Learning of Balancing Task

Alternatively, we ran an experiment for learning the balancing task in the one-leg mechanism. This experiment aims to determine the performance of the output response of the policy π obtained through the AHN-based reinforcement learning (AHN-RL) approach.

The balancing task was set up as follows. A state of the mechanism considers the distance R and the angle θ such that $\mathbf{s}_t = (R_t, \theta_{3_t})$. The goal state is $\mathbf{s}_{goal} = (R, \theta_3)$ and the cost function $c(\mathbf{s}_t, \mathbf{a}_t)$ is expressed as (19).

$$c(\mathbf{s}_t, \mathbf{a}_t) = \sqrt{\sum_i (\mathbf{s}_{goal,i} - \mathbf{s}_{t,i})^2} \tag{19}$$

Initial dataset $\mathcal{D}$ collected 10 rollouts with random initial states $s_{initial} = (0.21, 90 \pm 15)$ for coverage. Five iterations were fixed for training AHN. We set the AHN with 10 molecules and learning rate $\eta = 0.01$. We preprocessed the training data by normalizing it to be mean 0 and standard deviation 1 ensuring that the loss function weights all parts of the state equally. For the policy search, we set $K = 100$ candidate sequences and $H = 3$ horizon steps. After training, we performed an online policy search π via MPC with $K = 3$ and $H = 3$ and the usage of the trained AHN-RL. Figure 9 shows the output response of the performed π found by AHN-RL. Table 3 summarizes the performance indices for this experiment.

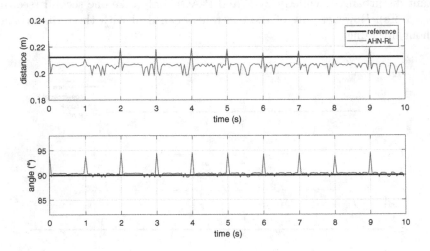

Fig. 9. Comparative output response for the balancing task using AHN-RL.

From Fig. 9, it can be seen that the output response did not reach the distance reference in steady-state. However, it also presented settling times 60% better than any of the control responses, and no over-shooting. From Table 3, AHN-RL handles the balancing task similarly to the other controllers based on the performance indices.

5   Conclusions

This paper presented a force-balanced one-leg mechanism for easy controlling. The proposed design was modeled and simulated. In addition, we proposed to use artificial hydrocarbon networks for controlling and learning the balancing task in the mechanism. Experimental results provided information to conclude that the proposed mechanism is easily controlled by simple P-control strategies. In addition, both P-AOC control and AHN-RL learning approaches performed well when the mechanism was subjected to disturbances.

For future work, we will investigate controllability and learnability of more complex robotic tasks such as walking, jumping and climbing stairs using this mechanism in legged robots. Also, we will characterize the dynamics of the one-leg mechanism in detail.

References

1. Wieber, P.-B., Tedrake, R., Kuindersma, S.: Modeling and control of legged robots. In: Siciliano, B., Khatib, O. (eds.) Springer Handbook of Robotics, 2nd edn, pp. 1203–1234. Springer, Cham (2016). https://doi.org/10.1007/978-3-319-32552-1_48
2. Rebula, J.R., Neuhaus, P.D., Bonnlander, B.V., Johnson, M.J., Pratt, J.E.: A controller for the LittleDog quadruped walking on rough terrain. In: Proceedings of the IEEE International Conference on Robotics and Automation, pp. 1467–1473 (2007)
3. Yoshida, E., Kanoun, O., Esteves, C., Laumond, J.P.: Task-driven support polygon reshaping for humanoids. In: Proceedings of the 2006 6th IEEE-RAS International Conference on Humanoid Robots, HUMANOIDS, pp. 208–213 (2006)
4. Del Prete, A., Tonneau, S., Mansard, N.: Fast algorithms to test static equilibrium for legged robots. In: IEEE International Conference on Robotics and Automation (ICRA), Stockholm, Sweden, May 2016, pp. 1601–1607 (2016)
5. Najafi, E., Lopes, G.A.D., Babuska, R.: Balancing a legged robot using state-dependent Riccati equation control. IFAC Proc. Volumes **47**(3), 2177–2182 (2014)
6. Grasser, F., D'Arrigo, A., Colombi, S., Rufer, A.C.: JOE: a mobile, inverted pendulum. IEEE Trans. Ind. Electron. **49**(1), 107–114 (2002)
7. Kashki, M., Zoghzoghy, J., Hurmuzlu, Y.: Adaptive control of inertially actuated bouncing robot. IEEE Trans. Robot. **33**(3), 509–522 (2017)
8. Wensing, P., Wang, A., Seok, S., Otten, D., Lang, J., Kim, S.: Proprioceptive actuator design in the MIT Cheetah: impact mitigation and high-bandwidth physical interaction for dynamic legged robots. IEEE/ASME Trans. Mechatron. **22**(5), 2196–2207 (2017)
9. Komoda, K., Wagatsuma, H.: Energy-efficacy comparisons and multibody dynamics analyses of legged robots with different closed-loop mechanisms. Multibody Syst. Dyn. **40**, 123–153 (2017)
10. van der Wijk, V., Herder, J.L.: Synthesis of dynamically balanced mechanisms by using counter-rotary countermass balanced double pendula. ASME J. Mech. Des. **131**(11), 111003-1–111003-8 (2009). https://doi.org/10.1115/1.3179150
11. García de Jalón, J., Bayo, E.: Kinematic and Dynamic Simulation of Multibody Systems: The Real-Time Challenge. Springer, New York (1994). https://doi.org/10.1007/978-1-4612-2600-0
12. Munoz de Cote, E., Garcia, E.O., Morales, E.F.: Transfer learning by prototype generation in continuous spaces. Adapt. Behav. **24**(6), 464–478 (2016)
13. Molina, A., Ponce, H., Ponce, P., Tello, G., Ramirez, M.: Artificial hydrocarbon networks fuzzy inference systems for CNC machines position controller. Int. J. Adv. Manuf. Technol. **72**(9–12), 1465–1479 (2014)
14. Nagabandi, A., Yang, G., Asmar, T., Kahn, G., Levine, S., Fearing, R.S.: Neural network dynamics models for control of under-actuated legged millirobots. arXiv:1711.05253 (2017)
15. Sung-Kwun, O., Pedrycz, W., Rho, S.-B., Ahn, T.-C.: Parameter estimation of fuzzy controlle and its application to inverted pendulum. Eng. Appl. Artif. Intell. **17**(1), 37–60 (2004)

16. Ponce, H., Ponce, P.: Artificial organic networks. In: In 2011 IEEE Conference on Electronics, Robotics and Automotive Mechanics, pp. 29–34. IEEE, Cuernavaca (2011)
17. Ponce-Espinosa, H., Ponce-Cruz, P., Molina, A.: Artificial Organic Networks. SCI, vol. 521. Springer, Cham (2014). https://doi.org/10.1007/978-3-319-02472-1
18. Ponce, H., Ponce, P., Molina, A.: The development of an artificial organic networks toolkit for LabVIEW. J. Comput. Chem. 36(7), 478–492 (2015)
19. Ponce, H.: A novel artificial hydrocarbon networks based value function approximation in hierarchical reinforcement learning. In: Pichardo-Lagunas, O., Miranda-Jiménez, S. (eds.) MICAI 2016. LNCS (LNAI), vol. 10062, pp. 211–225. Springer, Cham (2017). https://doi.org/10.1007/978-3-319-62428-0_18
20. Ponce, H., Ponce, P., Molina, A.: Artificial hydrocarbon networks fuzzy inference system. Math. Probl. Eng. 1–13, 2013 (2013)
21. Ponce, H., Ponce, P., Molina, A.: A novel robust liquid level controller for coupled-tanks systems using artificial hydrocarbon networks. Expert Syst. Appl. 42(22), 8858–8867 (2015)
22. Rao, A.: A survey of numerical method for optimal control. Adv. Astronaut. Sci. 135, 497–528 (2009)

An Adaptive Robotic Assistance Platform for Neurorehabilitation Therapy of Upper Limb

José Daniel Meneses-González[1], Omar Arturo Domínguez-Ramírez[1(✉)],
Luis Enrique Ramos-Velasco[2], Félix Agustín Castro-Espinoza[1],
and Vicente Parra-Vega[3]

[1] Research Center in Information Technologies and Systems,
Autonomous University of Hidalgo State (UAEH), Pachuca, Mexico
dants_300@hotmail.com, {omar,fcastro}@uaeh.edu.mx
[2] Aerospace Engineering Department,
Metropolitan Polytechnic University of Hidalgo (UPMH), Tolcayuca, Mexico
lramos@upmh.edu.mx
[3] Center for Research and Advanced Studies (CINVESTAV), Saltillo, Mexico
vparra@cinvestav.mx

Abstract. There are many human-robot physical interaction methods
for physical therapy in patients of upper limbs disabilities. The use of
haptic devices for this purpose is abundant, as are the different proposals
for motion control in haptic guidance, as part of a clinical protocol with
the patient in the loop. A conclusive result of these interaction platforms
is the need to modify elements of the control strategy and the motion
planning, this for each patient. In this paper, we propose a new approach
to the control of human-robot physical interaction systems. To guarantee the bilateral energy flow between the robotic system and the patient
under stable conditions and, without modifying the interaction platform;
we propose an adaptive control structure, free of the dynamic model. The
control scheme is called PID Wavenet, and identifies the dynamics using
a radial basis neural network with daughter RASP1 wavelets activation
function; its output is in cascaded with an infinite impulse response (IIR)
filter toprune irrelevant signals and nodes as well as to recover a canonical form. Then, online adaptive of a discrete PID regulator is proposed,
whose closed-loop guarantees global regulation for nonlinear dynamical
plants, in our case a haptic device with the human in the loop. Effectiveness of the proposed method is verified by the real-time experiments on
a Geomagic Touch haptic interface.

Keywords: Human robot interaction · Haptic interface
Wavelet neural network control · Rehabilitation robotics

© Springer Nature Switzerland AG 2018
I. Batyrshin et al. (Eds.): MICAI 2018, LNAI 11289, pp. 291–303, 2018.
https://doi.org/10.1007/978-3-030-04497-8_24

1 Introduction

1.1 Background

Haptic interaction with a virtual object establishes a kinesthetic sensation to the user. It is well accepted that to better perceive a given virtual object, it is relevant to yield some surface properties of the object, for instance the shape through the normal contact force, the roughness by the sliding friction, and the texture as a combination of both of them. Based on a novel formulation of the computation of the contact force of haptic interfaces, a new paradigm for haptic guidance is proposed. Guided kinesthetic feedback is provided to improve and effectively train the user with PID Wavenet robot control. The system introduces a training path using potential fields, which can be tuned according to the handicap score of the user, to gradually improve the motor skills of the user. To this end, we present a haptic guidance platform to support physical rehabilitation of neuromuscular disabilities, providing a solution to the problem of the increasing demand of neuromuscular therapy in overcrowded facilities with deficit of rehabilitation professionals. The platform, characterized by a portable modular architecture, can be configured according to the treatment suggested by the physician, for instance, a Local Haptic Guidance configuration for patients requiring a continuous repetition of coordinated movements to recover, or improve, motor skills, and, a Remotely-Assisted Haptic Guidance for a direct intervention of the therapist, evaluating and stimulating, simultaneously, the neuromuscular condition of the patient. To this goal, a novel global PID control scheme for nonlinear MIMO systems is proposed and synthesized for a haptic interface. The identification process (human-robot physical interaction) is used for online tuning of the discrete linear PID feedback gains. Inverse dynamics identification is based on radial basis neural network with daughter RASP1 wavelets activation functions in cascaded with an infinite impulse response (IIR) filter in the output to prune irrelevant signals and nodes. The closed-loop system guarantees global regulation for a class of Euler-Lagrange systems, convenient for instance in plants whose dynamics are rather uncertain or unknown, such as in haptic devices.

1.2 Contribution

In this paper, a novel wavenet control based on closed-loop identification scheme of the human-robot interaction as a full contact is proposed, and representative experiments are presented to purpose in rehabilitation of upper limb. The input of the wavelet neural network is the human contact force as altogether with the position and velocity of the haptic device, such that approximate identification of the coupled dynamics is achieved. The adaptive wavenet adapts to the time-varying nature of human arm dynamics coupled to the haptic device such that a multiresolution wavenet scheme synthesizes a so-called model-free $PIDWavenet$ controller. The proposed methodology has been implemented in a Geomagic Touch haptic robot system.

2   The Problem to Solve

Consider a haptic device as an interaction platform; the nonlinear dynamic model is similar to a robot manipulator, as follows:

$$H(q)\ddot{q} + C(q,\dot{q})\dot{q} + G(q) = \tau - \tau_f, \tag{1}$$

where $q \in R^n, \dot{q} \in R^n$ are the generalized position and velocity joint coordinates, respectively, $H(q) \in R^{n \times n}$ denotes a symmetric positive definite inertial matrix, $C(q,\dot{q}) \in R^{n \times n}$ represents the Coriolis and centripetal forces, $G(q) \in R^n$ models the gravity forces, and $\tau \in R^n$ stands for the torque input. Where it only compensates the device dynamics. Term $\tau_f = f_b\dot{q} + f_c \tanh(\gamma\dot{q})$ stands for joint friction, for f_b, f_c, γ are positive definite $n \times n$ matrices modelling viscous damping and the dry friction and its coefficient, respectively. However, the control techniques that allow global regulation to be solved a perfect tracking of trajectories have been worked by the control community, [9]. When considering the human operator in the loop of the device, the dynamics change remarkably; the nonlinear dynamic model is:

$$H(q)\ddot{q} + C(q,\dot{q})\dot{q} + G(q) = \tau - \tau_f + \tau_h, \tag{2}$$

Where the disturbance torque τ_h is assumed a differentiable bounded small time-varying function as the bounded persistent disturbance term, and represent the human in the loop. Now, in our application case, the human operator is a patient with disability in upper limb. Previous work, integrated the solution with non-linear control structures with limited performance, [10]. The problem is the modification of the control gains for each patient, see Fig. 1.

The conclusion of [10] is the necessity of the adaptability of the control strategy, being the purpose of this preliminary work to induce rehabilitation conditions. In the Fig. 1, the patient is guided in a trajectory based on a clinical protocol.

2.1   Experimental Platform

Consider a *Geomagic Touch* haptic interface, whose haptic device is a joint nonlinear robot of three degrees of freedom, see Fig. 2, which although shows purposely low apparent inertia at its end-effector, there arises strong nonlinear coupling and it is subject to joint friction and gravitational torques. The experimental platform runs on a workstation iCore7 at 3.6 GHz Intel Core with 16 Gb of RAM. Software features is under OS Windows 10, running a compiler Simulink under Matlab 2014. Experiments are run at $[h = 1\,\text{ms}]$ or a sampling frequency of 1 KHz.

3   Adaptive Human Robot Physical Interaction

3.1   Intelligent Control Design

The wavenet PID controller scheme is based on an identification of inverse dynamics and a IRR filter to tune PID feedback gains (k_p, k_i and k_d), and guarantees global regulation, see Fig. 3. The following variables are used: $\mathbf{y}_{ref}(k)$ is

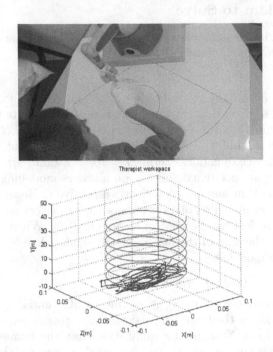

Fig. 1. Auto-tuning feedback gains of the PDH controller.

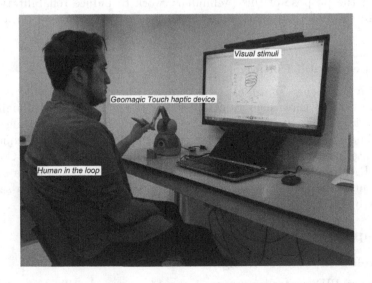

Fig. 2. Experimental platform.

the reference signal, $\epsilon(k)$ stands for the error signal, the control input is $\mathbf{u}(k)$, $r(k)$ models the noise signal, $\mathbf{y}(k)$ is the HRpI (human patient in the haptic loop) output with $\hat{\mathbf{y}}(k)$ its estimate, and $\mathbf{e}(k)$ the error estimated, finally, $v(k)$ stands for the persistence signal.

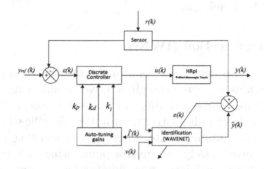

Fig. 3. Scheme of the PID Wavenet controller.

Haptic Device Dynamics. Consider the nonlinear dynamic model of a rigid serial n-link robot manipulator as follows, in the continuous domain. HRpI based on human patient in the loop and haptic interface as the Eq. 2, can be interpreted as a general nonlinear MIMO dynamical system, which can be represented by a general discrete state equation, [6]:

$$x(k+1) = f[\mathbf{x}(k), \mathbf{u}(k), k] \tag{3}$$
$$\mathbf{y}(k) = g[\mathbf{x}(k), k] \tag{4}$$

where $\mathbf{x} \in R^n$, $\mathbf{u}, \mathbf{y} \in R^p$ and

$$f : R^n \times R^p \longrightarrow R^n \tag{5}$$
$$g : R^n \longrightarrow R^p \tag{6}$$

are unknown smooth functions. Robot friction and disturbances are considered affine and state dependent, then those are represented in (5)–(6). Notice that input $\mathbf{u}(k)$ and system output $\mathbf{y}(k)$ are the only data available, and since the linearized system of (2) is observable around the equilibrium point, then there exists an input-output representation that can be reconstructed with a basis, [6]. That is, consider the following canonical realization:

$$\mathbf{y}(k+1) = \beta[\mathbf{Y}(k), \mathbf{U}(k)] \tag{7}$$

where

$$\mathbf{Y}(k) = [\mathbf{y}(k)\ \mathbf{y}(k-1),\ \cdots,\ \mathbf{y}(k-n+1)] \tag{8}$$
$$\mathbf{U}(k) = [\mathbf{u}(k)\ \mathbf{u}(k-1),\ \cdots,\ \mathbf{u}(k-n+1)] \tag{9}$$

Then, there exists a function β that maps the output $y(k)$, input $u(k)$ and their $n-1$ past values in $y(k+1)$, [6]. Thus, [1] establishes that there exists a wavenet neural network $\hat{\beta}$ that can be trained to converge locally to β. In this paper, we exploit this property of wavenets, but additionally we consider IIR filter in the output layer to prune irrelevant signals to build an efficient identification scheme useful to tune PID feedback gains.

3.2   Wavenet Identification (IWNN)

It is proposed radial basis neural network for the identification process, in which the activation functions $\psi(\tau)$ are daughter wavelet functions $\psi_j(\tau)$ of RASP1 type. This incorporates three IIR filter in cascade whose function is to filter neurons that have little or null contribution in the identification process, allowing a reduction in the number of iterations in the learning process, [3]. This scheme identifies approximately the inverse plant using as few neurons as possible, which stands for an efficient approximator for practical applications due to its reduced computational load. The general interconnection and signal propagation is presented in Fig. 4, where $\tau_l = \frac{\|\mathbf{u}(k)-\mathbf{b}_l\|}{\mathbf{a}_l}$. Infinite impulse response (IIR) recurrent structure, in cascading structure, yields double improving speed of learning by pruning those nodes with insignificant relevant information from the cross contribution summation of daughters wavelets, located in the third layer. The inner structure of the IIR filter is shown in Fig. 5, notice the forward delayed structure modulated by the input and the feedback loop modulated by the persistent signal to allows swapping a range of frequency. The mother wavelet function $\psi(k)$ generates daughter wavelets $\psi_{a,b}(\tau)$ by its property of expansion or contraction and translation, represented as, [1]:

$$\psi_l(\tau_l) = \frac{1}{\sqrt{a}} \, \psi(\tau_l) \tag{10}$$

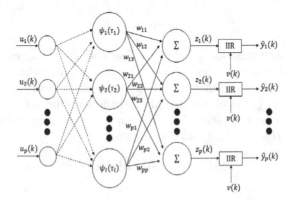

Fig. 4. Diagram of a wavenet neural network with an IIR filter in cascade.

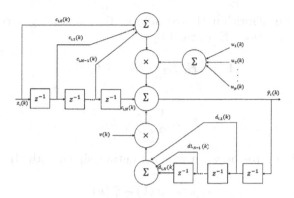

Fig. 5. IIR filter structure.

with $a \neq 0$; $a, b \in R$ and

$$\tau_{l_j} = \left(\sum_{j=1}^{p} (u_j - b_{l,j})^2 \right)^{1/2} / a_{l_j} \qquad (11)$$

The j scale variable, a_{l_j} allows expansion and contraction, and $b_{l,j}$ stands for the (l, j) translation variable at k, in the classical role of RBF, with the advantage of dealing with more refinement through daughters wavelets $\psi_l(\tau_l)$. This last feature is essential in the present algorithm together with the pruning capability of the IIR filter. As suggested in [1], the mathematical representation of wavelet RASP1 is a singularity-free normalization of the argument of the wavelet

$$RASP1 = \frac{\tau}{(\tau^2 + 1)^2} \qquad (12)$$

whose partial derivative with respect to $b_{i,j}$ is

$$\frac{\partial \tau}{\partial b_{i,j}} = \frac{1}{a} \frac{3\tau^2 - 1}{(\tau^2 + 1)^3} \qquad (13)$$

In this way, the i wavenet approximation signal with IIR filter can be calculated as:

$$\hat{y}_i(k) = \sum_{q=1}^{p} \sum_{l=0}^{M} c_{i,l} z_i(k - l) u_p(k) + \sum_{j=1}^{N} d_{i,j} \hat{y}_i(k - j) v(k)$$

$$z_i(k) = \sum_{l=1}^{L} w_{i,l} \psi_l(k) \qquad (14)$$

where L stands for the number of daughter wavelets, w_l the weights of each neuron in the wavenet, c_i and d_j are the coefficients of forward and backward IIR filter, respectively, and M and N the coefficients number of forward and backward IIR filter, respectively. The wavenet parameters are optimized by a least mean square algorithm (LMS) subject to minimizing a convex radially unbounded cost functions **E**, defined by

$$\mathbf{E} = \begin{bmatrix} E_1 \; E_2 \; \cdots \; E_i \; \cdots \; E_p \end{bmatrix}^T \tag{15}$$

where

$$E_i = \frac{1}{2} \sum_{k=1}^{T} e_i^2(k) \tag{16}$$

Let the estimation error between wavenet output signal with IIR filter and system output be

$$e_i(k) = y_i(k) - \hat{y}_i(k) \tag{17}$$

To minimize **E**, the steepest gradient-descent method is considered. To this end, notice that partial derivatives of **E** wrt **A**(k), **B**(k), **W**(k), **C**(k),**D**(k) are required to update the incremental changes of each parameter along its negative gradient direction. That is,

$$\Delta\mathbf{W}(k) = -\frac{\partial\mathbf{E}}{\partial\mathbf{W}(k)} \tag{18}$$

$$\Delta\mathbf{A}(k) = -\frac{\partial\mathbf{E}}{\partial\mathbf{A}(k)} \tag{19}$$

$$\Delta\mathbf{B}(k) = -\frac{\partial\mathbf{E}}{\partial\mathbf{B}(k)} \tag{20}$$

$$\Delta\mathbf{C}(k) = -\frac{\partial\mathbf{E}}{\partial\mathbf{C}(k)} \tag{21}$$

$$\Delta\mathbf{D}(k) = -\frac{\partial\mathbf{E}}{\partial\mathbf{D}(k)} \tag{22}$$

then, the tuning update parameter becomes:

$$\mathbf{W}(k+1) = \mathbf{W}(k) + \mu_{\mathbf{W}}\Delta\mathbf{W}(k) \tag{23}$$
$$\mathbf{A}(k+1) = \mathbf{A}(k) + \mu_{\mathbf{A}}\Delta\mathbf{A}(k) \tag{24}$$
$$\mathbf{B}(k+1) = \mathbf{B}(k) + \mu_{\mathbf{B}}\Delta\mathbf{B}(k) \tag{25}$$
$$\mathbf{C}(k+1) = \mathbf{C}(k) + \mu_{\mathbf{C}}\Delta\mathbf{C}(k) \tag{26}$$
$$\mathbf{D}(k+1) = \mathbf{D}(k) + \mu_{\mathbf{D}}\Delta\mathbf{D}(k) \tag{27}$$

The update parameters follows the next rule:

$$\Delta\theta(k) = -\mu\frac{\partial\mathbf{E}}{\partial\theta(k)} \tag{28}$$

$$\theta(k+1) = \theta(k) + \Delta\theta(k) \tag{29}$$

where θ can be any of the parameters set $\mathbf{W}(k)$, $\mathbf{A}(k)$, $\mathbf{B}(k)$, $\mathbf{C}(k)$ o $\mathbf{D}(k)$. The value of $\mu \in R^m$ represents the learning rate for each of the parameters.

3.3   Discrete PID Controller

Canonical observable realization of the discrete domain equations of the robot suggests, in view of previous subsection, that the following models the unknown robot, which is useful to derive a wavenet schemes as follows. Consider

$$\mathbf{y}(k+1) = \Phi[\mathbf{Y}(k), \mathbf{U}(k)] + \Gamma[\mathbf{Y}(k), \mathbf{U}(k)] \cdot \mathbf{u}(k) \tag{30}$$

when terms Φ and Γ are exactly known, computed torque or inverse dynamics algorithms provides the following controller $u(k)$ that ensured convergence to a desired output $\mathbf{y}_{ref}(k+1)$ as follows:

$$\mathbf{u}(k) = \Gamma^{-1}[\mathbf{Y}(k), \mathbf{U}(k)](\mathbf{y}_{ref}(k+1) - \Phi[\mathbf{Y}(k), \mathbf{U}(k)]) \tag{31}$$

Thus, the closed-loop Eqs. (30)–(31) produces, ideally $\mathbf{y}(k+1) = \mathbf{y}_{ref}(k+1)$. However, it is assumed that such realization (30) is not known, but observability of robot dynamics implies that there exist basis function that approximates (30). Then, (30) is estimated as follows

$$\hat{\mathbf{y}}(k+1) = \hat{\Phi}[\mathbf{y}(k), \Theta_\Phi] + \hat{\Gamma}[\mathbf{y}(k), \Theta_\Gamma] \cdot \mathbf{u}(k) \tag{32}$$

System (32) is estimated by two wavenet functions as follows

$$\hat{\Phi}_i[\mathbf{y}(k), \Theta_\Phi] = \sum_{j=1}^{N} d_{i,j}\hat{y}(k-j)v(k) \tag{33}$$

$$\hat{\Gamma}_{i,q}[\mathbf{y}(k), \Theta_\Gamma] = \sum_{q=1}^{p}\sum_{l=0}^{M} c_{i,l}z_i(k-i)u_q \tag{34}$$

$$z_i(k) = \sum_{l=1}^{L} w_{i,l}\psi_l(k) \tag{35}$$

with adjustable parameters Θ_Φ and Θ_Γ, for function $\hat{\Phi}_i$ and $\hat{\Gamma}_{i,q}$ representing the i component of $\hat{\Phi}$ and the (i,q)-element of the matrix $\hat{\Gamma}$, respectively. Therefore, since nonlinear functions wavenet functions $\hat{\Phi}(k)$ and $\hat{\Gamma}(k)$ estimate $\Phi(k)$ and $\Gamma(k)$, as $k \to \infty$, then error $e_i(k) = y_i(k) \to \hat{y}_i(k)$ can be used as a Lebesgue measure useful to tune feedback gains. The following discrete PID controller is proposed:

$$\begin{aligned} u_\sigma(k+1) = u_\sigma(k) &+ k_{p_\sigma}(k)[\varepsilon_\sigma(k) - \varepsilon_\sigma(k-1)] + \\ k_{d_\sigma}(k)&[\varepsilon_\sigma(k) - 2\varepsilon_\sigma(k-1) + \varepsilon_\sigma(k-2)] + \\ k_{i_\sigma}(k)&\varepsilon_\sigma(k) \end{aligned} \tag{36}$$

where $k_{p_\sigma}(k)$, $k_{i_\sigma}(k)$ and $k_{d_\sigma}(k)$ stand for strictly positive definite proportional, integral and derivative feedback gains, respectively; $u(k)$ is the controller at instant k, and error is defined as

$$\varepsilon_\sigma(k) = y_{ref_\sigma}(k) - y_\sigma(k) \tag{37}$$

for $\sigma = 1, 2, 3, \ldots, p$. Each feedback gain is tuned according to the corresponding error they affect in (36) and modulated by $\hat{\Gamma}$, the input matrix of (32).

3.4 Auto-tuning PID Gains

Due to the gains $k_{p_\sigma}, k_{i_\sigma}$ are considered within the cost function (16), those can be updated similar to (23)–(27). Let

$$k_{p_\sigma}(k) = k_{p_\sigma}(k-1) + \mu_p e_\sigma(k)\hat{\Gamma}_{i,q}(k)[\varepsilon_\sigma(k) - \varepsilon_\sigma(k-1)]$$
$$k_{i_\sigma}(k) = k_{i_\sigma}(k-1) + \mu_i e_\sigma(k)\hat{\Gamma}_{i,q}(k)\varepsilon_\sigma(k)$$
$$k_{d_\sigma}(k) = k_{d_\sigma}(k-1) + \mu_d e_\sigma(k)\hat{\Gamma}_{i,q}(k)$$
$$[\varepsilon_\sigma(k) - 2\varepsilon_\sigma(k-1) + \varepsilon_\sigma(k-2)]$$

where $\hat{\Gamma}$ is defined by (34), for $0 < \mu < 1$ the learning rate of the PID controller gains. Notice that learning rates μ are designer parameters.

3.5 The Medical Task

The Haptic Guidance Platform is a modular system designed to assist rehabilitation treatments in patients with neuromuscular affections. It offers the possibility of being configured according to the therapy for a wide class of disabilities. The platform is composed of three main functional blocks: *Rehabilitation Therapy* concerning medical aspects in a medical protocol to determine the treatment according to the condition of the patient and suggests the visual and kinaesthetic stimuli for his/her neuromuscular system. *Haptic Interface* including hardware and software of the haptic device itself, including drivers, close loop controllers and a visual display for virtual reality environments, and *Computational Platform* containing the path planning algorithms, communication interfaces and protocols, and software interfaces required to connect the host computer to the external devices. The whole design, obtained under a robotic approach, respects biomedical specifications to allow its use with patients, i.e., anthropometry, ergonomics, and safety.

3.6 Proposal for Rehabilitation

Hemiparesis is relatively common condition in children, of congenital or acquired aetiology. It is characterized by a unilateral decrement of force and precision of movements, due to injuries in the nervous system. The system in Local Haptic Guidance configuration, assists the child in coordination tasks tracking flat patterns indicated visually. The guidance action corrects his movements by an attractive force towards the desired trajectory producing a kinaesthetic stimulus that promotes the synchrony and the correct sequence of movements. This configuration was implemented using a Geomagic Touch haptic device (Fig. 2) and is proposed for a therapy session in patients with hemiparesia. The close loop controller was designed under a Intelligent control based on a Wavenet controller with high performance.

4   The Experimental Results

The experiment consists of applying the adaptive control strategy based on a wavenet network and the self-tuning of a PID control. The graphs of the signals derived from each experiment correspond to: (i) the workspace (x vs y vs z), (ii) the variation in time of the operational coordinates, (iii) the operational error signals, (iv) the joint control signals, (v) the identification error by degree of freedom, (vi) the dynamic weights and parameters of the neural network, and (vii) the self-tuning of discrete PID control gains. Two experimental phases are presented: (a) experiment without human in the loop, in this case the control adapts the gains according to the admissible configurations in the tracking task and the compensation of the dynamics without uncertainty in the loop; and (b) experiment with human in the loop, with the same signals as in the previous case, but with the human operator in the loop. In which there is presence of artificially induced disturbances to show the response of adaptability.

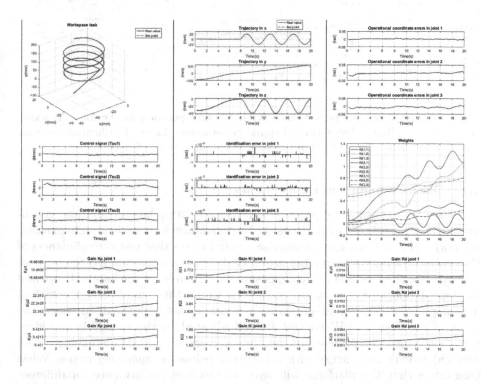

Fig. 6. Adaptive control performance **without** human operation in the loop. Tracking in Workspace. Haptic device control signal performance. Error identification for each degree of freedom. The weights of the wavelet neural network. Control gains.

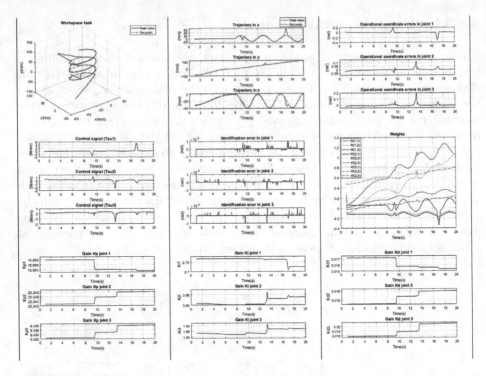

Fig. 7. Adaptive control performance **with** human operation in the loop. Tracking in Workspace. Haptic device control signal performance. Error identification for each degree of freedom. The weights of the wavelet neural network. Control gains.

4.1 Discussions of the Experiments

Although the experimental practice is developed with a human operator in conditions of physical and mental health, the emulation of a patient that modifies the trajectory involuntarily or exercises spatial forces that test the robustness of the platform, a high performance is observed. The control signal has the purpose to guarantee the tracking and compensate the dynamics of the haptic device and the human operator: $\tau = \tau_{haptic\_device} - \tau_h$. Under these conditions, the task of haptic guidance is given satisfactorily; it is visible that with the disturbance, the gains of the discrete PID control are self-adjusted by the change in the identification error. The presence of the human operator, and its disturbing action demand an excess of energy with magnitudes below the operating values. What guarantees that the platform will have an excellent performance for different users.

5 Conclusion

In this paper, a novel identification and control scheme for nonlinear MIMO systems based in wavenet with IIR filter, as pruning of irrelevant nodes, and RASP1 daughters wavelets is proposed for an efficient inverse dynamics approximation scheme. Such approximation capabilities and the basis of the input matrix is used to design a time-varying feedback gains such that recurrent discrete tuning of PID feedback gains occurs without any a priori knowledge of the haptic device. Contrary to most neural network controllers, approximation of inverse dynamics are not used for stabilization but for tuning feedback gains of a simple PID controller. Global convergence is obtained as iteration increases. Experiments show the viability of the proposed scheme for practical implementation, where typically the exact model is now known, in particular, it captures the essential full and real nonlinear dynamics, without any linearization nor idealization of non-linear robotics. Experiments on a nonlinear 3D highly coupled robotic system in real-time shows the relevance of the proposed control scheme. Under these conditions, the physical interaction platform is ideal for the application of clinical protocols through a task tracking with the patient in the control loop.

References

1. Daubechies, I.: Ten Lectures on Wavelets. SIAM, Philadelphia (1992)
2. Fu, M.J., Cavusoglu, M.C.: Human-arm-and-hand-dynamic model with variability analyses for a stylus-based haptic interface. IEEE Trans. Syst., Man, Cybern., Part B (Cybern.) **42**, 1633–1644 (2012)
3. Haykin, S.: Kalman Filtering and Neural Networks. Wiley, Hoboken (2001)
4. Jarillo-Silva, A., Domínguez-Ramírez, O.A., Parra-Vega, V., Ordaz-Oliver, J.P.: PHANToM OMNI haptic device: kinematic and manipulability. In: IEEE, Electronics, Robotics and Automotive Mechanics Conference 2009, CERMA 2009, pp. 193–198 (2016). https://doi.org/10.1109/CERMA.2009.55
5. Krebs, H.I., Hogan, N., Aisen, M.L., Volpe, B.T.: Robot-aided neurorehabilitation. IEEE Trans. Rehabil. Eng. **6**, 75–87 (1998)
6. Levin, A., Narendra, K.: Control of nonlinear dynamical systems using neural networks: controllability and stabilization. IEEE Trans. Newral Netw. **4**, 192–206 (1993)
7. Ramos-Velasco, L.E., Domínguez-Ramírez, O.A., Parra-Vega, V.: Wavenet fuzzy PID controller for nonlinear MIMO systems: experimental validation on a high-end haptic robotic interface. Appl. Soft Comput. **40**, 199–205 (2016)
8. Suleman, K., Andersson, K., Wikander, J.: Dynamic based control strategy for haptic devices. In: IEEE, World Haptics Conference, pp. 131–136 (2011)
9. Spong, M.W., Vidyasagar, M.: Robot Dynamics and Control. Wiley, Hoboken (1989)
10. Turijan-Rivera, J.A., Ruiz-Sanchez, F.J., Domínguez-Ramírez, O.A., Parra-Vega, V.: Modular platform for haptic guidance in paediatric rehabilitation of upper limb neuromuscular disabilities. In: Pons, J., Torricelli, D., Pajaro, M. (eds.) Converging Clinical and Engineering Research on Neurorehabilitation. BIOSYSROB, vol. 1, pp. 923–928. Springer, Heidelberg (2013). https://doi.org/10.1007/978-3-642-34546-3_150

ROBMMOR: An Experimental Robotic Manipulator for Motor Rehabilitation of Knee

Gabriel A. Navarrete[1(✉)], Yolanda R. Baca[2(✉)],
Daniel Villanueva[2,3(✉)], and Daniel Martínez[1(✉)]

[1] Universidad Tecnológica del Centro de Veracruz, Av., Universidad no. 350,
Carretera Federal Cuitláhuac - La Tinaja, Dos Caminos,
Cuitláhuac, Ver., Mexico
{gabriel.navarrete,daniel.martinez}@utcv.edu.mx
[2] INFOTEC: Centro de Investigación e Innovación en Tecnologías de la
Información y Comunicación, Circuito Tecnopolo Sur 112,
20313 Aguascalientes, Ags., Mexico
{yolanda.baca,daniel.villanueva}@infotec.mx
[3] CONACyT Consejo Nacional de Ciencia y Tecnología, Dirección de Cátedras,
Insurgentes Sur 1582, Crédito Constructor, 03940 Mexico City, Mexico

Abstract. Nowadays, the role of robotics in patients' rehabilitation it is an area of interest for science and technological development. Besides, the motor rehabilitation has had great success in subjects with disability problems which require an intensive and specific therapeutic approach for each task through robots. Budgetary constraints limit to hand-to-hand therapy approach, so machines can offer a solution to further promote patient recovery and to better understand the rehabilitation process. This article presents a ROBMMOR: an experimental robotic manipulator of knee rehabilitation. The robot is capable of performing passive exercises in patients with motor movement problems in the knee. The robot's system helps the patient in the process in a personalized way through the positions of speed and strength required. Finally, *ROBMMOR* obtains data and generates an evaluation of the progress of the patient's rehabilitation that helps the therapist for future analysis.

Keywords: Robotic · Knee · Control system · Motor rehabilitation

1 Introduction

At present, rehabilitation in society has caused a great impact and technological development that has covered practically all areas of the human being. Successful motor rehabilitation in patients with accidents and traumatic injuries requires an intensive and specific therapeutic approach for each task. Smart machines can offer a solution to promote motor recovery and obtain a better understanding in rehabilitation [1].

This new field of automated or robot-assisted motor rehabilitation has emerged since the 1990s [2]. Also, robotics has had an important acceptance and successful results around robot-assisted rehabilitation. Therefore, it allows generating new alternatives for the different forms of therapies that currently exist and, as a consequence,

© Springer Nature Switzerland AG 2018
I. Batyrshin et al. (Eds.): MICAI 2018, LNAI 11289, pp. 304–317, 2018.
https://doi.org/10.1007/978-3-030-04497-8_25

increases the effectiveness of the exercise and diminishes the work of the expert in the domain [3].

According to the International Classification of Functioning, Disability and Health, people with disabilities "are those who have one or more physical, mental, intellectual or sensory deficiencies and who, when interacting with different environments of the social environment, can prevent their full and effective participation in equal conditions to others" [4].

Also, according to the WHO (World Health Organization), in 1969 rehabilitation is defined as: "part of the medical care in charge of developing the functional and psychological capabilities of the individual and activate their compensation mechanisms, in order to allow them to carry an autonomous and dynamic existence" [5].

Therefore, rehabilitation seeks that the person affected or with a disability, through therapies, gradually restore their abilities, trying to reach a normal state, or at least to a state where the individual can be having autonomy, that is, to be able to fend for himself. Some of the main problems in human disabilities are caused in legs and arms causing immobility.

For example, muscles weakened by old age, accidents and also neuronal problems. In order to recover the mobility of the body, the muscles have to be strengthened, through specialized exercises, for which the patient must undergo therapies carried out by expert physiotherapists who supervise and exercise the muscles of handicapped people in a manual way and, through repetitive and routine movements of damaged limbs [6].

This paper presents an experimental **ROB**otic Manipulator **MO**tor **R**ehabilitation of knee (*ROBMMOR*). An experimental robot capable of performing passive exercises in patients with motor problem in the knee, its system helps in the rehabilitation process. Therefore, it replaces the work and physical effort of the therapist, in addition accelerates the rehabilitation process in a personalized way through the positions of speed and strength necessary for each patient. Finally, ROBMMOR obtains data and generates an evaluation of the progress of the patient's rehabilitation that helps the expert therapist for future analysis.

2 State of the Art

2.1 Systems Motor Rehabilitation

Within the development of specialized robots for motor rehabilitation, specifically knee, there are important advances, as well as devices that are currently on the market as an attractive option for patients with disabilities [7]. For example, the case of the Optiflex 3, a unit of passive motor rehabilitation for the knee, which implements a progressive range of knee flexion, has a remote control, as well as a system that allows adjusting to the anthropometry of each patient [8]. Also, the commercial passive knee rehabilitation device is the Artromot k3, within its exclusive programming features include: heating mode settings, dual speed adjustment, patient operating time, total device operating time, features of load reversal, pause, extension and bending, making it a versatile device [9]. Within the motor rehabilitation there are several methods that physiotherapists carry out for the restoration of patients [10], within these methods we find the rehabilitation through the passive movement of the damaged limbs, which consists of repetitive movements of the joints and rehabilitation by means of active movements, where the patient performs certain exercises by himself [11].

2.2 An Spatial Augmented Reality Rehab System for Post-stroke Hand Rehabilitation

This work presents a Spatial Augmented Reality system for rehabilitation of hand and arm movement. The table-top home-based system tracks a subject's hand and creates a virtual audio-visual interface for performing rehabilitation-related tasks that involve wrist, elbow and shoulder movements. It measures range, speed and smoothness of movements locally, and is capable to send real time photos and data to the clinic for further assessment. By developing an AR rehab system, the coordination system of user's real world is unified with the virtual world. Thus, the patients feel that the assistive virtual objects that are displayed to help them carry out their exercises, are present and belongs to real world rather to be apart in a separate screen. Also, a vision-based system was developed to locate and track the hand of the subject using color marker and motion information. The system is capable to quantify the motion captured by camera using computing vision methods. Augmented Reality technology has the potential to impact on traditional rehabilitation techniques and it could generalize to real life settings to a greater extent than other computer-based approaches [12].

2.3 Towards Engaging Upper Extremity Motor Dysfunction Assessment Using Augmented Reality Games

This work presents an exploration of the potential of Augmented Reality (AR) using free hand and body tracking to develop engaging games for a uniform, cost-effective and objective evaluation of upper extremity motor dysfunction in different patient groups. Based on the insights from a study with 20 patients (10 Parkinson's disease patients and 10 stroke patients) who performed hand/arm movement tasks in AR, a set of different augmented reality games for upper extremity motor dysfunction assessment were created. In the set of games, virtual hand visual feedback was provided to help patients interact with the virtual content. Also, visual cues were provided as 3D lines between the center of the view of the HMD (head-mounted device) and the virtual object of interest when this was located outside the view of the HMD. Depending on the hand chosen to solve the AR tasks, the whole virtual scene is mirrored. The goal of the games is to provide an objective and quantitative measurement of human motor function in a controlled and engaging environment that offers the possibility to perform a variety of movements [13].

2.4 AR-REHAB: An Augmented Reality Framework for Poststroke-Patient Rehabilitation

This work proposes a framework based on Augmented Reality (AR) technologies that can increase a stroke patient's involvement in the rehabilitation process. The approach takes advantage of virtual-reality technologies and provides natural-force interaction with the daily environment by adopting a tangible-object concept. In the framework, the patient manipulates during the treatment session a tangible object that is tracked to measure her/his performance without the direct supervision of an occupational therapist. The core architecture of the framework and its subsystems that provide more convenience to patients and therapists, were introduced. Also, two exercises are presented: a shelf exercise and a cup exercise, as examples and perform preliminary usability study. In addition, some assessment measurements such as task-completion time, compactness of task, and speed of hand movement by capturing the patients' hand movements with the tangible object were introduced. Motivating virtual objects are overlaid on top of the real scene so patients are efficiently encouraged to repeat boring and tedious rehabilitation procedures in a more pleasant way. The preliminary usability study has shown that these two key advantages are fulfilled quite well in the implemented framework [14].

2.5   Robot-Assisted Upper-Limb Rehabilitation Platform

This work presents a robotic platform for upper-limb rehabilitation robotics. It integrates devices for human multisensorial feedback for engaging and immersive therapies. Its modular software design and architecture allows the implementation of advanced control algorithms for effective and customized rehabilitations. A flexible communication infrastructure allows straightforward devices integration and system expandability. The platform is mainly composed by a robot arm, an open PC based controller, various devices to allow human-robot interaction at different levels, and a safety system which guarantee the reliability of the whole system. In the last section conclusions and future developments are drawn. In addition, the platform incorporates a stereo-vision system with a twofold purpose: completely track upper-limb and trunk movements for clinical and control feedbacks; track robot movements for safety feedback (refer to the next section). A graphic user interface has been implemented to properly control and monitor the robotic platform functionalities. It has been designed taking into account physiotherapist ergonomics using a touch-screen device [15].

2.6   Challenges and Opportunities in Exoskeleton-Based Rehabilitation

Robotic systems are increasingly used in rehabilitation to provide high intensity training for patients with motor impairment. The results of controlled trials involving human subjects confirm the effectiveness of robot enhanced methods and prove them to be marginally superior over standard manual therapy in some cases. Although very promising, this line of research is still in its infancy and further studies are required to fully understand the potential benefits of using robotic devices such as exoskeletons. Exoskeletons have been widely studied due to their capability in providing more control over paretic limb as well as the complexities involved in their design and control. This work briefly discusses the main challenges in development of rehabilitation exoskeletons and elaborates more on how some of these issues are addressed in the design of CLEVERarm, a recently developed upper limb rehabilitation exoskeleton. The mechanical design of the exoskeleton is centered on reducing the weight and bulkiness of the whole structure. Robots could allow patients to start therapy in the very early stages of recovery, without having to deal with the hassles of frequent and long visits to clinics. In the comfort of their own homes, people could get specific training at the appropriate level of intensity [16].

3   ROBMMOR Methodology

This section describes the methodology used for the construction of the prototype. In Fig. 1, the *ROBMMOR* methodology is shown, which allows the design and construction of the robot. The methodology is based on recurrent engineering. On the one hand, it involves mechanical engineering for the design and construction of the robotic base. On the other hand, it involves the electronic engineer for the design and construction of the control and power plates. Finally, computer engineering for the development of the graphic interface and system programming.

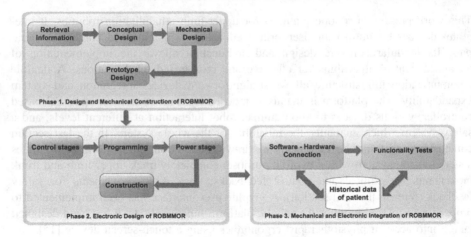

Fig. 1.   ROBMMOR methodology

Phase 1: Design and Mechanical Construction

In this phase, the development of the mechanical design that includes the recommendations of the specialist on the exercises and positions of the body that allow to move the knee is shown. Therefore, a solution is generated through the development of *ROBMMOR* for the exercise and rehabilitation of the patient. Once the information is obtained, freehand sketches are made and with industrial drawing techniques to determine the mechanical structure, obtained the base, the computer design is made through the SolidWorks software, which allows to computationally design a mechanical sketch of the Robotic platform.

Phase 2: Electronic Design

In this phase, we start to work with the electronic part, which consists in determining the actuators that will integrate the robot, considering the necessary torques and the angular movements for the knee flexion. Also, the control system and the electronic power stage are designed. Through a learning tool on the design and simulation of circuits, the designs of the plates are generated for later, in order to carry out the construction and tests of their operation. In addition, in this stage the programming algorithms for the microcontroller and the communication with the interface implemented in a computer are determined.

Phase 3: Electronic Design

In this phase the different developments of phase 1 and phase 2 are integrated. On the one hand, the mechanical, electronic design, actuators and sensors are integrated in the mechanical platform. On the other hand, the integration of the microcontroller, the control cards and the electronic power cards, is performed. Likewise, communication with the computer is developed through an interface that allows the analysis of the data. Finally, the construction of the prototype for its experimentation is carried out.

4 Design of the Platform ROBMMOR for Motor Rehabilitation

4.1 Anthropometric Study

To carry out the conceptual and mechanical design of *ROBMMOR,* determining factors were considered, on which the correct functioning of the system depends. The system works for different ages, consequently, the dimensions of the lower limb vary according to each factor. Therefore, it is necessary to make a study based on scientific research that serves to know the anthropometric measures on three cases of study, adults, seniors and children of the Mexican population. The measurements define the variations in the physical dimensions in which an individual belongs. Figure 2, shows the anthropometric measurements of the human body.

In the case of the study on adults, the data were taken from the references of the anthropometric measurements of the body, specifically measure 12, whose study consisted of an evaluation of 210 workers whose average age was 40 years.

- Height of the floor at the knee (Measurement reference 7) - 51.84 cm.
- Height of the floor to the back of the knee (Measurement reference 18) - 43.41 cm.
- Length of the back of the knee to the back of the chair (Measurement reference 21) - 49.83 cm.
- Knee length at the back of the chair (Measurement reference 22) - 60.60 cm.

For the case study of the elderly according to measurement reference 13, for a population with an average age of 66.9 for men and 67.3 years for women, with a sample of 508 individuals, the results are the following:

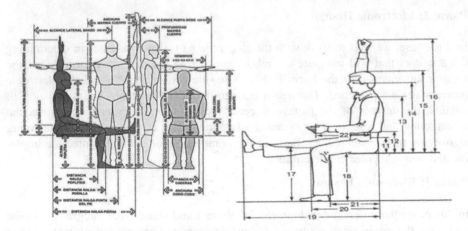

Fig. 2. Anthropometric measurements of the body.

- Men's knee height: 52.04 cm.
- Women's knee height: 47.94 cm.

In the case of children's anthropometric study, the data were taken from the measurement reference 14 for an average age of 10 years:

- Floor height at the knee - 39 cm.
- Knee length at the back of the chair - 48.3 cm.

4.2   Analysis of the Strength of the Knee Joint

The strength of the knee joint used to determine the necessary torque needed to move a person's knee and, in this way, determine the power and type of physical motor to be used within the system. According to the study carried out in measurement reference 15 of Fig. 2, which consisted in determining the strength of the knee joint in both flexion and extension, A 40 people sample was selected, which was composed by 20 men and 20 women, with the help of a Biodex 3 dynamometer team, which will help determine the angles and average strength. The selected people with an average age of 30.7 years, average height equal to 1.74 m in men and 1.57 in women, and with an average weight of 81.7 kg in the case of men and women. 57 kg in the women's [17]. Tables 1, 2, 3 and 4, show the results of the case study.

The data show that the highest value is the extension force of a man's right leg, the magnitude is 189.156 Nm of torque. Therefore, the motor must provide a greater force to consider, not only the weight of the knee, but also the support that raises the leg. Also, it is necessary to consider that in the passive movement the knee must be moved even though the patient puts some resistance. However, care must be taken in order to avoid damaging the patient's joint.

Table 1. Torque of the flexion of a man's leg

Leg	Degrees	Average torque in flexion
Right	60°	107.806 Nm
	90°	98.456 Nm
	120°	98.722 Nm
Left	60°	101.622 Nm
	90°	96.706 Nm
	120°	89.350 Nm

Table 2. Payment management torque of the spread of a man's leg

Leg	Degrees	Average torque in extension
Right	60°	189.156 Nm
	90°	182.861 Nm
	120°	153.694 Nm
Left	60°	181.911 Nm
	90°	164.694 Nm
	120°	152.622 Nm

Table 3. Flexion torque of a woman's leg

Leg	Degrees	Average torque in flexion
Right	60°	56.040 Nm
	90°	50.360 Nm
	120°	49.180 Nm
Left	60°	54.920 Nm
	90°	48.615 Nm
	120°	46.940 Nm

Table 4. Torque of the extension of a woman's leg.

Leg	Degrees	Average torque in extension
Right	60°	107.020 Nm
	90°	89.580 Nm
	120°	83.705 Nm
Left	60°	102.860 Nm
	90°	89.335 Nm
	120°	81.620 Nm

4.3   Physical Design of the Robot

The most appropriate way to perform knee rehabilitation exercises is with the patient sitting, So that, we decided build the mechatronic system based on the design of a chair or exercise bench, since this device is comfortable; once the bank was physically developed a sketch of the mechatronic system was designed. In Fig. 3, the different stages of design are shown, from the idea to using freehand drawings of a bank of exercises, considering that the patient needs to be comfortable, and to have a support to be able to sit, until the computer design which allows to obtain simulations of the movements of the mechanisms involved. Finally, the *ROBMMOR* prototype is shown.

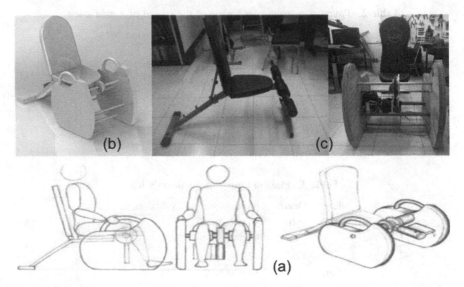

Fig. 3. Stages of design (a) Design on freehand drawings, (b) Computer design (c) ROBMMOR prototype.

4.4   Robot Control Design

This section describes the development of the *ROBMMOR* system for knee rehabilitation. The electronic design is responsible for driving the motor, controlling the speed and controlling the force applied by the device to the patient's leg. In addition, the electronic design can control over the position to determine the degree of flexion and extension. Finally, the strength that each patient applies through a communication between the computer and the prototype is recorded. In the speed control of the prototype, a converter cycle was designed and implemented to control the engine speed. Therefore, it allows to control the speed with which the device flexes the knee of the patient through the electric motor who provides the necessary power to do perfom it. The motor to be controlled is single-phase induction with capacitor start, its nominal current is 1.5 A, its supply voltage is 120 V at 60 Hz with a nominal speed is 1750 rpm and a power of 1/20 HP. This engine within the system is coupled to a gearbox with which its speed is reduced and its torque is increased, the reduction ratio is 200:1. Also, an incremental quadrature encoder is used for position control, the encoder has two main channels called A and B and a third channel that helps find the initial position called Index and, this encoder inside the prototype is installed in the speed reducer shaft. (see Fig. 3).

In Fig. 4, the block diagram of the electronic control system of *ROBMMOR* is shown. The diagram starts from a CA (Alternating Current) and, it goes to a rectification stage controlled by SCRs (Silisium Controlled Rectifiers). The obtained signal is filtered and taken to an inverter circuit designed with Mosfets. Then, the inverter that is controlled by the microcontroller using the sinusoidal PWM technique, generates an output signal with variation in frequency, this signal goes directly to the motor, which has a speed reducer and an incremental encoder installed, the signal of the sensor is sent to the microcontroller, the setPoint of the system is sent by the user interface to the microcontroller which finally sends the signals.

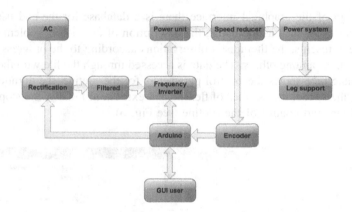

Fig. 4. Block diagram of electronic system.

In the case of the robotic system to be able to control the speed of the single-phase induction motor, the current of the AC supply must first be converted to CD, the DC potential level must be regulated, and then converted from CD to AC, but the AC frequency must be modulated, which results in a CA to AC cycloconverter. Which provides to the output a regulated potential and a modulated work frequency.

5   Discussion and Evaluation

5.1   Modulation with Sinusoidal PWM Technique

Modulation with sinusoidal technique consists of emulating the behavior of a sine wave through the control of pulse width in time intervals. The technique uses a signal of nine pulses per quarter of cycle, each pulse must increase in duration until it reaches the quarter of a cycle where, the next quarter of a cycle, the pulses decrease. To apply this signal to the inverse of full bridge it is necessary to generate two signals but with a phase shift. Figure 5, shows the modulation of *ROBMMOR* with the sinusoidal technique through a microcontroller and the simulation of its outputs.

Fig. 5. Waveform of the two sinusoidal PWM signals of the microcontroller.

5.2   Graphic Interface

For the design of the graphical interface there is a database for the end user. At the base, data is stored, such as the name, the description of the motor problem presented by the patient. resents, the therapist's observations according to the progress shown in the rehabilitation, among others. The data is accessed through the file with the patient's name. The interface shows the control parameters for the routine, the duration time of the routine, the speed, the degrees of flexion and extension. Finally, the stop and start control and the start control of the routine (see Fig. 6).

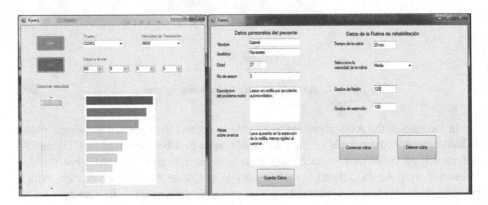

Fig. 6.   Graphic interface ROBMMOR

According to the calculations obtained through the anthropometric analysis, the correct functioning of *ROBMMOR* is ensured. Therefore, a torque capable of lifting a 10 kg leg is obtained. By graphs obtained from the behavior of the current and voltage, and by means of position and speed control, the pertinent routines can be programmed for therapy. In addition, varying the frequency and power supply of the motor, it can be determined if the robot can lift the desired weight. Figure 7, shows the graphs and the behavior parameters for one of the experimental tests performed with *ROBMMOR*.

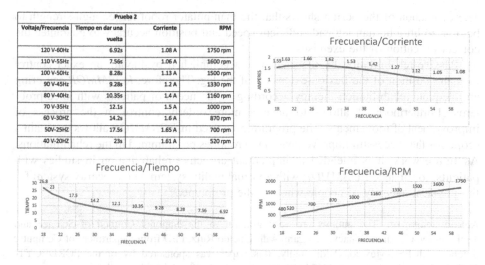

Fig. 7. Experimental test of frequencies with ROBMMOR.

The sequences and control strategies for the execution of *ROBMMOR* are programmed from the graphic interface. Through the interface, the patient's history and the data of each routine, can be stored. The data stored provides a personalized record of the progress of the exercises carried out. Subsequently, the records can be used to analyze the results and monitor the patient's rehabilitation progress, and also, it can generate an alternative for the expert. Also, *ROBMMOR* ensures that the torque is enough to lift a human leg, the experimentation was done with different weights of 8 kg, 9 kg, 10 kg and a maximum 12 kg to check the *ROBMMOR* positions.

6 Conclusions and Future Works

By means of the design and construction of *ROBMMOR* it is possible to verify that the electronics, robotics, computer science and the techniques for the development involved in the robot, allow the experimental manipulation in the technological development for the motor rehabilitation of the knee. There is an area of opportunity that can be addressed with the present development, or the case of people who are disabled by a motor problem or a disease that damages some of their body members. Likewise, the technological development of *ROBMMOR* benefits people with a motor disability. The robot is a complementary tool for therapists in the treatment and diagnosis of patients. Besides rehabilitation centers are benefited with this type of technological advances as part of a possible clinical application for users. On the one hand, by providing a robot that allows them to save physical effort and work time. On the other hand, to provide better control in motor rehabilitation therapies of the knee. The data stored provides a personalized record of the progress of the exercises carried out. Subsequently, the records can be used to analyze the results and monitor the patient's rehabilitation progress, and also, it can generate an alternative for the expert.

The evaluation of the design shows that the manipulator robot has enough strength to be able to lift a human leg and sufficient speed and position accuracy to guarantee the correct execution of the exercises.

Passive rehabilitation by the patient is accomplished by having an exact control over the position of the flexion-extension of the leg. Therefore, *ROBMMOR* can bring an adequate rehabilitation therapy according to the needs of patients with knee problems. Furthermore, it allows to have a more orderly control of the process of improvement of movements. The programming used in *ROBMMOR* allows to obtain a scope for the successful improvement in the routines programmed in the rehabilitations. As future work, it is intended to implement an active rehabilitation in studies with patients, to enhance *ROBMMOR* with a virtual reality system and a security system for the physical integrity of people who use the prototype.

Acknowledgment. The authors are very grateful to the National Council of Science and Technology, Cátedras Conacyt Program, with Ubaldo Ruiz, PhD from Department of Computer Science, CICESE-México. Additionally, this paper was sponsored by of the FORDECYT-CONACYT with the project 296737 "Consortium in Artificial Intelligence".

References

1. Kwakkel, G., Wagenaar, R., Twisk, J., Lankhorst, G.: Intensity of leg and arm training after primary middle-cerebral-artery stroke: a randomised trial. Lancet **354**, 191–196 (1999)
2. Hesse, S., Schmidt, H., Werner, C.: Upper and lower extremity robotic devices for rehabilitation and for studying motor control. Curr. Opin. Neurol. 16, 705–710 (2003). journals.lww.com
3. Lünenburger, L., Colombo, G.: Biofeedback for robotic gait rehabilitation. J. Neuroeng. Rehab. **4**, 1 (2007). jneuroengrehab.biomedcentral.com
4. Ayuso-Mateos, J., Nieto-Moreno, M., Sánchez-Moreno, J.: Clasificación Internacional del Funcionamiento, la Discapacidad y la Salud (CIF): aplicabilidad y utilidad en la práctica clínica. Med. Clin. (Barc) **126**, 461–466 (2006). sid.usal.es
5. García, C., Sánchez, A.: Clasificaciones de la OMS sobre discapacidad. Boletín del RPD **50**, 15–30 (2001). referato.com
6. Akdoğan, E., Taçgın, E., Adli, M.: Knee rehabilitation using an intelligent robotic system. J. Intell. Manuf. **20**, 195 (2009)
7. Roy, A., Krebs, H., Williams, D., Bever, C.: Robot-aided neurorehabilitation: a novel robot for ankle rehabilitation. IEEE Trans. Robot. **25**, 569–582 (2009). ieeexplore.ieee.org
8. García, D., Gabriel, P.: Diseño de un rehabilitador de rodilla (2012)
9. Ocaña, S., Edmundo, W., Amaguaña, S., Adolfo, M.: Diseño y construcción de un banco de cuádriceps automático, para rehabilitación física de rodilla en pacientes del Patronato de Amparo Social Niño de Isinchi–Pujilí (2017)
10. Camacho, J., Chamorro, C.: Implementation of a service-oriented architecture for applications in physical rehabilitation. Facultad de Ingeniería 26, 113–121 (2017). revistas.uptc.edu.co
11. Ceballos, E., Díaz-Rodriguez, M., Paredes, P., Vargas, P.: Development of a passive Rehabilitation Robot for the wrist joint through the implementation of an Arduino UNO microcontroller, June 2017

12. Hondori, H.M., Khademi, M., Dodakian, L., Cramer, S.C., Lopes, C.V.: A spatial augmented reality rehab system for post-stroke hand rehabilitation. Stud. Health Technol. Inf. **184**, 279–285 (2013)
13. Cidota, M., Lukosch, S., Bank, P.: Towards engaging upper extremity motor dysfunction assessment using augmented reality games. In: 2017 IEEE International Symposium on Mixed and Augmented Reality, ISMAR-Adjunct (2017). ieeexplore.ieee.org
14. Alamri, A., Cha, J., El Saddik, A.: AR-REHAB: an augmented reality framework for poststroke-patient rehabilitation. IEEE Trans. Instrum. Meas. **59**, 2554–2563 (2010). ieeexplore.ieee.org
15. Malosio, M., Pedrocchi, N., Molinari Tosatti, L.: Robot-assisted upper-limb rehabilitation platform. In: Proceedings of the 5th ACM/IEEE International Conference on Human-Robot Interaction (2010). dl.acm.org
16. Soltani-Zarrin, R., Zeiaee, A., Langari, R.: Challenges and opportunities in exoskeleton-based rehabilitation. arXiv preprint arXiv:1711.09523 (2017). arxiv.org
17. Mercado, P., Zarco, R., Arias, D.: Relation between proprioception and muscle strength of knee in asymptomatic subjects. Revista mexicana de medicina física y rehabilitación **15**, 17–23 (2003). academia.edu

A Bio-Inspired Cybersecurity Scheme
to Protect a Swarm of Robots

Alejandro Hernández-Herrera[✉], Elsa Rubio Espino,
and Ponciano Jorge Escamilla Ambrosio

Centro de Investigación en Computación - Instituto Politécnico Nacional,
07738 Mexico City, Mexico
b160516@sagitario.cic.ipn.mx, {erubio,pescamilla}@cic.ipn.mx,
http://www.cic.ipn.mx

Abstract. Swarm robotics describes a multi-robot system characterized by the simplicity of its agents, homogeneous architecture, limited
communication skills, local detection, execution of parallel tasks, robustness, scalability, flexibility and decentralized control. However, being a
technology in development, the security and vulnerability of the swarm
of robots against possible cybernetic attacks have been commonly overlooked. This is of major concern when executing mission-critical activities that inherently require an adequate management of security. In this
work, a bio-inspired security mechanism applied to a swarm of robots is
proposed. Through computer simulations, it is observed how the mechanism, when executing a homing towards a stationary landmark, allows
the swarm of robots to identify abnormal behaviors, caused by a certain
cyberattack; subsequently, it establishes a certain tolerance to it and
allows to improve the level of availability that is required to continue
executing the task at hand.

Keywords: Swarm robotics · Bio-inspired cyber security scheme
Cybersecurity · Cyberattack

1 Introduction

Currently, the field of swarm-based robotics is mainly exploratory, although some
practical applications have been built that extend or surpass the capabilities of
current systems (see [8,34]). Very few tasks of practical interest have been carried out using the swarm-based methodology. However, many of the capabilities
of agents in the natural world are amazing ([13,14]) and it is interesting to
understand how they occur. Once they are understood, it is possible that their
extension to real cases of engineering allows the solution of problems for which
the complexity of the solution is a serious prejudice for its construction.

Swarm robotics is a relatively new technology, whose potential is being
explored for use in a variety of different applications and environments, such
as surveillance, search and rescue, and medicine to name a few. Emerging technologies have often overlooked the security component implied in information

© Springer Nature Switzerland AG 2018
I. Batyrshin et al. (Eds.): MICAI 2018, LNAI 11289, pp. 318–331, 2018.
https://doi.org/10.1007/978-3-030-04497-8_26

management and it is until later stages of development, when security has had to be adapted in a forced, inconsistent and sometimes costly manner. Swarm robotics, due to its unique technical characteristics, does not allow the existing security mechanisms to be applied uniformly, this deficiency could have a significant impact on critical mission applications that require information management that provides integrity, continuity and availability, current concepts used in the security of systems and information. In recent years, there has been great interest in the distributed systems of multiple robots whose members act on the basis of information acquired through local detection and/or communication with other robots in their spatial vicinity. When these local interactions result in global collective behaviors (e.g. cohesion or dispersion), the system is known as swarm robotics [1,22].

The swarms of robots are composed of a large number of agents that are homogeneous. In addition, the swarms are robust for the action of adding or subtracting agents, which gives them the scalability and robustness properties for the individual failure of a robot. However, these same characteristics also make the swarm vulnerable to manipulation by agents that could be inserted into the swarm by an adversary for the purpose of altering the performance of the swarm; likewise, when a swarm of robots unfolds in an unknown and hostile environment to fulfill a mission, there are many problems that affect the system, such as random failures of the robots, communication problems caused by the robots themselves or by some external factors (unstable network, dirty sensors, etc.). Each system has its own characteristics from the point of view of security, so it is important to find a unique approach and consider the characteristics of each system.

The article is structured in the following way, in Sect. 2 the state of the art regarding the security in the robot swarm systems is reviewed, we continue with Sect. 3 that is related to the risk analysis and the identification of the requirements of security, Sect. 4 presents, in a general way, the solution of resilience against cyber attacks followed by the experimental simulation and results in Sect. 5, and finally, in Sect. 6 the conclusions of this work.

2 Related Work

Zikratova et al. in [41] consider the problem of building protection mechanisms to protect the robotic multi-agent system against malicious robots. Feng et al. they describe in [10] their approach based on reputation and trust. They propose two algorithms to calculate the level of trust between the nodes. The nodes form an estimation vector and generate a member estimation matrix. These estimates are formed according to the validity of the information and the distance to the objective. As a result, the reputation function is described by the Weibull-Gnedenko distribution. The drawback of this approach is the fact that trust is calculated at a particular time and for a particular problem. In addition to that, only one parameter is used, called the distance to the target. If an intruder carries out several different attacks, which are not related to the accuracy and precision of the data, the system will not be able to detect it.

Vikshin et al. [40] propose another approach for estimating the stability of algorithms based on trust and reputation. The authors use the following parameters for the estimation: interaction radius, number of agents and percentage of adversaries in the total number of agents. The location of the robots was chosen at random in each experiment. The results reflect the percentage of legitimate agents and malicious agents correctly recognized throughout the network and within the interaction radius. As a result of the experiments, the authors confirmed that the trust model can successfully oppose intruders in any amount. The main drawbacks of this approach for estimating network security is the insufficient amount of estimation parameters. Only the type of intruder can be recognized from the presented parameters. The intruder model is described only for a certain type of security breach.

In [9], the authors develop an approach for the detection of malicious node according to their behavior. In this approach, a group of robots follows a set of rules in fulfilling their mission. Then, robots observe the behavior of others and exchange relevant information. The behavior of the robots is described by a formal model proposed by the authors. The approach is based on the principle of using a protocol that allows robots to detect the abnormal behavior of others under certain conditions and modify their own behavior to protect other robots from malicious activities. The protocol has two main components, the first is a monitor of the behavior of the neighbor that compares them with the agent's own behavior; the second is the algorithm that allows to combine "opinions" of different monitors by communication. The authors carry out experiments that show that at least six robots must control the situation to detect intruders properly. The disadvantage of this method can be summarized as follows, when the system detects an intruder, it is based on the behavior of a node, which is characterized by its movements along a trajectory, that is, no others are detected types of attacks. In order to correctly track an intruder, the system must have special monitor nodes, which follow the behavior of the network, and its quantity must be at least six. This fact brings additional limitations to the network, in addition, the authors do not analyze the effects of the variability in the number of intruders and the maximum number of intruders diverted.

Due to the lack of prior knowledge of the environment in which the task will be executed, the pre-planning based approaches [19,32] will not work, since subtasks created dynamically during the scan can not be handled. Due to the hostile environment, the coordination and communication of robots can be very challenging. Approaches based on collaboration through the physical medium (stigmergy) [39] require leaving traces or marking the environment, and may be difficult to achieve in unfamiliar and hostile environments, such as the ocean floor, asteroids or other planets, etc. Approaches based on auction [7,18,39] are difficult due to the high demand for communication. In addition, the robots that win the auction may malfunction or fail in the process and result in a high failure rate and require even more communication resources and bandwidth to recover faults. Potentially, maintaining communication through the sending of messages [4,31] may not be possible and may cause the problem of clustering

[27]. The approach based on the meeting point [27,30,31] requires robots to be in a preestablished location and communicate. It is a concept taken from the protocol used in human rescue teams under limited communication. The meeting point, particularly the fixed meeting point, is reliable but inefficient.

Meng, et al. in [27], also propose a framework type reference solution, partially addressing the problems in the coordination of robots under extreme conditions with limited communication through the use of visual cues as alternatives. However, using visual cues is not a viable option if there are heavy rains or dust or sand storms. In addition, its model considers a hierarchical system, which generally requires greater coordination from higher level planning robots to lower level robots. In addition, failures of leading robots can cause more coordination problems for the recovery and location of a new leader.

Recently in [37], an approach was proposed for swarms of robots based on self-regeneration and inspired by the process of formation of granulomas, which is a process of containment and repair that is found in the immune system of living beings , and that allows the recovery of certain failure modes during the operation of the swarm. The disadvantage is that to contain the fault, a certain number of robots are assigned in a coordinated way to separate the affected robots from the rest of the swarm, which causes more resources to be used for the containment of the problem.

Using an industrial approach, in [20] Khaldi et al. propose a detection of faults by means of control charts, using a multivariate analysis method known as PCA (Principal Component Analysis) for the variation parameters and a control model called VVC (Virtual Viscoelastic Control) to keep the robots ordered in a uniform way at a distance from each other. The drawback of this last study is that you must previously have the control charts for the specific problem you want to detect, and thus be able to build a reference model at the time of detection. In turn, the swarm of robots must remain with some geometric structure, due to its control VVC, so that what is limited to only work under certain conditions and in scenarios without many obstacles.

3   Risk Analysis and Identification of Security Requirements

Because swarm robotics presents particular security challenges that do not exist in other technologies, it is necessary to perform a risk analysis of particular characteristics that adapt to the swarm of robots and allow to identify their vulnerabilities and threats. However, as the security analysis for these systems is still in its infancy and there are only a few works that address the security requirements, we only consider those for which the swarm of robots maintains the availability for the execution of the assigned task. Those security requirements that have to do with aspects such as integrity or confidentiality from the point of view of information security are not considered.

In the brief study of the literature on swarms of robots and their security requirements, we observe the following points:

- Security requirements are often implicit and specific to a particular system.
- The security standards, for example, the Common Criteria for the evaluation of information technology security [36] are too general to apply directly to formal analysis.
- The security requirements are specified in different levels of details with different assumptions.
- The security requirements are expressed in several formats. Therefore, it is difficult to combine existing requirements from different sources to make a complete set of security requirements.

Therefore, our objective is to identify a set of explicit security requirements for swarms of general robots, specifically oriented to maintain availability to perform the assigned task and specify them in such a way as to provide information on vulnerability and threat relationships.

3.1   Security Requirements Identification

Robotic swarming systems are, on the one hand complex, since they consist of a large number of autonomous and, on the other hand, dynamic subcomponents, since the subcomponents can vary according to the unpredictable environment. To study the security requirements of a general robotic swarm system, we first identify the *subcomponents* that can be potentially vulnerable and, as a consequence, affect the behavior of the swarm. For each subcomponent, we identify, both the objective of security, that for our study is the availability of execution of the task by the swarm; as well as the attacks that can be applied to that subcomponent, assuming that the environment of the swarm of robots can be malicious

The corresponding relationship between the attack and the security properties are maintained, with the intuition that the security objective persists during the attack, therefore in summary, the security requirements are composed of: *(i) Identification of vulnerable subcomponents.* The following subcomponents were identified: Actuators (motors), power supply (battery), sensors (proximity), communications and control software. *(ii) Identification of the security objective.* It was determined that the security objective or security service as specified in [15] is related to availability, which in this study, should ensure that the swarm of robots is available to execute the assigned task and *(iii) Identification of attack scenarios.* Unlike security objectives, security attacks are events that have a negative effect on system security due to unauthorized access and modification of data, information, services, networks and systems devices. Different subcomponents have different vulnerabilities and, therefore, are targets of different attacks. For each subcomponent, we collect and identify a set of possible attacks [12, 17, 21, 25, 26, 35].

3.2   Identified Subcomponents and Cyber Attacks

In this section we carry out a Failure Mode and Effect Analysis (FMEA) for a robot swarm connected wirelessly. The methodology is simple, see [6]. We try to identify all the possible dangers, which can be attacks/failures in robots or subsystems of robots. Then, in each case, we analyze the effect of the danger in each of the global behaviors of the swarm. Therefore, we constructed an image of the tolerance of the swarm to both types of danger and began to understand which hazards are the most serious in terms of the general behavior of the swarm. FMEA is, in this stage, essentially qualitative. In this study, we consider only internal risks. External risks (i.e., communications noise) are extensively investigated in [29]. Since, in our case, the robots in the swarm are all identical, the (internal) risk analysis requires that we consider only the faults that could occur in one or more individual robots, and then consider their effect on the global behavior of the swarm. In Table 1, we present in summary the results of each step in the process of identification of security requirements. The identified sub-components are presented, as potentially vulnerable to an attack that compromises the swarming behavior of robots, as well as possible attacks that could affect them.

Table 1. Summary of security requirements.

Attacks	Subcomponent				
	Actuator	Sensor prox.	Sensor loc.	Comm.	Software
Jamming		Apply	Apply	Apply	
Flooding		Apply	Apply	Apply	
Denial of service	Apply	Apply	Apply	Apply	Apply
Buffer overflow				Apply	Apply
Fuzzing attack		Apply	Apply		Apply
Man in the middle (Injection or modification of code)	Apply	Apply	Apply	Apply	Apply
Suppression of physical component warning	Apply	Apply	Apply		
Security objective	Availability in the execution of tasks				

4 Bio-Inspired Resilience Against Cyber Attacks

An interesting analogy can be made between a system that has the capacity to withstand and recover from attacks, that is, a resilient system, and the adaptive immune system of the human body. The adaptive immune system must allow the cells and tissues of the body to function normally, while mounting an immune response or attacking what may be abnormal cells or tissues (e.g., infected cells and cancer cells) [16]. Our proposed solution is based on the Cross-Regulation Model or CRM [3,23,24], which is a mathematical model whose strength is to capture the characteristic of the immune system known as *body maintenance* [5,38], defined as the ability of the immune system to maintain its host despite the unpredictable events that you will undoubtedly encounter in the course of your life.

The idea of body maintenance in biological systems is comparable to resilience in engineering systems; both continue to operate even in unusual or unexpected circumstances. With this idea, the MRC model allows the system to discriminate between antigens (for this research, normal and abnormal behaviors) based solely on its density and persistence in the environment. The model is able to simulate the tolerance of body antigens (the molecular components of body tissues) that are characteristically persistent and abundant and, at the same time, perform an immune response to foreign pathogens, which are characterized by not being persistent or abundant. The research presents three general contributions:

1. A distributed behavior observation model for swarms of spatially distributed large-scale robots that allow a robot to characterize the behavior of its neighboring robots using limited detection capabilities and sporadic observations.
2. Detection of attacks that can be parameterized to balance the compensation between latency in the identification of defective robots in the swarm, and the number of false positive incidents; and
3. A distributed swarm coalition algorithm to consolidate the decisions taken by individual robots on their neighbors to a robust swarm-level decision on the normal/abnormal state of the robot, crucial to provide collective strategies for recovering attacks for the swarm.

4.1 Behavior Detection

The detection process is divided into the following three phases:

1. The robots observe and characterize the behavior of their neighbors over a period of time and estimate the corresponding behavioral characteristics (Fig. 1a).
2. Each robot executes the MRC and classifies the behaviors observed as normal or abnormal, where the abnormalities are consistent with attacks on the robot (Fig. 1b).
3. The robots form voting coalitions to consolidate their decisions at the individual level on the abnormalities of behavior detected (Fig. 1c).

4.2 Robot Behavoir Characterization

Robot behavoir characterization according to the result of the risk analysis, is divided into three general classes, (i) the immediate environment of the robot (sensors), (ii) Robot movements (actuators) and (iii) the response of the robot to events (sensorimotor interactions). The behavior characteristics of each class are used to identify the behavior of an observed robot, coding each characteristic in binary string (present $= 1$, absent $= 0$) to form a binary chain that we call the characteristics vector (VC). In the simulations, a VC comprises the concatenation of six functions (F_1, F_2, F_3, F_4, F_5, F_6), with two functions for each class of behavior characterization grouped as follows: ***Characteristics in the***

 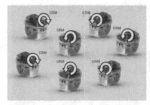 

(a) Behavior observation   (b) Individual execution of the CRM   (c) Coalition to detect abnormal behavior

Fig. 1. (a) The robots observe the behavior of the neighbors, (b) Each robot executes its copy of the CRM individually and (c) The coalition is carried out to detect the abnormal behaviors.

immediate environment of the observed robot. The first two characteristics $F1_j^i(\tau)$ and $F2_j^i(\tau)$ at time τ correspond to the number of neighbors of the robot r_j as observed by the robot r_i. ***Characteristics of the action of the observed robot.*** The following two characteristics $F3_j^i(\tau)$ y $F4_j^i(\tau)$, belong to the motor actions of the robot r_j observed by r_i. ***Characteristics of the sensorimotor interactions of the observed robot*** The two final characteristics, $F5_j^i(\tau)$ and $F6_j^i(\tau)$, belong to the sensorimotor interactions of the robot r_j, as observed r_i.

4.3   Attack Detection

In the next phase of detection of the attacks, an instance of the MRC is executed in each robot to classify the observed behavioral characteristics vectors as normal or abnormal. Behaviors in the swarm that are persistent and abundant (performed by most robots) should be treated as normal. Conversely, rare behaviors (exhibited by fewer robots) must be classified as abnormal.

5   Experimental Setup and Results

We use a multirobot simulator based on physics and discrete time called ARGoS [33], designed to realistically simulate complex experiments involving large swarms of robots. We simulate a swarm of robots composed of 20 e-puck [28] robots located in an environment with a size of $3 \times 3\,\text{m}^2$ (see Fig. 2). The e-puck robot has a diameter of $7.5\,\text{cm}$, a maximum speed of $5\,\text{cm/s}$, and a control cycle of $0.1\,\text{s}$. In our experiments, the e-puck robot model is equipped with eight infrared proximity sensors to avoid obstacles and two actuators that control the speed and direction of movement of the robots. Each robot is also equipped with an extension plate to measure range and heading (location) [11], which allows a robot to estimate the relative location and orientation of neighboring robots.

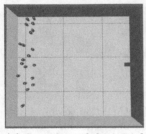 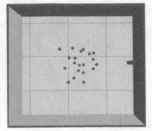 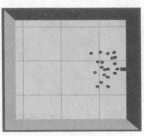

(a) Snapshot of the initial position of the robots

(b) Snapshot of progress in search of the objective

(c) Snapshot of the final position of the swarm when locating the target

Fig. 2. The robots are established in an initial random position (a) and advance in a group cooperatively looking for the objective (b) until its location (c).

5.1   Behavior Simulated by the Robotic Swarm

In the homogeneous behavior of the swarm, all the robots of the swarm execute an identical behavior throughout the duration of the simulation. The homogeneous behavior to be simulated will be that of *search of a source or objective*, where the robots move towards a single pre-established objective, which serves as a point of reference, this behavior also implies that they move away if they are too close to the target or other robots. The position of the objective is fixed at the beginning of the experiment. At the beginning of the experiment, the robots are placed randomly in the first 1.5 m^2, so that from there they can move towards the target. Robot behaviors are implemented using a substraction architecture [2].

5.2   Simulated Attack Behavior

In the experiments, the attacks are simulated by means of the fault signals directly in the sensors and actuators of the robot. Consequently, the resulting defective behavior of the robot corresponds to the actions performed by its controller, which is provided with the input of defective sensors or whose commands to the actuators are not executed correctly by the underlying hardware. The simulated attacks in the experiments are represented by seven different scenarios involving permanent attacks on the sensors and actuators of the e-puck robots.

Infrared Proximity Sensors. For infrared proximity sensors, sensor control attacks represent scenarios involving intentionally disconnected proximity sensors (SMIN), obstructions (simulating, for example, a portion of dust) stuck in proximity sensors (SMAX), and scenarios between these two extremes such as an obstruction in the proximity sensor that temporarily and partially develops as the robot moves through the environment (SRND). *Actuators.* Attacks on the control of the actuators of the e-puck robots occur when one or both of the motors appears to be malfunctioning, or if the tire on the rim of the robot is worn. This attack simulation prevents the robot from turning the affected wheel (left wheel IACT, right wheel DACT, both wheels AACT).

5.3 Results

In the Table 2, the average distances that were obtained in the experiments are shown. In the experiments with the swarm of 20 e-puck robots, simulating a search behavior of a source or objective, 70% performed the normal behavior and the remaining robots, the abnormal behavior indicated.

Table 2. Summary of results

Experiment	Average swarm distance to target
Normal	23.8 cm
DoS-AACT (s/MRC)	>150 cm
DoS-AACT (c/MRC)	39.92 cm
MiM-SMIN (c/MRC)	24.87 cm
MiM-SMAX (c/MRC)	27.94 cm
MiM-SRDN (c/MRC)	30.52 cm

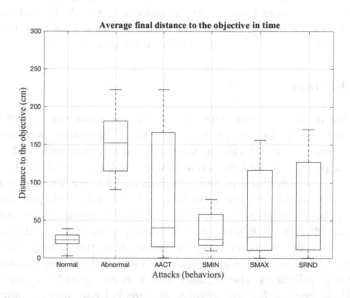

Fig. 3. Final average distance to the objective for all experiments

In the Fig. 3, by means of a box diagram, a comparison of the final average distances to the objective that were obtained for each experiment is shown. As expected, in the *Normal* experiment, the swarm reached the target in a very compact manner and was located at the shortest distance from all the experiments. On the contrary, the experiment called *Abnormal* (DoS-AACT without MRC), does not have a large dispersion but at no time approaches the target, because the swarm of robots is "anchored" to the distance where the attack occurred.

Now, when implementing the scheme based on the MRC model, we can see that, even though there is more dispersion in the experiments *AACT* (DoS-AACT with MRC), *SMAX* (MiM-SMAX with MRC) and *SRND* (MiM-SRND with MRC), due to the affected robots that are left behind, the trend in the average distance of the swarm approaches an acceptable value very close to the target, so it can be assumed that the swarm ended or the task will eventually end. On the other hand, the experiment *SMIN* (MiM-SMIN with MRC) is the one that comes closest to the normal experiment, both in the dispersion and in the average distance, however, in this experiment some collisions arise due to the alteration in the sensors of proximity, this in some cases could be tolerable or not, depending on the type of application in which the swarm of robots is deployed.

Although our proposal is limited, on the one hand, to a limited number of robots that form the swarm and, on the other hand, to the type of attacks executed, the most important advantages of our model are that, in a generic way, it allows us to observe the Robot behavior in a distributed manner, considering the limited detection capabilities and periodic observations. Likewise, the resulting distributed coalition algorithm allows to consolidate the decisions taken by the individual robots, providing a robust decision of swarm level that is crucial to provide collective recovery strategies before other types of cyber attacks.

6   Conclusions

The detection of attacks and tolerance to failures represent two of the most important problems in the field of swarm robotics. The swarms of robots in many real-world scenarios will operate in unstructured environments and, therefore, will require a protection and detection system that can adapt to the temporal variations of robot behavior and perturbations of the robot. environment. As far as we know, this is the first thesis work that has studied this problem. So, for the security scheme, we tried to formalize it, design it and test it experimentally to present and discuss its efficiency as a solution proposal. In this work it has been shown that by characterizing the behavior of the robot as a vector of characteristics, certain types of attacks can be identified towards the swarm, with a generally high reliability. Having said that, it is recognized that the diagnosis of attacks that generate failures in robot swarms is by no means complete.

There are still a number of problems to be addressed, such as how a system can reliably diagnose the types of attacks that can manifest and be generated in different ways, and even how they can be diagnosed in real time for very changing environments. Finally, from a reliability perspective, future work should focus on expanding the methods for the mathematical modeling of robotic swarm systems, working on the more detailed analysis of "security" at both the swarm level and at the level of a robot. individual and develop methodologies and practices for engineering tests in robot envelopes. It is clear that it is necessary to do a lot of work before the swarms of robots can become a reliable reality from the point of view of engineering.

Acknowledgments. This work was supported by IPN under Projects SIP-1894 and SIP-20180460. The first author gratefully acknowledge the support from the Mexican National Council for Science and Technology (CONACyT) and IPN-PIFI BEIFI grant to carry out this work.

References

1. Brambilla, M., Ferrante, E., Birattari, M., Dorigo, M.: Swarm robotics: a review from the swarm engineering perspective. Swarm Intell. **7**(1), 1–41 (2013)
2. Brooks, R.: A robust layered control system for a mobile robot. IEEE J. Robot. Autom. **2**(1), 14–23 (1986)
3. Carneiro, J., et al.: When three is not a crowd: a crossregulation model of the dynamics and repertoire selection of regulatory CD4+T cells. Immunol. Rev. **216**(1), 48–68 (2007)
4. Chen, X.: Enabling disconnected transitive communication in mobile ad hoc networks. In: Proceedings of Workshop on Principles of Mobile Computing, Colocated with PODC 2001, pp. 21–27 (2001)
5. Cohen, I.: Tending Adam's Garden: Evolving the Cognitive Immune Self. Elsevier Science (2000)
6. Dailey, K.: The FMEA Pocket Handbook: Failure Mode and Effects Analysis. DW Publishing (2004)
7. Dias, M.B., Zinck, M., Zlot, R., Stentz, A.: Robust multirobot coordination in dynamic environments. In: 2004 IEEE International Conference on Robotics and Automation, 2004, Proceedings, ICRA 2004, vol. 4, pp. 3435–3442, April 2004. https://doi.org/10.1109/ROBOT.2004.1308785
8. Dorigo, M., Maniezzo, V., Colorni, A.: Ant system: optimization by a colony of cooperating agents. Trans. Syst. Man Cybern. Part B **26**(1), 29–41 (1996)
9. Fagiolini, A., Dini, G., Bicchi, A.: Distributed intrusion detection for the security of industrial cooperative robotic systems. IFAC Proc. Vol. **47**(3), 7610–7615 (2014). https://doi.org/10.3182/20140824-6-ZA-1003.02666
10. Feng, R., Xu, X., Zhou, X., Wan, J.: A trust evaluation algorithm for wireless sensor networks based on node behaviors and D-S evidence theory. Sensors **11**(2), 1345–1360 (2011)
11. Gutiérrez, A., Campo, A., Dorigo, M., Amor, D., Magdalena, L., Monasterio-Huelin, F.: An open localization and local communication embodied sensor. Sensors **8**(11), 7545–7563 (2008). https://doi.org/10.3390/s8117545
12. Hartmann, K., Steup, C.: The vulnerability of UAVs to cyber attacks - an approach to the risk assessment. In: 2013 5th International Conference on Cyber Conflict (CYCON 2013), pp. 1–23, June 2013
13. Holldobler, B., Wilson, E.O.: The Ants. The Belknap Press of Harvard University Press, Cambridge (1990)

14. Holldobler, B., Wilson, E.O.: Journey to the Ants. The Belknap Press of Harvard University Press, Cambridge (1994)
15. (ISO), I.S.O.: ISO/IEC 7498-2:1989. Information processing systems - Open Systems Interconnection - Basic Reference Model - Part 2: Security Architecture. Iso7498-2:1989, pp. 1–32 (1989)
16. Janeway, C.A.J., Travers, P., Walport, M., Shlomchik, M.J.: Immunobiology: The Immune System in Health and Disease, 5th edn. Garland Science, New York (2001)
17. Javaid, A.Y., Sun, W., Devabhaktuni, V.K., Alam, M.: Cyber security threat analysis and modeling of an unmanned aerial vehicle system. In: 2012 IEEE Conference on Technologies for Homeland Security (HST), pp. 585–590, November 2012
18. Jones, E.G., Browning, B., Dias, M.B., Argall, B., Veloso, M., Stentz, A.: Dynamically formed heterogeneous robot teams performing tightly-coordinated tasks. In: Proceedings 2006 IEEE International Conference on Robotics and Automation, ICRA 2006, pp. 570–575, May 2006
19. Kalech, M., Kaminka, G.A., Meisels, A., Elmaliach, Y.: Diagnosis of multi-robot coordination failures using distributed CSP algorithms. In: American Association for Artificial Intelligence (AAAI 2006). AAAI Press (2006)
20. Khaldi, B., Harrou, F., Cherif, F., Sun, Y.: Monitoring a robot swarm using a data-driven fault detection approach. Robot. Auton. Syst. **97**, 193–203 (2017). https://doi.org/10.1016/j.robot.2017.06.002
21. Kim, A., Wampler, B., Goppert, J., Hwang, I., Aldridge, H.: Cyber attack vulnerabilities analysis for unmanned aerial vehicles. In: Infotech@Aerospace (2012)
22. Kolling, A., Walker, P., Chakraborty, N., Sycara, K., Lewis, M.: Human interaction with robot swarms: a survey. IEEE Trans. Hum.-Mach. Syst. **46**(1), 9–26 (2016). https://doi.org/10.1109/THMS.2015.2480801
23. León, K., Lage, A., Carneiro, J.: Tolerance and immunity in a mathematical model of T-cell mediated suppression. J. Theor. Biol. **225**(1), 107–126 (2003). https://doi.org/10.1016/S0022-5193(03)00226-1
24. León, K., Peréz, R., Lage, A., Carneiro, J.: Modelling T-cell-mediated suppression dependent on interactions in multicellular conjugates. J. Theor. Biol. **207**(2), 231–254 (2000). https://doi.org/10.1006/jtbi.2000.2169
25. Mansfield, K., Eveleigh, T., Holzer, T.H., Sarkani, S.: Unmanned aerial vehicle smart device ground control station cyber security threat model. In: 2013 IEEE International Conference on Technologies for Homeland Security (HST), pp. 722–728, November 2013. https://doi.org/10.1109/THS.2013.6699093
26. Mansfield, K., Eveleigh, T., Holzer, T.H., Sarkani, S.: DoD comprehensive military unmanned aerial vehicle smart device ground control station threat model. Publ. Def. Acquis. Univ. **23**(2), 240–273 (2015)
27. Meng, Y., Nickerson, J.V., Gan, J.: Multi-robot aggregation strategies with limited communication. In: 2006 IEEE/RSJ International Conference on Intelligent Robots and Systems, pp. 2691–2696, October 2006
28. Mondada, F., et al.: The e-puck, a robot designed for education in engineering. In: Proceedings of the 9th Conference on Autonomous Robot Systems and Competitions, vol. 1, no. 1, pp. 59–65 (2009). ISBN 978-972-99143-8-6
29. Nembrini, J., University of the West of England: Minimalist coherent swarming of wireless networked autonomous mobile robots. Ph.D. thesis, University of the West of England (2005)
30. Nickerson, J.V.: A concept of communication distance and its application to six situations in mobile environments. IEEE Trans. Mob. Comput. **4**(5), 409–419 (2005). https://doi.org/10.1109/TMC.2005.60

31. Nickerson, J.V., Olariu, S.: A measure for integration and its application to sensor networks. In: Workshop on Information Technologies and Systems (WITS), August 2005

32. Parker, L.E.: Alliance: an architecture for fault tolerant multirobot cooperation. IEEE Trans. Robot. Autom. **14**(2), 220–240 (1998)

33. Pinciroli, C., et al.: ARGoS: a modular, parallel, multi-engine simulator for multi-robot systems. Swarm Intell. **6**(4), 271–295 (2012)

34. Schoonderwoerd, R., Holland, O.E., Bruten, J.L., Rothkrantz, L.J.M.: Ant-based load balancing in telecommunications networks. Adapt. Behav. **5**(2), 169–207 (1997)

35. Strohmeier, M., Lenders, V., Martinovic, I.: On the security of the automatic dependent surveillance-broadcast protocol. IEEE Commun. Surv. Tutor. **17**(2), 1066–1087 (2015)

36. The Common Criteria: (22 de Junio de 2018). https://www.commoncriteriaportal.org

37. Timmis, J., Ismail, A., Bjerknes, J., Winfield, A.: An immune-inspired swarm aggregation algorithm for self-healing swarm robotic systems. Biosystems **146**, 60–76 (2016). https://doi.org/10.1016/j.biosystems.2016.04.001. Information Processing in Cells and Tissues

38. Timmis, J., Andrews, P., Hart, E.: On artificial immune systems and swarm intelligence. Swarm Intell. **4**(4), 247–273 (2010)

39. Vain, J., Tammet, T., Kuusik, A., Juurik, S.: Towards scalable proofs of robot swarm dependability. In: 2008 11th International Biennial Baltic Electronics Conference, pp. 199–202, October 2008. https://doi.org/10.1109/BEC.2008.4657513

40. Viksnin, I., Iureva, R., Komarov, I., Drannik, A.: Assessment of stability of algorithms based on trust and reputation model. In: Proceedings of the 18th Conference of Open Innovations Association FRUCT, FRUCT 2018, pp. 364–369. FRUCT Oy, Helsinki, Finland (2016). https://doi.org/10.1109/FRUCT-ISPIT.2016.7561551

41. Zikratov, I.A., Lebedev, I.S., Gurtov, A.V.: Trust and reputation mechanisms for multi-agent robotic systems. In: Balandin, S., Andreev, S., Koucheryavy, Y. (eds.) NEW2AN 2014. LNCS, vol. 8638, pp. 106–120. Springer, Cham (2014). https://doi.org/10.1007/978-3-319-10353-2_10

Chaos Optimization Applied to a Beamforming Algorithm for Source Location

Karla I. Fernandez-Ramirez[✉] and Arturo Baltazar[✉]

Robotics and Advanced Manufacturing Program,
Centro de Investigación y de Estudios Avanzados del Instituto Politécnico
Nacional–Unidad Saltillo, Ramos Arizpe, Mexico
k.ivonne.fernandez.rmz@gmail.com,
arturo.baltazar@cinvestav.edu.mx

Abstract. In this work, the delay and sum (DAS) beamforming algorithm commonly used in several areas of engineering and robotics is modified and implemented for the identification and localization of acoustic sources. Its classical approach uses a systematic scanning of all points in a given space domain to localize a disturbance source. DAS is efficient when the searching area is small, but it becomes time consuming when the area increases, or when its topology is unknown. Here, an algorithm that uses beamforming information and a chaotic search scheme for optimal target localization is proposed. The algorithm is performed in two stages: first, the entire work area is mapped using chaotic sequences to determine a vector with the locations with a high probability of finding a source. The second stage initiates a search for a global optimum using chaotic walk trajectories. The proposed algorithm is tested with known analytical functions and then implemented using a time domain simulation of acoustic field and an array of sensors. The algorithm performance for the synthetic signals was compared with the traditional systematic scan. The results showed a reduction in searching time of 90% with similar localization accuracy as the typical beamforming.

Keywords: Optimization · Chaos · Scanning

1 Introduction

The theory of how to find a strategy for the search of a target or a perturbation source, fixed or in movement has been studied by scientists for more than 50 years. In search theory, a systematic scan is commonly used as a basic searching technique [1]. Many applications of the search theory can be found in various areas of engineering and science i.e., robotics, astronomy, mechanical engineering, etc., and for applications in the industry, medicine, biology and energy exploration.

Beamforming is a technique commonly used for detection of a target to estimate the direction of arrival of a signal or to intensify a selected signal that is mixed with noise, reverberations or other signals from other sources [2]. The algorithm is a spatial filter that combines the signals received by an array of sensors to generate a beam in a particular direction, making it possible to observe the desired signal and attenuate

I. Batyrshin et al. (Eds.): MICAI 2018, LNAI 11289, pp. 332–343, 2018.
https://doi.org/10.1007/978-3-030-04497-8_27

signals from other directions. The beamforming algorithm has been studied and used in the area of robotics, seismology, communications, radar, sonar, to mention a few. To improve the performance of classical time domain beamforming, applications of frequency-domain analysis as well as the use of adaptive algorithms that allow modifying the weight and delay based on the characteristics of the received signals has been studied [3].

Recently, in the area of Structural Health Monitoring (SHM), beamforming algorithms were used to process the data received from ultrasonic sensors to identify the location, shape, size of structural damage when exciting an ultrasonic wave and measuring the reflected waves of damage using an array of transducers [2].

For a systematic scanning the value of the probability of finding an unknown source tends to one when the searching step resolution is small, and the workspace is well defined. However, such a scheme can be time consuming and computationally intensive. The problem of searching for a stationary or non-stationary target by a mobile robot has also gained considerable attention from the scientific community for its potential application in searching tasks such as in military, surveillance, or rescue operations [4]. Chaos dynamics has been recently reported as a tool to explore vigilance zones with a complete and fast coverage of the workspace using robotics systems. The problem of searching for a target can be interpreted as an optimization problem where the optimum on a map distribution defined by the cost function needs to be found.

The stochastic optimization algorithms are based on random generation of sampled points for the implementation of non-linear local optimization search procedures. The main disadvantage of these algorithms is their premature converge which often leads to a local instead of the global optimum [5]. Recently, some researchers investigated the use of hybrid algorithms, especially stochastic algorithms combined with chaos optimization algorithms (COA) [6]. These use chaotic sequences to produce a global search variable and have been studied using numerical sequences generated by a chaotic map [7]. The advantage of the hybrid algorithms is that they can easily escape from local minima when compared with classical stochastic optimization algorithms [5]. Unfortunately, the statistical properties of the chaotic maps might affect their performance, requiring further studies.

In this work, a beamforming scheme based on chaotic optimization that implements sounds sensors to detect a target is discussed. The objectives are: first, to study the effect of using chaotic and uniformly random maps on optimization; second, to develop a modified COA using chaotic walk [8] to reduce detection time of a target; and third, to implement the proposed algorithm using beamforming with a time domain simulation of an acoustic field and an array of acoustic sensors.

The paper is organized as follows: Sect. 2 introduces basic theory on chaos mapping and their statistical properties. Section 3 describes the beamforming technique applied for acoustic sources. In Sect. 4 we discussed the proposed COA algorithm with chaotic walk. Finally, the numerical results and conclusions are presented in Sects. 5 and 6 respectively.

2 Stochastic and Chaotic Optimization

Stochastic optimization algorithms do not depend on a strict mathematical property of the optimization problem, such as continuity, differentiability, or an accurate mathematical description of objective functions and constraint conditions, and therefore their optimization procedures are easy to implement [6]. However, these algorithms usually exhibit premature convergence and weak exploitation capability, leading to a local optimum instead of global optimum and thus slow convergence. Alternatively, chaos has been proposed as an optimization algorithm due to its topological and statistical properties. Chaos occurs in many nonlinear systems, where their evolution on time is unpredictable. To classify the behavior of a system as chaotic, this must have certain properties: sensitivity to initial conditions, pseudo-randomness and ergodicity that implies that a chaotic sequence can visit all the state of a strange attractor [9]. These properties found in chaos optimization algorithm could have a better performance than stochastic algorithms [7].

Figure 1 shows the distribution sequence of a random uniform variable and its respective distribution for the first 500 values. The results for this small sample make evident the nature of the uniform distribution.

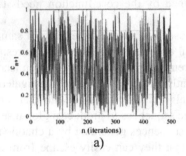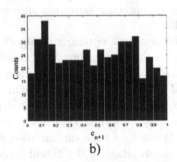

Fig. 1. (a) Uniform random distribution sequences; (b) corresponding histogram.

Now, we turn to the chaotic dynamical system. A one dimensional simple map that can generate a chaotic sequence is the logistic map. This is a polynomial map that can lead to complicated phenomena of chaotic dynamics and can be represented as:

$$c_{n+1} = \mu c_n(1 - c_n), 0 < \mu \le 4, c_n \in (0, 1), \tag{1}$$

where c_n is the n-th chaotic number, with n denoting the number of iteration and μ a control parameter. The system (1) has different dynamics characteristic for different value of μ. When the value of $\mu = 4$ the trajectory is chaotic between values (0, 1). The simulation results using $\mu = 4$ are shown in Fig. 2(a). The resulting probability density function (PDF) is a Chebyshev distribution [7], where the center of the distribution show a quasi-uniform distribution, as can be seen in the histogram (Fig. 2(b)).

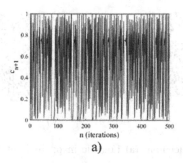 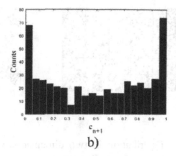

a)   b)

Fig. 2. (a) Chaotic trajectory generated by the logistic map with $\mu = 4$; (b) histogram of the chaotic map

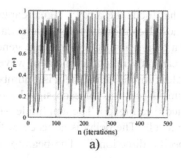 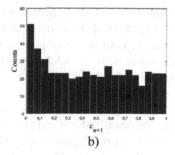

a)   b)

Fig. 3. (a) Chaotic trajectory generated by the Kent map; (b) histogram of the chaotic sequence.

It has been reported that a logistic map can have some limitation when used for chaos optimization due to its non-uniform distribution [7]. An alternative is the Kent map which has a theoretical uniform PDF distribution and it is defined as:

$$c_{n+1} = \begin{cases} \frac{c_n}{\beta} & 0 < c_n \leq \beta \\ \frac{1-x_n}{1-\beta} & \beta < c_n \leq 1 \end{cases}, \qquad (2)$$

where β is a control parameter with interval $0 < \beta < 1$, and the sequence is bounded in $(0,1)$. Figure 3 shows an example for $\beta = .8$ and its corresponding histogram. The plotted distribution exhibits a pattern that resemblances a uniform distribution as n increases.

The chaotic sequences in a 2D workspace for these two chaotic maps are compared with a uniform distribution. Figure 4 show results for 1000 iterations. Clearly the distributions obtained using chaotic maps provide an almost uniform distribution with the advantage that according to topological properties of chaos theory all open set in the map can be visited. We then tested these maps with selected functions to study their performance for global maximum search.

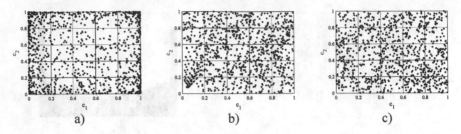

Fig. 4. Distribution in two dimensions with 1000 iteration; (a) Logistic map; (b) Kent map; (c) Random uniform distribution.

3 Delay and Sum Beamforming Theory

Beamforming is an array processing algorithm, that focus the array's signal-capturing abilities in a defined direction of a working space. Beamforming algorithms can be used to detect both active or passive sources. An active acoustic source can generate acoustic waves actively, while a passive source can only reflect an incident wave.

The idea of the delay and sum beamforming algorithm is that if there is a number of signal measurements $(y_m, m \in 1 \ldots M)$ taken from M sensors, for each signal a delay (Δ_m) and an amplitude weight (w_m) can be applied (see Fig. 5). The amplitude weight enhances the beam's shape and reduces sidelobe levels. The delay is adjusted to focus the array's beam on signals propagating in a specific direction $\bar{\gamma}^0$. The beamformer's output signal can then be expressed as:

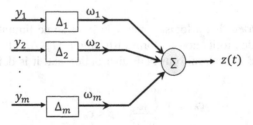

Fig. 5. The output of the beamformer $z(t)$ is obtained by adding together the sensor's output after a delay and amplitude weight has been applied to each. The delay can be chosen to maximize the array's sensitivity to waves propagating from a specific direction.

$$Z(t) = \sum_{m=0}^{M-1} w_m y_m (t - \Delta_m). \tag{3}$$

At each point in the working space there will be a representation of Eq. (3); thus, the beamforming energy (or power) at this scanning location becomes a global maximum when the measurement's location is coincident with the location of a "real" acoustic source. By repeating the same procedure at all tested points, the beamforming energy map in the space can be reconstructed. Finally, the target (global maximum) on the reconstructed beamforming power map can be identified.

In this work, instead of the typical beamforming with a systematic search, we are proposing to approach the search as an optimization problem using the beamforming energy information as the cost function.

4 Proposed Optimization Algorithm Based on Chaos Theory

Optimization algorithms depends on the initial configuration to obtain the global minimum and to avoid being trapped in a local minimum. The proposed chaos optimization algorithm will include two stages as described in Fig. 6. The first stage of the algorithm maps the entire area aiming to find areas where is more likely to find the global optimum. Thus, it is possible that the second stage could begin near a local optimum, this might allow to assure a rapid convergence.

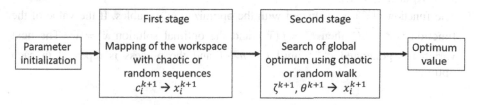

Fig. 6. Two-stage proposed chaos optimization scheme.

The proposed optimization algorithm is described below:

First stage:

1. To generate the chaotic or random sequences, initial values $c_i^{k=1} (k = 1, \ldots n)$ are set and initialization of parameters is defined to include: number of iterations n, lower limit L_i and upper U_i of the sequence; where i represents the dimensions $(i = 1, 2\, \text{or}\, 3)$ of the function to be optimized. Also, an optimal value threshold $f_{th}^* = f^o$ is established.

2. The chaotic variables c_i^k (bounded by $(0, 1)$) are mapped in the range of the set limits (optimization variables x_i^k) by Eq. (4)

$$x_i^k = L_i + (U_i - L_i)c_i^k \tag{4}$$

3. The function is evaluated with the optimization variables $f(\vec{x}^k)$. If the value of the function $f(\vec{x}^k) < f_{th}^*$, the value is saved to form a local optima vector $\vec{f}_j^* = f(\vec{x}^k)$ where j is the number of local optima found and the vector of optimal solution $\vec{x}_j^* = [x_i^k]$. By iteration, the next variable is generated c_i^{k+1} up to $k = n+1$. This process will generate a data matrix $(D = [\vec{x}_j^* \vec{f}_j^*])$ with column vectors of the

optimum vector position ($\vec{x}_j^*$) and the corresponding value of the cost function evaluated at that position ($\vec{f}_j^*$).

Second stage:

4. The number of iterations n and the initial values $\theta^{k=1}$ and $\zeta^{k=1}$ are defined.
5. A modified Pearson's equation is used to implement optimization using chaotic walk. The Pearson equation is defined by:

$$x_i^k = \sum_{k=1}^{n} \zeta^k e^{i\theta^k}, \tag{5}$$

where x_i^k is the new position, which is given as a sum of n two-dimensional vectors with a step size (ζ^k) of the iteration k and orientation (θ^k). The value of ζ^k y θ^k are given by the chaotic map. The values of θ^k are mapped within the range $(0, 2\pi)$ using Eq. (5).

6. The optimization variables outside the limits are adjusted by: $x_i^k < L_i$ then $x_i^k = L_i$; $x_i^k > U_i$ then $x_i^k = U_i$.
7. The function $f(\vec{x}^*)$ is evaluated with the optimization variables. If the value of the function $f(\vec{x}^*) < \vec{f}_j^*$ then $\vec{f}_j^* = f(\vec{x}^*)$ and the optimal solution $\vec{x}^* = \vec{x}^k$. The next variable is generated c_i^{k+1} until $k = n+1$ and the process is repeated in each optimal.

5   Numerical Results

5.1   Test with Benchmark Function

To test the efficiency and performance of the proposed algorithm these were evaluated using the Himmelblau's function (Fig. 7). The Himmelblau's function has four local minimum (without a global) all with optimum value $f^* = 0$ in the positions $x_1^* = (3, 2)$, $x_2^* = (-2.8, 3.1)$, $x_3^* = (-3.7, -3.2)$ and $x_4^* = (3.5, -1.8)$.

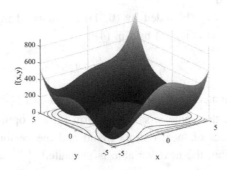

Fig. 7.   Himmelblau's function $f(x, y) = (x^2 + y - 11)^2 + (x + y^2 - 7)^2$.

Figure 8(a) shows a schematic of the result of the general search carried out for the first stage of the algorithm in the Himmelblaus function. The gray points indicate the location where the function was evaluated and the black dots are the optimal. To reduce redundancy, a search area (rectangle) around the lowest optimum found is set. In this example, we leave four possible candidates as initial points for the second stage. In each of the candidates the second stage is performed (Fig. 8(b)). The second search will end when the convergence criteria is satisfied (number of iteration) and the optimal value is found (asterisk).

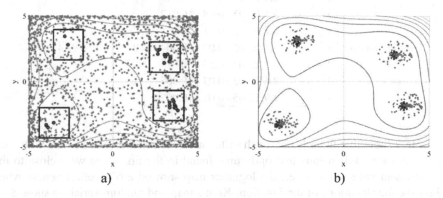

a)   b)

Fig. 8. Himmelblau's function projection (X–Y plane); (a) Fist stage; (b) Second stage.

Table 1 shows the positions and values found with our proposed algorithm and are compared with the actual values of the function. The total computational time (for both stages) is also given. The location of the optimum location (1–4) is shown.

Table 1. Simulation results of Himmelblau's with the logistic map.

# Solution	Found position		Function value	Time (sec.)	Actual position		Function value
	x	y			x	y	
1	2.9	2	.02	3.4	3	2	0
2	−2.7	3.1	.06		−2.8	3.1	0
3	−3.7	−3.2	.008		−3.7	−3.2	0
4	3.5	−1.8	8.3e−4		3.5	−1.8	0

Table 2 the results of executing the algorithm 30 times on the Himmelblaus's function are given. To study the effect of the search point distribution on the optimization algorithm performance several repeated tests (30 times) using the logistic map, Kent map and uniform random distributions were carried out. A total of 1100 search points were used in each case.

Table 2. Simulation results of Himmelblau's with the proposed algorithms.

Cases	Optimum	Optimum value			Time (sec)		
		Best	Worst	Average	Best	Worst	Average
Logistic map	1	3.1 E−06	0.04	0.006	1.8	3.4	2.5
	2	2 E−04	0.03	0.004			
	3	9 E−06	0.02	0.004			
	4	5 E−06	0.03	0.006			
Kent map	1	6.7 E−06	0.02	0.005	2.2	3.8	2.6
	2	2.4 E−05	0.01	0.003			
	3	5.1 E−05	0.18	0.017			
	4	6.7 E−05	0.01	0.004			
Uniform random	1	9.7 E−05	0.03	0.004	2	3.1	2.4
	2	2.1 E−05	0.13	0.013			
	3	2.6 E−06	0.07	0.005			
	4	1.9 E−05	0.01	0.002			

Table 2 show that there is not much difference between the values found with each of the sequences, this means that optimums found in the first stage were close to the final optimum values. However, the logistics map showed a 67% effectiveness when finding the four locations of the function. Kent's map and random variables show 87% effectiveness.

5.2   Test with Beamforming and Acoustic Signal

To test the efficiency of the algorithm in the detection of one or several sources and the possible reduction of time in the search by applying the DAS algorithm, simulation of the active sources and the generation of the received signals by the sensor array was used. The time domain acoustic and ultrasound simulations was carried out using the open source k-Wave toolbox for Matlab [10]. The program solves the wave acoustic equation using finite difference approach. The acoustic problem is described in Fig. 9 (a). An search area of 50×35 cm with a spacing with an spatial resolution of $0.2 \times 0.2 \, cm^2$. A linear array of 9 sensors was located on the $(0,0)$ coordinate axis of the rectangular area and a spacing between $d = .4$ cm sensors was used. The source (target) was placed in $(10, 14)$ position. A tone bust of 3 cycles of $f = 200$ kHz was used as the acoustic disturbance coming from the source. Figure 9 show the simulation of the wave propagation using the k-Wave. An example of the received signal by the odd sensors of the array is show in Fig. 9(c). The dot line shows how the signal is received with certain delay which depend the position of the sensor and the origin of the source.

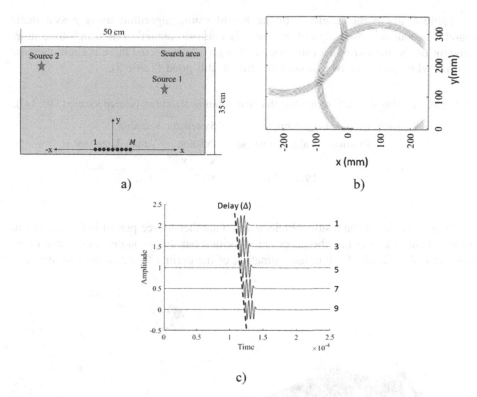

a)

b)

c)

Fig. 9. (a) Simulation design where the propagation velocity is $c = 150,000$ cm/s; (b) Source signal; (c) received signal example from the first three sensors.

Figure 10(a) shows the reconstruction of the area's energy field using Eq. (3) by performing a systematic scanning. Figure 10(b) shows the results of applying the proposed optimization algorithm. Based on the received signals, an energy is estimated to define the threshold. To define the limits of the discretization area, an $\delta = 5$ was used.

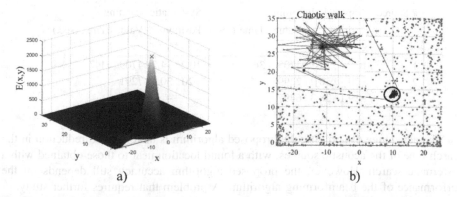

a)

b)

Fig. 10. (a) Energy map reconstruction using DAS beamforming algorithms in 3D; (b) Source search using the proposed algorithms.

Table 4 shows the results with the beamforming algorithm using a systematic search and our proposed algorithm. There is dramatic reduction in search time independent of the number of acoustic sources (target). As expected there is slight increase in time when you have two sources but this is still good (Table 3).

Table 3. Simulation results of applied the beamforming algorithm (source location (10, 14)).

# Source	Proposed algorithm				Systematic scanning			
	Position		Value	Time (sec.)	Position		Value	Time (sec.)
	x	y			x	y		
1	9.8	13.9	1974	30	10.2	14.2	2051	311

Figure 11 shows the results obtained when another source placed in (−20, 30) was added. Table 4 shows that the proposed algorithm has three sources, this is due to the selection of the limits for the discrimination of the optima found in the first stage.

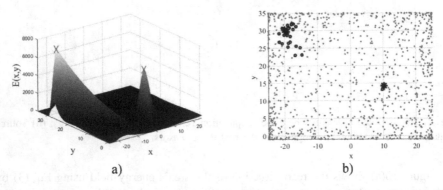

a)   b)

Fig. 11. (a) Energy map reconstruction using DAS beamforming algorithms in 3D with two source; (b) Source search using the proposed algorithms.

Table 4. Simulation results of applied the beamforming algorithm (source 1 location = (10, 14) and source 2 location (−20, 30)).

# Source	Proposed algorithm				Systematic scanning			
	Position		Value	Time (sec.)	Position		Value	Time (sec.)
	x	y			x	Y		
1	10.2	14.1	4708	36	10.2	14.2	4709	447
2	−20	30.2	7299		−20	30.2	7291	
	−19.1	28.3	6497		−	−	−	

The advantage of applying the proposed algorithm is the observed reduction in the search time of the acoustic sources, with a found location near to those obtained with a systematic search. However, the proposed algorithm accuracy still depends on the performance of the beamforming algorithm. A problem that requires further study.

6   Conclusions

A beamforming algorithm to detect acoustic sources was proposed. The algorithm uses a quasi-stochastic optimization technique using chaotic variables. The algorithm methodology have to basic stages: first, a search of the possible optimal values (maximum or minimum) is carried out; second a chaotic walk to find optimum values with initial values found in first stage is performed. The proposed methodology was compared with different schemes, combining random and chaotic sequences. The logistic map and the map of Kent were used to generate the chaotic sequences. It was observed for known functions that the best performance for the sequence with better uniform distribution. A time domain acoustic problem with an acoustic (or multiple) sources and an array of sensors was modeled using finite different method. The optimization algorithm was then applied for the detection of one and two acoustic sources using a modified beamforming algorithm. The results showed a large reduction in the search time when compare with beamforming using systematic scanning. As part of the future work, it is the application of the algorithm in robotics sensing equipped with an ultrasonic system for searching and identification of acoustic sources.

Acknowledgements. The authors thank CONACYT for providing financial support through the project CB-286907.

References

1. Gertner, I., Zeevi, Y.Y.: Scanning strategies for target detection. In: Proceeding SPIE, Data Structure and Target Classification, Orlando, Florida, vol. 1470 (1991)
2. Benesty, J., Chen, J., Huang, Y.: Microphone Array Signal Processing, 1st edn. Springer, Berlin (2008)
3. Van Trees, H.L.: Optimum Array Processing Part IV of Detection Estimation, and Modulation Theory, 1st edn. Wiley, Canadá (2002)
4. Benkoski, S., Monticino, M., Weisinger, J.: A survey of the search theory literature. Naval Res. Logist. **38**(4), 469–494 (1991)
5. Liu, B., Wang, L., Jin, Y.H., Tang, F., Huang, D.X.: Improved particle swarm optimization combined with chaos. Chaos Solitons Fractals **25**, 1261–1271 (2005)
6. Alatas, B.: Chaotic bee colony algorithms for global numerical optimization. Expert Syst. Appl. **37**, 5682–5687 (2010)
7. Yang, D., Liu, Z., Zhou, J.: Chaos optimization algorithms based on chaotic maps with different probability distribution and search speed for global optimization. Commun. Nonlinear Sci. Numer. Simul. **19**(4), 1229–1246 (2014)
8. Baltazar, A., Fernandez-Ramirez, K.I., Aranda-Sanchez, J.I.: A study of chaotic searching for their application in an ultrasonic scanner. Eng. Appl. Artif. Intell. **74**, 271–279 (2018)
9. Baker, G., Gollud, J.: Chaotic Dynamics: An Introduction, 2nd edn. Cambridge University Press, New York (1996)
10. K-Wave. http://www.k-wave.org/. Accessed 21 May 2018

Data Augmentation in Deep Learning-Based Obstacle Detection System for Autonomous Navigation on Aquatic Surfaces

Ingrid Navarro[✉], Alberto Herrera, Itzel Hernández, and Leonardo Garrido

Department of Computer Science, Tecnológico de Monterrey,
Av. Eugenio Garza Sada 2501, Monterrey, Nuevo León, Mexico
{A00569236,A01196165,A01128828,leonardo.garrido}@itesm.mx

Abstract. Deep learning-based frameworks have been widely used in object recognition, perception and autonomous navigation tasks, showing outstanding feature extraction capabilities. Nevertheless, the effectiveness of such detectors usually depends on large amounts of training data. For specific object-recognition tasks, it is often difficult and time-consuming to gather enough valuable data [10]. Data Augmentation has been broadly adopted to overcome these difficulties, as it allows to increase the training data and introduce variation in qualitative elements like color, illumination, distortion and orientation. In this paper, we leverage on the object detection framework YOLOv2 [12] to evaluate the behavior of an obstacle detection system for an autonomous boat designed for the International RoboBoat Competition. We are focused on how the overall performance of a model changes with different augmentation techniques. Thus, we analyze the features that the network learns by using geometric and pixel-wise transformations to augment our data. Our instances of interest are *buoys* and *sea markers*, thus to generate training data comprising these classes, we simulated the aquatic surface of the boat and collected data from the COCO dataset [8]. Finally, we discuss that significant generalization is achieved in the learning process of our experiments using different augmentation techniques.

Keywords: Data augmentation · Synthesized images · Deep learning
Object detection · Computer vision

1 Introduction

Recent Deep Learning-based object recognition frameworks have shown that in general, the effectiveness of object recognition tasks improves as the volume of data increases. Nevertheless, training data is often not enough to produce robust detection and classification systems. Another aspect that has an important impact on model performance is feature extraction. If a model is able to

© Springer Nature Switzerland AG 2018
I. Batyrshin et al. (Eds.): MICAI 2018, LNAI 11289, pp. 344–355, 2018.
https://doi.org/10.1007/978-3-030-04497-8_28

extract rich features from a dataset by reducing the amount of information needed to describe it then, it will most certainly be able to generalize well to new data. However, it is frequently difficult to adapt public datasets like COCO [8] or ImageNet [13] datasets to specific interests. Furthermore, collecting and annotating images could be time-consuming because it involves being able to find data that introduces sufficient variability to the training set to produce good models. Training with inadequate datasets might result in consequences like having models that do not generalize well to new samples, i.e. they overfit, or even models that are unable to fit the training data, i.e. they underfit.

These limitations have motivated using a widely known technique called Data Augmentation (DA) [1], which consists on extending the training dataset by generating synthetic samples to improve the robustness of the models. Previous work has been focused on demonstrating that DA can behave as a regularization technique to prevent over-fitting. Nonetheless, it is frequently difficult to determine which type of augmentations are useful to produce accurate models as it often depends on the morphology of the object being analyzed by the neural network. This means that there are augmentation methods that may not work for all kinds of objects because different appearance factors might be captured by the networks with each technique. For this reason, as stated in [10], Convolutional Neural Networks (CNNs) might learn contrasting features from images containing the same type of object when different DA techniques are applied.

In this paper, we are focused on improving the performance of a detection system for an autonomous vehicle that needs to navigate through an aquatic obstacle course for the *International Roboboat Competition*. This self-navigating boat attempts to mimic coastal surveillance tasks and other oceanographic operations that are currently being developed for autonomous maritime surveillance systems. Our obstacle detection system leverages on the CNN presented in [12] to perform detection and classification of our instances of interest, i.e. spherical *buoys* and *sea markers*. However, due to our small and highly biased dataset, we propose exploring Data Augmentation techniques to prevent overfitting and improve the performance of the system. Thus, in the following sections, we outline each of the data augmentations techniques used to generate training data, and we discuss the overall results of the experiments that were carried out in this work. Finally, to our knowledge, approaches that perform detection and classification of our classes of interest have not been explored, therefore, the results that we present are only compared against our experiments.

2   Related Work

This section gives an insight into the work focused on designing and evaluating Data Augmentation techniques. Different types of augmentation transformations can be grouped into several categories depending on the criteria being evaluated in each research work. In [9], two different categories for data augmentation are proposed: geometric transformations and pixel value changes. The former describes changes that must be applied to the image and its ground truth label:

translations, rotations, mirroring, crops and distortions. The latter produces changes only on the image without affecting its ground truth label: noise, blurring and changing color spaces. The methods used in the aforementioned approach were used for road segmentation with the KITTI [2] dataset. Their results, show that geometric transformations introduce higher variability thus, better segmentation. However, by using both types of methods together, they showed better generalization against changes in illumination, texture and perspective. Random Erasing (RE) is introduced in [16]. It is a complementary method to traditional DA techniques. During the training process, RE chooses a rectangle region in a given image and changes its pixels with random values. This method shows a reasonable improvement in various recognition tasks. For instance, it responds fairly well against occluded objects in images from the CIFAR-10 and CIFAR-100 datasets [6].

In [10], using the R-CNN detector [3], they evaluate in an empirical fashion the kind of transformations that result more valuable to increase the robustness of object recognition tasks. They explore creating synthetic images using CAD models of different rendering types. Additionally, they apply transformations such as cropping, blurring and color shifts. Also, they extend the dataset using the Pascal3D+ dataset [15]. They concluded that synthetic data by itself under-performs compared to using only real data. Nevertheless, they discussed that certain types of rendered images when used along with the non-augmented dataset add important information that improves accuracy and robustness.

Recent approaches have explored the use of Generative Adversarial Neural Networks (GANs) [4] to augment datasets. These are powerful learning techniques that use one network to generate synthetic data during the training process and another network to perform the recognition task. Using GANs, [7] proposes a sophisticated process called Smart Augmentation (SA). It attempts to learn the best augmentation strategies for a given class of input data. This is done by merging samples of the same class and used them to train the network. SA shows an improvement in model performance allowing the network to come up with unusual augmentation techniques that work even better when used along with traditional DA. An important aspect that was demonstrated in this work is that augmentation is a process that can be automated reducing the time needed to create augmented datasets.

Finally, the approach presented in [11], compares the validation accuracy of a model using both traditional DA transformations and styled transformations using GANs. The first approach includes shifts, flips, distortions and rotations, whereas the in second approach they translate the original dataset images into the styles of artists like Monet, Van Gogh and Cezanne [17]. They also proposed another experiment with Neural Augmentation. Using their original dataset, they took two random images from the same class and concatenated them to create an augmented image. In their results, they show that traditional methods have a higher accuracy than the use of GANs. Nevertheless, both experiments significantly improve the robustness of the model. They also noted that neural augmentation works in many experiments but has no effect in certain experiments, hypothesizing that it may be due to the simplicity of the features presented in some of the datasets they used.

3 Approach

The main goal of this paper, as we mentioned in the previous sections, is to demonstrate how the use of Data Augmentation transformations allow to improve the accuracy of a detection system used on an autonomous boat. Therefore, in the following subsections, we explain in detail the framework used to perform detection and classification. Then, we describe how the datasets were generated and the data augmentation techniques used and finally, the implementation details of our approach.

3.1 Object Detection System

This section provides details about YOLOv2 [12], the object detection framework used in our experiments. Compared to many state-of-the-art detection systems, YOLOv2 is significantly faster. This is because many frameworks depend on complex feature extractors, such as VGG-16 [14]. Although the aforementioned model is highly robust, it requires more than 30 billion floating point operations for a each 224×224 image. In contrast, YOLOv2 uses a network model called Darknet-19, composed of 19 convolutional layers and 5 maxpooling layers. Their network model requires 5.58 billion operations to process an image of the same resolution. This allows the system to be faster, albeit slightly less accurate. For our work, we consider a rapid response is an important aspect to acknowledge as the autonomous boat relies on the latency of detection predictions. However, there is still room for improvement in terms of the accuracy of the system. To address this, we propose using the data augmentation which will be further explained in Sect. 3.3.

3.2 Datasets

Prior to outlining the data augmentation techniques used in this work, we describe the three datasets that were used to train the network model from the previous section. As mentioned before, we are focused on detecting two different objects: *buoys* and *sea markers*. To build our datasets, we gather data from different surroundings containing objects that share similar characteristics with *buoys* and *sea markers*. With this dataset, we evaluate if overall generalization is achieved by introducing images from various surroundings. However, we also collected images from a simulation of the environment were the boat navigates. Finally, we build a dataset by combining both dataset 1 and 2. Below, we provide more details of each dataset.

Dataset 1. For this dataset we collected 456 images of objects that share characteristics with *buoys* and *sea markers*. These images were obtained from the

COCO dataset. A total of 300 images were selected from the *sports ball* class to represent the *buoys*. The remaining images were gathered from the *bottle* and *fire hydrant* categories to represent the shape of *sea markers*. The "sports ball" category includes predominantly baseballs, tennis balls and soccer balls, whereas the other two categories contain mainly plastic water bottles and fire hydrants. Examples of these objects are shown in Fig. 1.

It is important to mention that we chose these particular objects under the assumption that they will not be found on the real surroundings where the autonomous boat would navigate.

Fig. 1. Dataset 1: *sports ball* images (above) and *bottle* and *fire hydrant* images selected from COCO dataset (below).

Dataset 2. It consists of 229 manually collected and annotated images. Firstly, we simulated the aquatic environment where the boat navigates with simple arrangements of balls to represent the *buoys*. A total of 34 images were obtained from this setting. The remaining data was gathered from the *RoboBoat 2017* competition with real *buoys*, and *sea markers*. Some examples of the settings are shown in Fig. 2. This dataset offers less substantial variation. However, it provides a more realistic representation of the environment where the autonomous boat will be operating.

Fig. 2. Dataset 2: images simulating the aquatic surface.

Dataset 3. This final set of images was created by combining Dataset 1 and Dataset 2. The objective is to verify if the neural network is able extract good features by capturing the qualitative aspects that each dataset offers separately.

3.3   Data Augmentation

To produce a model that is able to accurately generalize to new data, we find two aspects to take into consideration. In one hand, the framework described in Sect. 3.1 trades off accuracy for fast detection. In the other hand, there is a need to extend the dataset used to perform detection. To address these problems, we explore the use of data augmentation. Following the approaches in [5,9,11], we analyze the two types of transformations described below.

Geometric Transformations. The first type of augmentation involves changes on the image, as well as the ground truth annotations. This means that the transformations perform changes on the position and orientation of the object.

- *Flipping:* Mirrors the image about the x-axis or the y-axis.
- *Rotation:* Generates an arbitrary angle between the bottom side of the image and the x-axis.

In Fig. 3, we show examples of this type of transformations used in out tests. It is clear that these augmentations allow to respond to changes in object position and orientation.

Fig. 3. Geometric transformations: horizontal flipping (above) and rotation (below).

Pixel-Wise Transformations. The second type of augmentation involves changes in the image that do not affect the position of the object, thus the annotations remain unchanged. These transformations have effects on the object illumination and color because they alter the color components on the image.

- *Channel Shifting:* This technique takes the image color channels, i.e. the red (R), green (G) and blue (B) components, and shifts them from right to left. In OpenCV, the default color format is BGR. This technique, for instance, shifts all the values in the G channel to the B channel. Following this shifting methodology for the remaining channels we end up with an image in an GRB color format.

- *Negative:* This method depends on how the image pixels are laid out in memory. The transformation follows the next simple formula:

$$N = M - V \qquad (1)$$

where N represents the value of the transformation, M is the maximum value for a given data layout (i.e. 8-bit, 16-bit or 32-bit data representations) and V is the current value of a channel in a particular pixel. A common image layout in OpenCV is 8UC3. This means that the images are represented as a 3-channel, 8-bit integer matrix. Using this data layout, each image value can be represented with a number from 0 to 255. Thus, if the B channel of a given pixel has a value of 50, then by applying the negative value we know that the resulting value (N) is 205.

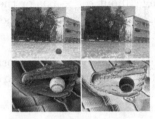

Fig. 4. Pixel-wise transformations: channel shifting (above) and negative (below). (Color figure online)

In Fig. 4, we show examples of this type of transformations used in our experiments. Using these techniques allowed our system to introduce color variation. Particularly, Dataset 1 from the previous section, contained several images of white and yellow sports balls. This predominance had an impact while detecting *buoys* of other colors. Thus, *channel shifting* was a key technique to vary the colors of the sports balls, which allowed our system to detect *buoys* of different colors. However, detecting black *buoys* and white *sea markers* was still a problem to address. Using the *negative* technique allowed to solve this problem because most of the balls had bright colors, whereas the fire hydrants and bottles were dark. This technique allowed to invert this color feature.

3.4 Implementation Details

For the training and validation process, we used a GeForce GTX 1080. Our network configuration included detection of the two classes stated in the previous sections as: *spherical buoys* and *sea markers*. The image scale for our training samples was 640 × 640 pixels with auto-scaling activated. We also trained with 64 pictures per batch and 8 subdivisions. Our tests were carried with a GTX 1050 Ti GPU. This computer is placed inside the autonomous boat to allow for independent control. The detection threshold is set to 0.25, which is the default value for YOLOv2. Video input comes from an USB web cam, attaining an average detection of 25 fps.

4   Experiments

As stated in the previous sections, our focus is to explore if the overall performance of a model increases as we apply different data augmentation techniques to our datasets, and what are the features that the network learns with such transformations. A set of 4 different experiments were conducted on each of the three datasets that were outlined in Sect. 3.2, leading to a total of 12 tests. Below, we describe in detail each one of the experiments.

1. **No transformation test:** In this experiment, we simply train YOLOv2 on each one of the datasets (i.e. Dataset 1, 2 or 3) without applying data augmentations. We further refer to this test as **NT**.
2. **Geometric transformation test:** In this experiment, we apply the *rotation transformation* described in Sect. 3.3 to the dataset before training. We refer to this test as **GT**.
3. **Pixel-wise transformations test:** We augment the dataset using the *channel shift transformation* and *negative transformation* transformations described in Sect. 3.3. We further denote this test as **PT**.
4. **Combined transformations test:** In this experiment, we employ geometric transformations, as well as pixel-wise transformations to generate new samples, and then we train YOLOv2 on the augmented dataset. We further refer to this test as **CT**.

In our experiments, Dataset 1 and 2 are randomly split into training (80%) and validation (20%) sets, whereas Dataset 3 combines the training set from Dataset 1 and 2, and likewise for the validation set. Also, when applying data augmentation to an experiment (i.e., experiments 2, 3, and 4), we randomly take 40% of the images contained in the training samples and apply a transformation. These new synthetic pictures are added back to the set before starting the training process. Validation images are not modified, since the intent is to verify how augmentations improve the model. The objective to evaluate each dataset with the same four experiments is to analyze whether a given augmentation technique is effective across different input samples and identify the one with best performance.

5   Results

In this section we present the results that were obtained from the tests described in Sect. 4. Validation for every experiment is made against a control group, containing representative images from all used datasets. In total, we show the results of 12 different tests, i.e. 4 tests per dataset. Every table shows the *average precision* for each of the classes. We refer to the *buoys* class as *class 0* and the *sea markers* class as *class 1*. We also present the mean average precision *mAP*, and the *F1 scores* of the experiments.

Table 1 shows the results of the tests conducted on Dataset 1 from Sect. 3.2, where images where gathered from the COCO dataset. In Fig. 5, we show images from the NT test against the CT test, which had the best results. The bounding boxes on the images show the class of the object (i.e. *b* for the buoys and *p* for the sea markers), and the detection confidence in a range from 0 to 100.

From Table 1 and Fig. 5, we observe an overall improvement with the tests where data augmentation was applied. However, in the experiment with geometric augmentation, average precision for the *sea markers* class had the lowest results of the experiment. Nevertheless, neither of the tests were able to robustly detect this class. Also, we observed that most of the sports balls from COCO dataset were white or yellow. For this reason, on the NT test, the network predominantly detected *buoys* of these colors. Data Augmentation allowed to slightly mitigate the latter problems on this experiments.

Table 1. Tests conducted on Dataset 1.

Test	Class 0 (AP%)	Class 1 (AP%)	mAP (%)	F1 score
NT	49.00	22.11	36.30	0.41
GT	56.97	8.02	32.49	0.48
PT	**65.26**	26.92	46.09	0.50
CT	60.29	**28.97**	**46.61**	**0.55**

Fig. 5. Experiments conducted on Dataset 1. Detections from NT test (above) against detections from CT test (below)

Table 2 shows the results of the experiments conducted on Dataset 2. We observe that, although the *AP* for the *sea markers* improved, the network still under-performs when using geometric transformations. Particularly, we observed that the network detected false positives of the *sea markers* class when similar objects, but in a horizontal position, appeared on the images. Nevertheless, compared to the tests conducted on Dataset 1, the results obtained with this tests had an overall improvement. In Fig. 6 we show images from the NT test against the PT test, which had the best results on this experiment. We see that detection precision improved on the PT test, as we observe tighter and more accurate bounding boxes compared to the NT test. Also, on this experiment, the network was able to detect the sea markers, which did not occur on the experiments conducted on Dataset 1.

Table 2. Tests conducted on Dataset 2.

Test	Class 0 (AP%)	Class 1 (AP%)	mAP (%)	F1 score
NT	60.36	**37.32**	48.84	0.58
GT	30.98	5.71	18.19	0.33
PT	**66.44**	35.97	**51.21**	**0.63**
CT	62.70	30.51	46.61	0.62

Fig. 6. Experiments conducted on Dataset 2. Detections from NT test (above) against detections from PT test (below)

Table 3. Tests conducted on Dataset 3.

Test	Class 0 (AP%)	Class 1 (AP%)	mAP (%)	F1 score
NT	69.59	46.92	58.25	0.66
GT	53.10	22.15	37.67	0.49
PT	82.44	**48.45**	**65.44**	0.69
CT	**83.96**	41.80	62.88	**0.70**

In Table 3, we show the results of the four tests conducted on Dataset 3. We observe a significant improvement in precision scores for both classes in comparison to those shown on Tables 1 and 2. However, we note that the average precision remains lower for class 1. Tests PT and CT from this experiments show the best results among the entire set of tests. In Fig. 7 we show images from the

Fig. 7. Experiments conducted on Dataset 3. Detections from NT test (above) against detections from CT test (below)

NT test against the CT test. We observe less accurate bounding boxes on the results from the NT test, and predictions with less confidence against the results from the CT test.

6   Conclusion and Future Work

In this paper, we focused on implementing a system that detects spherical *buoys* and *sea markers*. We explored data augmentation to increase the robustness of the detection system. Our results show that significant improvement was achieved using these techniques and that the network was able to capture important features of our objects of interest. We observed that geometric transformations did not improve performance significantly when tested separately. Particularly, rotations increased the number of false positives with objects, held in a horizontal position, that share similar characteristics than sea markers. In future work, our interest is to explore different types of augmentation techniques to increase the average precision for the *sea markers*, e.g. using CAD models to generate synthetic data, and different kind of geometric transformation. Finally, we would like to extend the detection capabilities of our autonomous boat by adding more object classes to its repertoire like ship, vessels, other types of boats and people.

Acknowledgements. We would like to thank Tecnológico de Monterrey, WritingLabs and TecLabs for providing the equipment used in our experiments and financial support in the production of this work. Additionally, we extend our gratitude to VANTEC, the student group from ITESM that invited us to participate in the *International RoboBoat Competition*.

References

1. Fawzi, A., Samulowitz, H., Turaga, D., Frossard, P.: Adaptive data augmentation for image classification. In: 2016 IEEE International Conference on Image Processing (ICIP), pp. 3688–3692, September 2016. https://doi.org/10.1109/ICIP.2016.7533048

2. Geiger, A., Lenz, P., Stiller, C., Urtasun, R.: Vision meets robotics: the kitti dataset. Int. J. Robot. Res. **32**, 1231–1237 (2013)
3. Girshick, R.B., Donahue, J., Darrell, T., Malik, J., Berkeley, U.C.: Rich feature hierarchies for accurate object detection and semantic segmentation. Technical report (2013)
4. Goodfellow, I., et al.: Generative adversarial nets. In: Ghahramani, Z., Welling, M., Cortes, C., Lawrence, N.D., Weinberger, K.Q. (eds.) Advances in Neural Information Processing Systems 27, pp. 2672–2680. Curran Associates, Inc. (2014). http://papers.nips.cc/paper/5423-generative-adversarial-nets.pdf
5. Jo, H., Na, Y., Song, J.: Data augmentation using synthesized images for object detection. In: 2017 17th International Conference on Control, Automation and Systems (ICCAS), pp. 1035–1038, October 2017. https://doi.org/10.23919/ICCAS.2017.8204369
6. Krizhevsky, A., Hinton, G.: Learning multiple layers of features from tiny images, vol. 1, January 2009
7. Lemley, J., Bazrafkan, S., Corcoran, P.: Smart augmentation learning an optimal data augmentation strategy. IEEE Access **5**, 5858–5869 (2017). https://doi.org/10.1109/ACCESS.2017.2696121
8. Lin, T.-Y., et al.: Microsoft COCO: common objects in context. In: Fleet, D., Pajdla, T., Schiele, B., Tuytelaars, T. (eds.) ECCV 2014. LNCS, vol. 8693, pp. 740–755. Springer, Cham (2014). https://doi.org/10.1007/978-3-319-10602-1_48
9. Muoz-Bulnes, J., Fernandez, C., Parra, I., Fernndez-Llorca, D., Sotelo, M.A.: Deep fully convolutional networks with random data augmentation for enhanced generalization in road detection. In: 2017 IEEE 20th International Conference on Intelligent Transportation Systems (ITSC), pp. 366–371, October 2017. https://doi.org/10.1109/ITSC.2017.8317901
10. Pepik, B., Benenson, R., Ritschel, T., Schiele, B.: What is holding back convnets for detection? In: Gall, J., Gehler, P., Leibe, B. (eds.) GCPR 2015. LNCS, vol. 9358, pp. 517–528. Springer, Cham (2015). https://doi.org/10.1007/978-3-319-24947-6_43
11. Perez, L., Wang, J.: The effectiveness of data augmentation in image classification using deep learning, December 2017
12. Redmon, J., Farhadi, A.: YOLO9000: better, faster, stronger. arXiv preprint arXiv:1612.08242 (2016)
13. Russakovsky, O.: ImageNet large scale visual recognition challenge. Int. J. Comput. Vis. **115**(3), 211–252 (2015). https://doi.org/10.1007/s11263-015-0816-y
14. Simonyan, K., Zisserman, A.: Very deep convolutional networks for large-scale image recognition, September 2014
15. Xiang, Y., Mottaghi, R., Savarese, S.: Beyond PASCAL: a benchmark for 3D object detection in the wild. In: IEEE Winter Conference on Applications of Computer Vision, pp. 75–82, March 2014. https://doi.org/10.1109/WACV.2014.6836101
16. Zhong, Z., Zheng, L., Kang, G., Li, S., Yang, Y.: Random erasing data augmentation. arXiv preprint arXiv:1708.04896 (2017)
17. Zhu, J.Y., Park, T., Isola, P., Efros, A.A.: Unpaired image-to-image translation using cycle-consistent adversarial networks. In: 2017 IEEE International Conference on Computer Vision (ICCV) (2017)

Combining Deep Learning and RGBD SLAM for Monocular Indoor Autonomous Flight

J. Martinez-Carranza[✉], L. O. Rojas-Perez, A. A. Cabrera-Ponce,
and R. Munguia-Silva

The Computer Science Department of the Instituto Nacional de Astrofisica Optica y
Electronica, Puebla, Mexico
carranza@inaoep.mx

Abstract. We present a system that uses deep learning and visual SLAM for autonomous flight in indoor environments. In this spirit, we use a state-of-the-art CNN architecture to obtain depth estimates, on a frame-to-frame basis, of images obtained from the drone's onboard camera, and use them in a visual SLAM system to obtain both camera pose estimates with a metric that is further passed to a PID controller, responsible for the autonomous flight. However, because depth estimation and visual SLAM system are computationally intensive tasks, the processing is carried out off-board on a ground control station that receives online imagery and inertial data transmitted by the drone via a WiFi channel during the flight mission. Further, the metric pose estimates are used by the PID controller that communicates back to the vehicle with the caveat that synchronisation issues may arise in between the frame reception and the pose estimation output, typically with the frame reception running at 30 Hz, and the pose estimation at 15 Hz. As a consequence, the controller may also exhibit a delay in the control loop, provoking a flight off-track the trajectory set by the way-points. To mitigate this, we implemented a stochastic filter that estimates velocity and acceleration of the vehicle to predict pose estimates in those frames where no pose estimate is available yet, and when available, to compensate for the communication delay. We have evaluated the use of this methodology for indoor autonomous flight with promising results.

1   Introduction

Autonomous navigation of a drone in the indoor environment is a challenging task that has attracted the attention of the robotics community due to similarities in the problems faced by autonomous robots. One of the main issues to be addressed is that of robot/drone localisation given the lack of access to GPS or any other external positioning system. Motion capture systems can be set in the environment to enable reliable localisation. Another option is that of sticking markers to the environment such as fiducial markers assumed to be seen by a vision system onboard the drone. Although these solutions may enable accurate

© Springer Nature Switzerland AG 2018
I. Batyrshin et al. (Eds.): MICAI 2018, LNAI 11289, pp. 356–367, 2018.
https://doi.org/10.1007/978-3-030-04497-8_29

localisation that can be exploited by the drone's controller, the reality is that there exist several scenarios where it is desirable that a drone autonomously navigates in an unknown scenario, where placing any external positioning system or marking of the scene is not an option. For instance: cave, tunnel or pipe inspection, or indoor environments with hazardous materials for humans, among others. The above calls for a solution where the drone is capable of localising itself without depending on any external localisation system. Furthermore, truly autonomous navigation calls for a mechanism that enables the drone not only to localise itself within an unknown scene but to obtain a representation of the scene that can be exploited for path planning, obstacle avoidance among other autonomous competences.

Motivated by the above, in this work we address the problem of autonomous flight in unknown indoor environments by using a well-known technique in robotics and vision communities: the visual Simultaneous Localisation and Mapping (SLAM). In addition, we are motivated by the idea of achieving indoors autonomous flight by using the least set of sensors, this is, a monocular camera and an inertial measurement unit, which is attractive in terms of energy consumption efficiency, an incentive for the development of micro aerial unmanned vehicles.

Visual SLAM for monocular systems has proved feasible at moderate frame rates (30 Hz) [1], and the field has seen this technology to become robust and effective to be used for drones with onboard monocular cameras [2,3], with the caveat that monocular SLAM delivers pose and map estimates up to scale. Depending on the task, this may become drawback as the controller may require metric localisation in order to command the drone to a specific location in the scene or to maintain flight within an area of interest.

In our previous work [4], we have addressed the scale issue by assuming that the ground surface below the drone is a flat surface, meaning that a planarity constraint can be introduced such that with the right camera angle, a synthetic depth image can be generated, this image can be used then in a conventional RGB-D SLAM system, where depth is used to initialise map point with metric. Although this approach proved effective, a planar ground is a strong assumption that may not hold in several non-structured indoor scenes. Moreover, in this previous approach, the camera had to be pointed partially to the ground, which may not be suitable if the same a camera has to be used to observe the scene.

Encouraged by the idea of generating a depth image on a frame-to-frame basis, such that can be exploited by a visual SLAM system such as ORB-SLAM, we began looking at the recent work on depth estimation in a single image. Pioneered by the seminal work of Saxena [5], depth estimation in a single image has been leveraged by the use of Convolutional Neural Networks (CNN) of several layers and their implementation on Graphics Processing Units (GPU), this has been labelled as *deep learning*. Thus, we chose the method proposed in [6], claimed to be one of the best in the state of the art, to obtain depth estimates from a single image in indoor scenes.

Deep learning has been used in visual SLAM for leveraging 3D reconstruction with scale [7] and several works mentioned the possibility of using visual SLAM with depth estimation via deep learning. However, in the context of indoors autonomous navigation, we only found the work of [8] that uses the estimated depth image to generate trajectories free of the collision.

Therefore, we propose a methodology where we incorporate the use of a depth image estimation on a frame-to-frame basis and couple it with ORB SLAM in its RGB-D version, in order to obtain pose estimates in metres. Thus, the drone's position in metres is used by the controller responsible for the autonomous flight, which commands the drone to follow a trajectory set by way-points whose coordinates are also given in metres, see Fig. 1.

Given that the deep learning process employed in this work is an intensive and expensive computational task, we tested with a drone platform that transmits imagery and inertial data to a Ground Control Station (GCS) with the caveat that a delay in the control loop is introduced. This means that the pose estimate may not correspond to the actual drone's pose, but some position behind. Hence, we propose to use a stochastic filter that uses a dynamic model up to the level of acceleration of the drone. With the pose estimates from ORB-SLAM as measurements, we use the corrected states of the filter to predict the current drone's position, in particular in those frames where no pose estimate is available as ORB-SLAM may be busy calculating the estimates.

To present our proposed approach in detail, this paper has been organised as follows: Sect. 2 describes relevant related work; Sect. 3 describes our proposed methodology; Sect. 4 describes our experimental framework; finally, our conclusions are discussed in Sect. 5.

Fig. 1. We present a methodology to achieve indoors autonomous flight with a drone equipped with a monocular camera only. We use the deep learning approach presented in [6] to obtain a depth image in a frame-to-frame basis and pass it to ORB-SLAM [9] in its RGB-D version. Thus, camera pose and map estimates with metric are obtained. A video of this work for reviewing purposes is found at https://youtu.be/7cuJCwjvI-Q (Color figure online).

2   Related Work

In recent years, MAV's equipped with onboard cameras together with SLAM and visual odometry systems are capable of performing autonomous navigation. The metric estimation in a map 3D is a challenge to those MAV's with a monocular camera by not offering a depth of the surrounding environment. The majority of works focus on a SLAM system using RGB-D cameras or laser scans to obtain metric depth [10,11]. In broad terms, autonomous flight base on vision has included the use of conventional cameras, RGB-D cameras and event cameras, with the latter being a promising technology that will enable pose estimation for fast, agile flight. Thus, there has been a plethora of works exploiting onboard cameras to carry out some pose estimation based on visual measurements namely, visual SLAM, visual odometry [12] or even visual-inertial odometry [13,14].

Direct methods [15–17] work with pixel intensities rather than visual features, but in addition to camera pose and map estimates, the methods become useful in areas with low texture. However, illumination changes or blurred introduced by camera motion may affect their performance. Research on how to use event cameras with visual SLAM principles is under development [18]. The outstanding potential of event cameras is that a robust state estimation can be obtained for autonomous navigation of very agile systems [19], something that is a strong limitation in current systems based on conventional cameras. Depth estimation is a difficult task in visual SLAM and in visual-inertial odometry systems that rely on monocular cameras, in particular, if pose and map estimates are desired to be recovered with metric. In recent year, depth estimation has become a popular topic in the deep learning community, where depth from a single image can be estimated by using convolutional neural networks (CNN). In this sense, large datasets of indoor and outdoor scenes are fed to the CNN models aiming at generating a model that enables depth estimation given a single image [20–22].

In this context, some works have used depth estimation with CNN to estimate camera motion from a frame without having to process two or more views [23] scenarios. In [24], the authors present a place recognition algorithm based on CNN models. Similarly, a robust system for monocular relocalisation is performed by using a trained CNN model that regresses the camera pose from a single RGB image [25]. The work in [26] addresses the indoors relocalisation problems using an RGB-D camera to obtain colour images and depth images. The deep learning uses these images in a CNN model to learn localisation features from images, thus to estimate camera poses from these features, achieving a good relocalisation performance.

Regarding 3D reconstruction, CNN models have been used to leverage a visual SLAM system that refines the initially estimated depth to produce an accurate 3D model with scale [27]. In [7], measurements obtained from SLAM are fused together with dense depth maps obtained with a CNN model that predicts a depth map using two frames separated in view by a small baseline. Similar to our work, the authors in [28] incorporate depth prediction with a CNN model in ORB-SLAM, although their efforts were more oriented towards enhancing the quality of 3D reconstruction. Although not directly related to depth estimation, the work in [29] presents the use of CNN models to obtain semantic structures that are added to the 3D map built during exploration of the environment, in particular in those low-texture scenes. Regarding motion estimation, the work in [30] presents an unsupervised learning method to predict depth and camera motion estimation using a monocular camera.

3   Methodology

Our approach is based on four main components: (1) ORB-SLAM [9] as visual SLAM system for a monocular camera, but in its RGB-D version; (2) Depth estimation in a single image using the method of [6]; (3) a stochastic estimator based on the Extended Kalman Filter to predict the drone's pose; (4) and a PID controller.

ORB-SLAM has become one of the best and versatile visual SLAM systems to date. It is available for monocular cameras, stereo, and RGB-D cameras. In this work, we use a vehicle with a monocular camera onboard, however, the RGB image acquired with the onboard camera is passed to the depth estimation, thus generating a depth image, at low, resolution, that is coupled with the RGB image (and re-scaled), thus obtaining an RGB-D image pair that is sent to ORB SLAM. Hence, pose and map estimates are recovered with scale.

We implemented a PID controller to control yaw, forward motion in pitch and lateral motion in a roll. The controller uses the pose estimates from ORB-SLAM or the filter, according to the experiments, to calculate the error w.r.t. to the way-point concerning distance. Note that this error is in metres, which also helps to tune the gains and thresholds in terms of metres, rather than having to change them according to an arbitrary scale. This is one of the main advantages of having metric in the system.

3.1   CNN Architecture

Current CNN architectures decrease the input resolution image using a series of convolutions and discretising the information from the feature map of the prior layer through pooling operations. In [6], a fully convolutional network for depth prediction is proposed. The architecture relies upon the receptive field, rather than neurons being fully connected.

The architecture of the previously mentioned CNN is composed of three important parts. The first part of the network is based on ResNet-50 with pre-trained weights, it accepts a 304×228 pixel RGB image as input, see Fig. 2. As stated by the authors, the network can handle higher resolutions as input which compromises precision of the predicted depth map. The second part encompasses a set of up-sampling blocks, which guides the network into learning its up-scaling, this is achieved with a sequence of un-pooling and convolutional layers. The final part, consisting of a convolutional layer, yields to the predicted depth map, with a size of 160×128. Figure 3 shows a representation of the RGB inputs and the depth map outputs of the CNN. We make use of this CNN model to recover metric scale in the SLAM system.

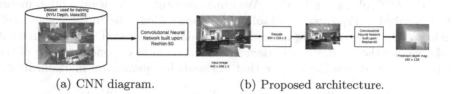

(a) CNN diagram.   (b) Proposed architecture.

Fig. 2. *a* Diagram of the architecture based on ResNet-50 [6]; *b* proposed architecture in the present work.

Fig. 3. Samples of different scenes taken as input for the CNN and their corresponding outputs. Notice that the predicted depth maps exhibit remarkable visual quality, and structure definition.

3.2 Filtering of Pose Estimates

The execution of the CNN model for depth estimation in a single image is an expensive computational process. Thus, for this work we decided to execute our system on a Ground Control Station with enough resources for the CNN model to deliver depth images at a frame rate of 30 Hz. To this end, we used a drone that transmits imagery and inertial data to the GCS via WiFi communication. The GCS process the images to obtain drone's pose estimates, passes it to the controller, and the controller communicates back to the vehicle the corresponding commands to execute autonomous flight.

Note that in the process described above, a delay is introduced in the control loop due to the WiFi communication has a natural lag in the transmission. Furthermore, ORB-SLAM runs at a slower rate than the camera rate (20 Hz). The main control loop runs within the image call back in the program that is executed every time an image arrives at the GCS. In this sense, due to the delay, some frames may not have an updated pose estimated, hence provoking delayed feedback in the controller that also leads to delay reaction of the drone, for instance, not reducing the speed in time when approaching a way-point.

To mitigate the delay, we implemented a stochastic estimation based on the Extended Kalman filter. Our state vector is given by position, orientation (represented by a quaternion vector), velocity and acceleration of the drone $\mathbf{X} = \left[r^W, q^{WR}, v^W, \omega^R, a^W, \alpha^R \right]^\top$ We use a constant acceleration motion model [31], described in Eq. 1, with variables A^W representing noise in the linear acceleration and Ψ^R representing noise in the angular acceleration. As measurement model we use the position and orientation, expressed as exponential map of the quaternion $exp(\cdot)$, see Eq. 2. Note that $\otimes$ stands for quaternion multiplication.

$$
\mathbf{X}_{new} = \begin{pmatrix} r^W_{new} \\ q^{WR}_{new} \\ v^W_{new} \\ \omega^R_{new} \\ a^W_{new} \\ \alpha^R_{new} \end{pmatrix} = \begin{pmatrix} r^W + v^W \Delta t + \frac{1}{2}(a^W + A^W)\Delta t^2 \\ q^{WR} \otimes q(\omega^R \Delta t + \frac{1}{2}(\alpha^R + \Psi^R)\Delta t^2) \\ v^W + (a^W + A^W)\Delta t \\ \omega^R + (\alpha^R + \Psi^R)\Delta t \\ a^W + A^W \\ \alpha^R + \Psi^R \end{pmatrix} \tag{1}
$$

$$
\begin{pmatrix} r^W_{new} \\ exp(q^{WR}_{new}) \end{pmatrix} = h\left(\mathbf{X}_{new} \right) \tag{2}
$$

Position measurements are obtained from ORB-SLAM, whereas orientation measurements are obtained as a weighted average of the orientation measured by the drone's IMU and the orientation estimated by ORB-SLAM, both aligned under the same coordinate system and converted to the exponential map representation. The average is a function of the frequency of the IMU, which is of 5 Hz and that of ORB-SLAM 15 Hz, giving more weight to the one with higher frequency.

4   Experiments

We present two sets of representative experiments where we evaluated the performance of our proposed methodology. In these experiments, the vehicle takes off and performs autonomous flight in an indoor scene by following a trajectory defined by a set of way-points.

For the first set of experiments, we evaluated the accuracy of the pose estimates obtained with ORB-SLAM whose estimates are returned in metres due to the use of the depth image predicted with the CNN model. We also compared the performance of the controller without the filter and with the filter. We used the motion capture system Vicon to obtain precise measurements of the drone's position in the scene. For the second set, we performed a different trajectory where the vehicle has to pass through a small gate. Note that for these experiments no previous map was built nor any assumption about the scene was made.

For our experiments, we used the Parrot Bebop 2.0 Drone. This vehicle can transmit inertially and visual data via WiFi. Visual data is captured from an onboard camera with an image resolution of 640 × 368 pixels transmitted at

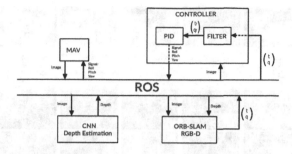

Fig. 4. Architecture implemented in this work based on the Robotic Operating System.

30 Hz; inertial data is captured with an onboard Inertial Measurement Unit (IMU) transmitting at 5 Hz; Communication and programming of control commands with the Bebop 2.0 is possible with the Software Development Kit (SDK) known as *bebop autonomy* available as a ROS package. This package is run on a Ground Control Station: an Alienware-Dell Laptop with Intel Core-i7, with 16 Gb in RAM and a GPU Nvidia GeForce GTX 1060. Linux version Ubuntu 16 LTS ran as the operating system. We used the Robotic Operating System (ROS), Kinetic version, for implementation of our approach and communication with the other programs running as nodes, namely ORB-SLAM, the Depth estimator, and our controller, Fig. 4, shows a scheme of our software architecture.

4.1   Autonomous Flight Following a Square Trajectory

In this experiment, we set a squared trajectory formed by four way-points. The square has dimensions of 3×3 m. The mapping is started right at the outset, with the drone sitting on the floor. After takeoff, the control takes over and begins to drive the drone towards the first way-point, see Fig. 5. Once the way-point is reached, the controller rotates the vehicle in the direction of the next way-point. This will continue until the drone returns to the origin, once in the origin the drone will land.

We tested our system with the controller using the ORB-SLAM pose estimates against the pose estimates predicted with the stochastic filter. We hypothesised that the filter would help to mitigate the delayed pose estimation induced by the transmission and the lag in ORB-SLAM. Figure 5 illustrates different instants of the autonomous flight. The first a row in the image shows an external view; map and tracked visual features by ORB-SLAM are shown in the second and the third figure; predicted depth image by the CNN model is shown in the fourth row; final row shows a schematic top view of the way-points and the drone depicted in white.

We carried out 10 runs for each configuration: with the controller using pose estimates from ORB-SLAM (CNN+ORB-SLAM); and with the controller using the filtered pose estimates (CNN+ORB-SLAM+Filter). Figures 6a–b show a top view of the 10 trajectories overlaid for the former and Figs. 6c–d show the 10

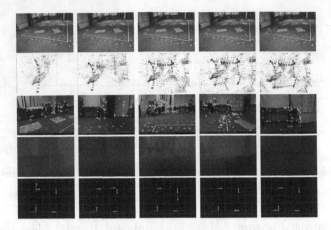

Fig. 5. Example run of the first experiment where the drone flew autonomously in an indoor environment following a squared trajectory of 3×3 m. For the sake of clarity in the first row, a red line is drawn under the drone in those images where it cannot be appreciated. The white line on the floor is a rectangle of 4×5 m placed as a reference to visually appreciate the drift of the pose estimation (Color figure online).

runs for the latter. Each pair figures show the Vicon output in black and the pose estimate in green according to accordingly.

Note that the estimates have a drift w.r.t. to the real position, which was expected since we observed that the predicted depth image was not accurate, at least in the scene where we tested it. Table 1 shows the average error of both approaches w.r.t. to the Vicon system, which is of 0.5 m roughly speaking, which represent an average of 2.6% of the traversed trajectory. Note that that the filter contributed to reduce the offset in the trajectory followed by the drone, see Fig. 6d, in contrast to the traversed trajectories were the delayed was not mitigated, see Fig. 6b.

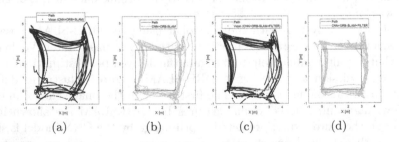

      (a)   (b)   (c)   (d)

Fig. 6. Comparison of the two approaches tested in the first experiment. The Vicon system was used to evaluate the accuracy of our method, for each approach, 10 runs were carried out.

Table 1. Accuracy of our system compared against the Vicon system

Method	Average error [m]	Std [m]	Average traversed distance [m]	Error in %
CNN + ORB-SLAM	0.5539	± **0.2566**	19.9024	2.7%
CNN + ORB-SLAM + FILTER	0.4955	± **0.2735**	19.7179	2.5 %

4.2 Autonomous Flight Through a Gate

Aware of the drift in the pose estimates of our system, we designed an experiment where such drift was considered when setting a number of way-points whose trajectory led the drone to fly through a small gate, see Fig. 7. We carried out 10 flights to evaluate the performance of our system under this configuration. The gate measured 1.5 m in height and 1.2 m wide. We use the CNN+ORB-SLAM+FILTER method in these experiments. To perform changes of rotation, the gate was placed 3 m to the right of the MAV and 3.5 m to the front. We should highlight that the drone successfully crossed the gate in 7 out of 10 runs. In the 3 failed cases, the drone hit the gate in one of the poles, however, it was very close to passing the gate. Finally, we should also highlight that in all our experiments the drone flew at the speed of 1 m/s, a fast speed for the reduced indoor workspace. The latter introduced some ground effect, however, the controller managed to drive the drone towards each way-point in all the runs.

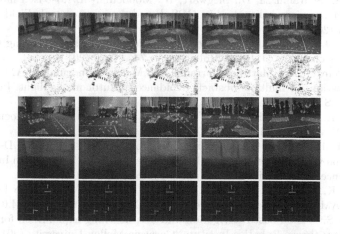

Fig. 7. Example run of the second experiment where the drone flew autonomously in an indoor environment where it had to fly through a gate. For the sake of clarity in the first row, a red line is drawn under the drone in those images where it can not be appreciated. In the fifth row, the position of the gate in the scene is indicated in orange. For reviewing purposes, a video of the experiments is found at https://youtu.be/7cuJCwjvI-Q.

5   Conclusions

We have presented a methodology where we combined depth estimation in a single image, on a frame-to-frame basis, with a visual SLAM system namely, ORB-SLAM, to obtain pose estimates with metric, which were used by a PID controller to drive the drone autonomously around a trajectory set by a number of way-points in an indoor scene. We also proposed to use a stochastic filter that models the drone's motion up to the level of acceleration, to predict the drone's pose and thus mitigate the delay in the control loop introduced by the transmission of image data from the drone to the Ground Control Station and the lag of the visual SLAM processing. We believe that the obtained results are promising and we will continue working on a faster and more accurate system that involves enhanced depth estimation in a single image, and faster visual SLAM operation such that these processes can run onboard the drone.

Acknowledgment. This work has also been partially funded by a CONACYT-INEGI fund with project no. 268528 and the Royal Society through the Newton Advanced Fellowship with reference NA140454.

References

1. Davison, A.J., Reid, I.D., Molton, N.D., Stasse, O.: MonoSLAM: real-time single camera SLAM. IEEE Trans. Pattern Anal. Mach. Intell. **29**, 1052–1067 (2007)
2. Weiss, S., Scaramuzza, D., Siegwart, R.: Monocular-SLAM-based navigation for autonomous micro helicopters in GPS-denied environments. J. Field Robot. **28**, 854–874 (2011)
3. Magree, D., Mooney, J.G., Johnson, E.N.: Monocular visual mapping for obstacle avoidance on UAVs. J. Intell. Robot. Syst. **74**, 17–26 (2014)
4. Rojas-Perez, L.O., Martinez-Carranza, J.: Metric monocular SLAM and colour segmentation for multiple obstacle avoidance in autonomous flight. In: IEEE 4th RED-UAS (2017)
5. Saxena, A., Chung, S.H., Ng, A.Y.: Learning depth from single monocular images. In: NIPS, vol. 18, MIT Press (2005)
6. Laina, I., Rupprecht, C., Belagiannis, V., Tombari, F., Navab, N.: Deeper depth prediction with fully convolutional residual networks. In: 2016 Fourth International Conference on 3D Vision (3DV), pp. 239–248, IEEE (2016)
7. Tateno, K., Tombari, F., Laina, I., Navab, N.: CNN-SLAM: real-time dense monocular SLAM with learned depth prediction. arXiv preprint arXiv:1704.03489 (2017)
8. Konam, S.: Vision-based navigation and deep-learning explanation for autonomy, in Masters thesis, Robotics Institute, Carnegie Mellon University (2017)
9. Mur-Artal, R., Montiel, J.M.M., Tardos, J.D.: ORB-SLAM: a versatile and accurate monocular SLAM system. IEEE Trans. Robot. **31**, 1147–1163 (2015)
10. Bi, Y., et al.: An MAV localization and mapping system based on dual realsense cameras. In: International Micro Air Vehicles, Conferences Competitions, National University of Singapore, Singapore (2016). Technical Report
11. Li, J., et al.: Real-time simultaneous localization and mapping for UAV: a survey. In: International Micro Air Vehicle Conference and Competition (IMAV) (2010)
12. Bloesch, M., Omari, S., Hutter, M., Siegwart, R.: Robust visual inertial odometry using a direct EKF-based approach. In: IROS (2015)

13. Teixeira, L., Alzugaray, I., Chli, M.: Autonomous aerial inspection using visual-inertial robust localization and mapping. In: Hutter, M., Siegwart, R. (eds.) Field and Service Robotics. SPAR, vol. 5, pp. 191–204. Springer, Cham (2018). https://doi.org/10.1007/978-3-319-67361-5_13
14. Lin, Y., et al.: Autonomous aerial navigation using monocular visual-inertial fusion. J. Field Robot. **35**(1), 23–51 (2018)
15. Xu, W., Choi, D., Wang, G.: Direct visual-inertial odometry with semi-dense mapping. Comput. Electr. Eng. (2018)
16. Mu, X., Chen, J., Zhou, Z., Leng, Z., Fan, L.: Accurate initial state estimation in a monocular visual-inertial SLAM system. Sensors **18**, 506 (2018)
17. Usenko, V., Engel, J., Stckler, J., Cremers, D.: Direct visual-inertial odometry with stereo cameras. In: 2016 IEEE International Conference on Robotics and Automation (ICRA), pp. 1885–1892 (2016)
18. Rebecq, H., Horstschaefer, T., Scaramuzza, D.: Real-time visualinertial odometry for event cameras using keyframe-based nonlinear optimization. In: British Machine Vision Conference (BMVC), vol. 3 (2017)
19. Vidal, A.R., Rebecq, H., Horstschaefer, T., Scaramuzza, D.: Hybrid, frame and event based visual inertial odometry for robust, autonomous navigation of quadrotors. arXiv preprint arXiv:1709.06310 (2017)
20. Mancini, M., Costante, G., Valigi, P., Ciarfuglia, T.A., Delmerico, J., Scaramuzza, D.: Toward domain independence for learning-based monocular depth estimation. IEEE Robot. Autom. Lett. **2**, 1778–1785 (2017)
21. Wang, P., Shen, X., Lin, Z., Cohen, S., Price, B., Yuille, A.L.: Towards unified depth and semantic prediction from a single image. In: Proceedings of the IEEE Conference on Computer Vision and Pattern Recognition, pp. 2800–2809 (2015)
22. Chakrabarti, A., Shao, J., Shakhnarovich, G.: Depth from a single image by harmonizing overcomplete local network predictions. In: Advances in Neural Information Processing Systems, pp. 2658–2666 (2016)
23. Godard, C., Mac Aodha, O., Brostow, G.J.: Unsupervised monocular depth estimation with left-right consistency. In: CVPR, vol. 2, p. 7 (2017)
24. Chen, Z., Lam, O., Jacobson, A., Milford, M.: Convolutional neural network-based place recognition. arXiv preprint arXiv:1411.1509 (2014)
25. Kendall, A., Grimes, M., Cipolla, R.: PoseNet: a convolutional network for real-time 6-DOF camera relocalization. In: IEEE International Conference on Computer Vision, pp. 2938–2946, IEEE (2015)
26. Li, R., Liu, Q., Gui, J., Gu, D., Hu, H.: Indoor relocalization in challenging environments with dual-stream convolutional neural networks. IEEE Trans. Autom. Sci. Eng. **15**(2), 651–662 (2017)
27. Weerasekera, C.S., Garg, R., Reid, I.: Learning deeply supervised visual descriptors for dense monocular reconstruction. arXiv preprint arXiv:1711.05919 (2017)
28. Mukasa, T., Xu, J., Stenger, B.: 3D scene mesh from CNN depth predictions and sparse monocular SLAM. In: Proceedings of the IEEE Conference on Computer Vision and Pattern Recognition, pp. 921–928 (2017)
29. Yang, S., Song, Y., Kaess, M., Scherer, S.: Pop-up SLAM: Semantic monocular plane SLAM for low-texture environments. In: 2016 IEEE/RSJ International Conference on Intelligent Robots and Systems (IROS), pp. 1222–1229. IEEE (2016)
30. Zhou, T., Brown, M., Snavely, N., Lowe, D.G.: Unsupervised learning of depth and ego-motion from video. In: CVPR, vol. 2, p. 7 (2017)
31. Gemeiner, P., Davison, A., Vincze, M.: Improving localization robustness in monocular SLAM using a high-speed camera. In: Robotics: Science and Systems (2008)

Author Index

Printed in the United States
By Bookmasters